International Society of Fire Service Instructors

# Fire Service Instructor

## Principles and Practice

JONES AND BARTLETT PUBLISHERS

*Sudbury, Massachusetts*

BOSTON TORONTO LONDON SINGAPORE

**Jones and Bartlett Publishers**
World Headquarters
40 Tall Pine Drive
Sudbury, MA 01776
978-443-5000
info@jbpub.com
www.jbpub.com

**Jones and Bartlett Publishers Canada**
6339 Ormindale Way
Mississauga, Ontario L5V 1J2
Canada

**Jones and Bartlett Publishers International**
Barb House, Barb Mews
London W6 7PA
United Kingdom

**National Fire Protection Association**
1 Batterymarch Park
Quincy, MA 02169-7471
www.nfpa.org

**International Association of Fire Chiefs**
4025 Fair Ridge Drive
Fairfax, VA 22033
www.iafc.org

**International Society of Fire Service Instructors**
2425 Highway 49 East
Pleasant View, TN 37146
www.isfsi.org

Jones and Bartlett's books and products are available through most bookstores and online booksellers. To contact Jones and Bartlett Publishers directly, call 800-832-0034, fax 978-443-8000, or visit our website, www.jbpub.com.



Substantial discounts on bulk quantities of Jones and Bartlett's publications are available to corporations, professional associations, and other qualified organizations. For details and specific discount information, contact the special sales department at Jones and Bartlett via the above contact information or send an email to specialsales@jbpub.com.



**Production Credits**

Chief Executive Officer: Clayton E. Jones
Chief Operating Officer: Donald W. Jones, Jr.
President, Higher Education and Professional Publishing: Robert W. Holland, Jr.
V.P., Sales and Marketing: William J. Kane
V.P., Production and Design: Anne Spencer
V.P., Manufacturing and Inventory Control: Therese Connell
Publisher, Public Safety Group: Kimberly Brophy
Senior Acquisitions Editor, Fire: William Larkin
Editor: Amanda J. Green
Production Manager: Jenny L. Corriveau
Associate Production Editor: Sarah Bayle
Photo Research Manager/Photographer: Kimberly Potvin
Director of Marketing: Alisha Weisman
Marketing Manager: Brian Rooney
Cover Image: Courtesy of the University of Nevada, Reno Fire Science Academy
Text Design: Anne Spencer
Cover Design: Kristin E. Ohlin
Composition: Shepherd, Inc.
Text Printing and Binding: Malloy, Inc.
Cover Printing: Courier Stoughton



**Copyright © 2009 by Jones and Bartlett Publishers, LLC, and the National Fire Protection Association.®**

All rights reserved. No part of the material protected by this copyright notice may be reproduced or utilized in any form, electronic or mechanical, including photocopying, recording, or by any information storage and retrieval system, without written permission from the copyright owner.

The procedures and protocols in this book are based on the most current recommendations of responsible sources. The International Association of Fire Chiefs (IAFC), National Fire Protection Association (NFPA®), International Society of Fire Service Instructors (ISFSI), and the publisher, however, make no guarantee as to, and assume no responsibility for, the correctness, sufficiency, or completeness of such information or recommendations. Other or additional safety measures may be required under particular circumstances.

Additional photographic credits appear on page 296, which constitutes a continuation of the copyright page.

Notice: The individuals described in "You are the Fire Service Instructor" and "Fire Service Instructor in Action" throughout the text are fictitious.



**Library of Congress Cataloging-in-Publication Data**

Fire service instructor : principles and practice / International Association of Fire Chiefs [and] National Fire Protection Association [and International Society of Fire Service Instructors].
p. cm.
ISBN-13: 978-0-7637-4910-1 (pbk.)
ISBN-10: 0-7637-4910-9 (pbk.)
1. Fire prevention—Study and teaching. 2. Fire extinction—Study and teaching. I. International Association of Fire Chiefs. II. National Fire Protection Association.
TH9120.F583 2008
628.9'2071—dc22

2008012181

6048

**Printed in the United States of America**
12 11 10 09 08 10 9 8 7 6 5 4 3 2 1

# Brief Contents

# Contents

# Resource Preview

## Fire Service Instructor: Principles and Practice

The National Fire Protection Association (NFPA®), International Association of Fire Chiefs (IAFC), and International Society of Fire Service Instructors (ISFSI) are pleased to bring you *Fire Service Instructor: Principles and Practice*. This resource provides you with the up-to-date information required to meet the modern job performance requirements for the Fire Service Instructor I and II as outlined by the National Fire Protection Association's 2007 edition of (NFPA) 1041, *Standard for Fire Service Instructor Professional Qualifications*.

*Fire Service Instructor: Principles and Practice* encourages critical thinking and is the core of an integrated teaching and learning system for Fire Instructor I and II level courses. The textbook is written in a clear, concise, and user-friendly writing style to simplify the material for students. *Fire Service Instructor: Principles and Practice* is not only designed to help you meet the requirements to become a fire service instructor, but also empowers you to become a great educator within the fire and emergency services.

*Fire Service Instructor: Principles and Practice* includes practical coverage of:

- Legal Issues
- The Learning Process
- Communication Skills
- Lesson Plan Development
- Safety During the Learning Process
- Instructor Management

## Chapter Resources

*Fire Service Instructor: Principles and Practice* serves as the core of a highly effective teaching and learning system. Its features reinforce and expand on essential information and make information retrieval a snap. These features include:

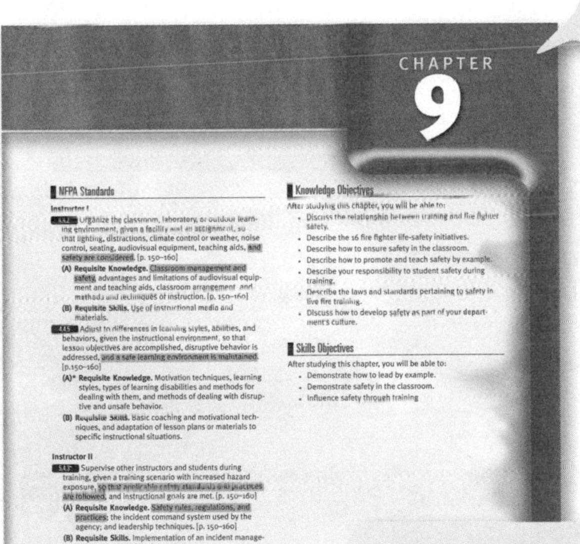
CHAPTER 9

**NFPA Standards**

**Instructor I**

Organize the classroom, laboratory, or outdoor learning environment, given a facility and an assignment, so that lighting, distractions, climate control or weather, noise control, seating, audiovisual equipment, teaching aids, and safety are considered. [p. 150–160]

(A) **Requisite Knowledge.** Classroom management and safety, advantages and limitations of audiovisual equipment and teaching aids, classroom arrangement, and methods and techniques of instruction. [p. 150–160]

(B) **Requisite Skills.** Use of instructional media and materials.

Adjust to differences in learning styles, abilities, and behaviors, given the instructional environment, so that lesson objectives are accomplished, disruptive behavior is addressed, and a safe learning environment is maintained. [p.150–160]

(A)* **Requisite Knowledge.** Motivation techniques, learning styles, types of learning disabilities and methods for dealing with them, and methods of dealing with disruptive and unsafe behavior.

(B) **Requisite Skills.** Basic coaching and motivational techniques, and adaptation of lesson plans or materials to specific instructional situations.

**Instructor II**

Supervise other instructors and students during training, given a training scenario with increased hazard exposure, so that applicable safety standards and practices are followed, and instructional goals are met. [p. 150–160]

(A) **Requisite Knowledge.** Safety rules, regulations, and practices; the incident command system used by the agency; and leadership techniques. [p. 150–160]

(B) **Requisite Skills.** Implementation of an incident management system used by the agency.

**Knowledge Objectives**

After studying this chapter, you will be able to:

- Discuss the relationship between training and fire fighter safety.
- Describe the 16 fire fighter life-safety initiatives.
- Describe how to ensure safety in the classroom.
- Describe how to promote and teach safety by example.
- Describe your responsibility to student safety during training.
- Describe the laws and standards pertaining to safety in live fire training.
- Discuss how to develop safety as part of your department's culture.

**Skills Objectives**

After studying this chapter, you will be able to:

- Demonstrate how to lead by example.
- Demonstrate safety in the classroom.
- Influence safety through training

**Chapter Objectives**

NFPA 1041 Standard, Knowledge Objectives, and Skills Objectives are listed at the beginning of each chapter.

- Portions of the NFPA 1041 Standard that are highlighted are applicable to the chapter.
- Fire Service Instructor Level I and Level II requirements are listed separately.
- Page references are included for quick reference to content.
- Knowledge and Skills Objectives outline the most important topics covered in the chapter.

**You Are the Fire Service Instructor**

Each chapter opens with a case study intended to stimulate classroom discussion, capture students' attention, and provide an overview for the chapter. An additional case study is provided in the end-of-chapter Wrap-Up material.

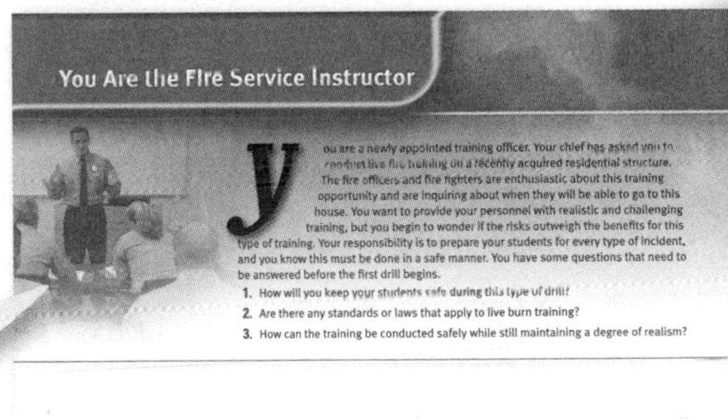
**You Are the Fire Service Instructor**

You are a newly appointed training officer. Your chief has asked you to conduct live fire training on a recently acquired residential structure. The fire officers and fire fighters are enthusiastic about this training opportunity and are inquiring about when they will be able to go to this house. You want to provide your personnel with realistic and challenging training, but you begin to wonder if the risks outweigh the benefits for this type of training. Your responsibility is to prepare your students for every type of incident, and you know this must be done in a safe manner. You have some questions that need to be answered before the first drill begins.

1. How will you keep your students safe during this type of drill?
2. Are there any standards or laws that apply to live burn training?
3. How can the training be conducted safely while still maintaining a degree of realism?

**Introduction**

Each year in the United States, countless fire fighters are injured during training. On average, 10 fire fighters are killed each year during training-related activities. This trend has accelerated over the past few years and is a growing concern among fire service leaders.

As a fire service instructor, you must take measures to stop the preventable injuries and deaths that occur during training. Of course, for training to be effective, it has to be realistic and therefore, a certain level of danger will exist in all training events. Live fire training tops the list of high-risk training events, but this same high risk can be found in many other training activities.

You must always place students' safety ahead of any performance expectations for the training session. This mandate includes everything from the assignment of safety officers to rehabilitation protocols to increasing instructor–student ratios to ensuring that enough trained eyes are watching out for the safety of all students. You have the standards, technology, and information needed to protect your personnel from training tragedies. You also need to apply relevant checklists, standards, and control measures to your training sessions correctly to keep your students safe. Fire fighters want meaningful drills that allow them to apply their skills, knowledge, and abilities. It is your responsibility to provide this training while keeping your students safe FIGURE 9.1.

**The Sixteen Fire Fighter Life-Safety Initiatives**

In 2004, fire service experts and leaders convened to discuss solutions to slow the number of line-of-duty deaths experienced by the fire service. Their work culminated in 16 initiatives that were given to the fire service with the goal of reducing line-of-duty deaths by 50 percent over a 10-year period TABLE 9.1. These initiatives should provide both guidance and motivation to you while you are developing and conducting any training. The initiatives highlight the importance of training as a strategy that will help reduce the number of fire fighter line-of-duty deaths. As a fire service instructor, you will be specifically charged with carrying out initiative 5 every time you instruct a course.

FIGURE 9.1 It is your responsibility to provide meaningful drills that allow fire fighters to apply their skills, knowledge, and abilities while keeping your students safe.

**Leading by Example**

As a fire service instructor, you will use many methods of delivery to teach. One of the most powerful methods is leading by example. Fire fighters, regardless of their rank, years on the job, or past experiences, will look to you and make mental notes on how the job *should* be done FIGURE 9.2. You

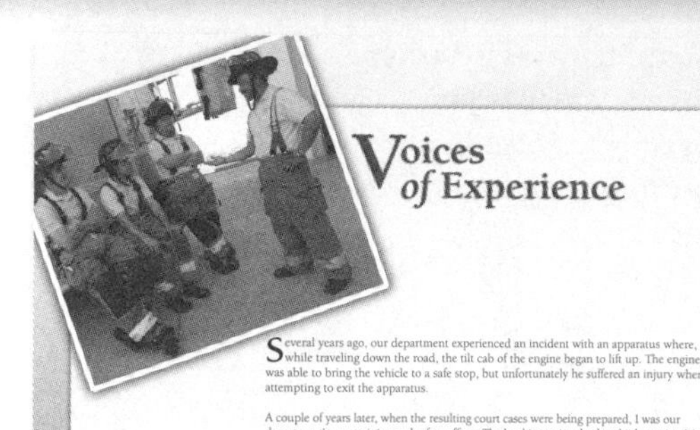

## Voices of Experience

Several years ago, our department experienced an incident with an apparatus where, while traveling down the road, the tilt cab of the engine began to lift up. The engineer was able to bring the vehicle to a safe stop, but unfortunately he suffered an injury when attempting to exit the apparatus.

A couple of years later, when the resulting court cases were being prepared, I was our department's new training and safety officer. The legal issues involved multiple parties. My involvement began when our department was subpoenaed to produce training records that could have a bearing on the issue. This material included the initial and ongoing training records for the driver as well as the training records for our department's mechanic personnel.

*"Because of our attention to maintaining accurate training records, we were able to produce the needed documents."*

Fortunately, the year before we had done an audit of our training records in conjunction with our state fire marshal's office. Because of our attention to maintaining accurate training records, we were able to produce the needed documents. This incident really emphasized for me the importance of tracking both initial training and ongoing training and documenting any certificates of completion and state certification.

Maintaining your training records is not as exciting as conducting a multiple-company drill, but it is just as important to your organization and your members. Many times training officers or instructors get some "grief" when we remind our personnel to "get the training reports done," but this incident was a good example of its importance. Remember—if it isn't properly documented, it didn't happen!

*Rudy Horist*
Elgin Fire Department
Elgin, Illinois

### Voices of Experience

In the Voices of Experience essays, veteran fire service instructors share their accounts of memorable incidents while offering advice and encouragement. These essays highlight what it is truly like to be a fire service instructor.

Chapter 3 Methods of Instruction 45

## Applying the JOB PERFORMANCE REQUIREMENTS (JPRs)

Effective training has many components. The audience, the learning environment, the instructor, and the expectations of the training process are all pieces of the puzzle that combine to make training effective. The methods of instruction chosen by the instructor are among the most vital parts of the puzzle, as they have to match the audience, the environment, and the expectations or outcomes of the training session. The wrong method may have disastrous effects on the outcome and lead to poor performance or even safety and survival issues on the fire ground. Students' motivation can be enhanced by varying approaches to instructional delivery and choosing the appropriate setting for training delivery.

| Instructor I | Instructor II |
| --- | --- |
| The Instructor I is responsible for delivering training from prepared lesson plans. Knowing how to identify the factors relating to the appropriate methods of instruction for the lesson plan will assist in the delivery of effective training. | Conducting an audience analysis and evaluating the learning environment help the Instructor II develop lesson plan materials that are effective for teaching students. The best learning process is interactive and student centered. Varied approaches to your training methods make this work. |
| **JPRs at Work** | **JPRs at Work** |
| Deliver the prepared lesson plan with the ability to adjust and adapt the lesson plan to the learning conditions presented by the students and the training location. Correct unproductive student behavior and control the delivery of training relating to the students performance. | Be able to adjust the delivery of a lesson plan by varying the methods of instruction to keep students motivated and interested while ensuring that all objectives prepared for the lesson plan are covered and the student performance is improved. |

**Bridging the Gap Between Instructor I and Instructor II**

The lesson plan content and the instructor's ability both play large roles in the method of instruction that is used to deliver a training session. Certain methods are required for specific types of objectives and intended outcomes. The Instructor II will prepare the lesson plan with a specific setting in mind for the training delivery. The Instructor I, based on his or her experience, comfort, and background, may vary the training delivery according to these factors.

**Instant Applications**

1. Identify the most effective and least effective training sessions you have attended. Determine which methods of instruction were appropriate or inappropriate and may have influenced your impressions of those training sessions.
2. Review a list of upcoming training sessions scheduled for your department and determine the best methods of instruction that should be used to achieve the best outcomes for them.
3. Rearrange your typical classroom seating arrangement to accommodate various methods of discussion such as a lecture, demonstration, or group discussion.

### Applying the Job Performance Requirements (JPRs)

This unique feature appears in every chapter and provides:

- An understanding of the relationship between Fire Service Instructor I and II.
- Identification of the specific responsibilities of the Fire Service Instructor I and II relating to how the Job Performance Requirements (JPRs) work.
- A discussion of how each level works together to achieve training goals.
- Instant application questions allow students an opportunity to review how the JPR works in their department.

### Teaching Tips

Teaching Tips offer important and insightful teaching strategies for new and experienced instructors.

Chapter 3 Methods of Instruction 53

**Teaching Tip**

In small groups, students should be encouraged to discuss what influences them to learn, including identifying their personal agendas and goals for the class. It certainly would be appropriate to have students present their group's results to their class. This exercise is a good way to encourage discussion and for you to learn more about the students.

## Lesson Plans

Ask yourself three basic questions before building your lesson plan:

1. *What are the goals of the lesson?* These goals are the purpose, aims, and rationale for the class period. You and your students will be working to achieve these goals.
2. *How will you achieve those goals?* Identify the objective of the lesson and focus on what the students will do to acquire the knowledge and skills of the lesson.
3. *How will you know when those goals are achieved?* Through some kind of assessment, you need to ensure that the students have learned, at a minimum, the desired information. Be sure to provide students with an opportunity to practice what you will be evaluating.

All students have special needs in the classroom that may require you to adjust some part of the lesson plan at some juncture. When you develop your plan, it should be detailed and complete enough to achieve the identified goals and desired outcomes, yet flexible enough that it can be adjusted to meet students' needs on the fly. Be prepared to repeat information in different ways or to perform demonstrations more than once in different ways. Make sure that you follow a detailed, sequential plan when demonstrating a skill; explain and outline the information to be explained; and discuss the results, including key questions to guide the discussion. These simple yet flexible components will enable you to develop an effective, yet concise lesson plan. Chapter 6, Lesson Plan Development, details this process.

**Safety Tip**

Always be aware of what is going on around you. As the instructor, it is your responsibility to ensure the safety of your class. Modifications to exits, changes in the environment, and unusual activities or commotions in the area should prompt a careful investigation to keep your learning environment safe. Items out of place on the floor, people not sitting with all of the legs of the chair on the floor, or furniture that is unstable may seem benign, but are actually dangers that can cause injury to the students in the classroom setting. Use common sense and practice safety.

### Safety Tips

Safety Tips reinforce safety-related concerns for both fire instructors and fire fighters.

### Ethics Tips

Ethics Tips are provided to help educate instructors on current ethical issues in the fire service.

### Theory into Practice Tips

Theory into Practice Tips assist instructors in conceptualizing important theoretical topics as they relate to today's fire service.

### Wrap-Up

End-of-chapter activities reinforce important concepts and improve students' comprehension. Additional instructor support and answers for all questions are available in the Instructor's Resource Manual.

### Fire Service Instructor in Action

This activity promotes critical thinking through the use of case studies and provides instructors with discussion points for the classroom presentation.

### Chief Concepts

Chief Concepts highlight critical information from the chapter and are provided in a bulleted format to help students prepare for exams.

### Hot Terms

Hot Terms are the key terms that the students must learn. They are easily identifiable within the chapter and the Hot Terms section at the end of the chapter defines these key terms. Finally, the Hot Term Explorer on www.fire.jbpub.com provides interactivities for students.

# Instructor Resources

A complete teaching and learning system developed by educators with an intimate knowledge of the obstacles that instructors face each day supports *Fire Service Instructor: Principles and Practice*. These resources provide practical, hands-on, time-saving tools such as PowerPoint presentations, customizable lecture outlines, test banks, and image/table banks to better support instructors and students. In addition, complete Fire Service Instructor I, Fire Service Instructor II, and combined Fire Service Instructor I and II curriculum packages have been created to meet every instructor's needs.

## Instructor's ToolKit CD-ROM

**ISBN: 978-0-7637-5971-1**

Preparing for class is easy with the resources on this CD-ROM. Instructors can choose resources for the Fire Service Instructor I, Fire Service Instructor II, or combined Fire Service Instructor I and II levels. The CD-ROM includes the following resources:

- **Adaptable PowerPoint Presentations.** Provides instructors with a powerful way to create presentations that are educational and engaging to their students. These slides can be modified and edited to meet instructors' specific needs.
- **Lecture Outlines.** Provides complete, ready-to-use lecture outlines that include all of the topics covered in the text. The lecture outlines can be modified and customized to fit any course.
- **Electronic Test Bank.** Contains multiple-choice questions and allows instructors to create tailor-made classroom tests and quizzes quickly and easily by selecting, editing, organizing, and printing a test along with an answer key, including page references to the text.
- **Sample Lesson Plans with Associated PowerPoint Presentations and Quizzes:** Twenty-four sample lesson plans and their associated PowerPoint Presentations and Quizzes are included to help instructors evaluate each student's unique presentation style and skills.
- **Skill Sheets:** Self Evaluation, Peer Evaluation, and Instructor Evaluation sheets are included. Additional course management forms and templates are also provided.
- **Image and Table Bank.** Offers a selection of the most important images and tables found in the text. Instructors can use these graphics to incorporate more images into the PowerPoint presentations, make handouts, or enlarge a specific image for further discussion.

## Instructor's Resource Manual CD-ROM

**ISBN: 978-0-7637-5972-8**

The Instructor's Resource Manual is the instructor's guide to the entire teaching and learning system. Instructors can choose resources appropriate to the Fire Service Instructor I, Fire Service Instructor II, or combined Fire Service Instructor I and II levels. This indispensable manual contains the following features for each chapter:

- Objectives with page references
- Support materials
- Enhancements
- Teaching tips
- Readings and preparation
- Presentation overview with suggested teaching times
- Lesson plans with corresponding PowerPoint slide text
- Answers to all end-of-chapter student questions found in the text
- Student quiz with instructor answer key
- Activities and assignments

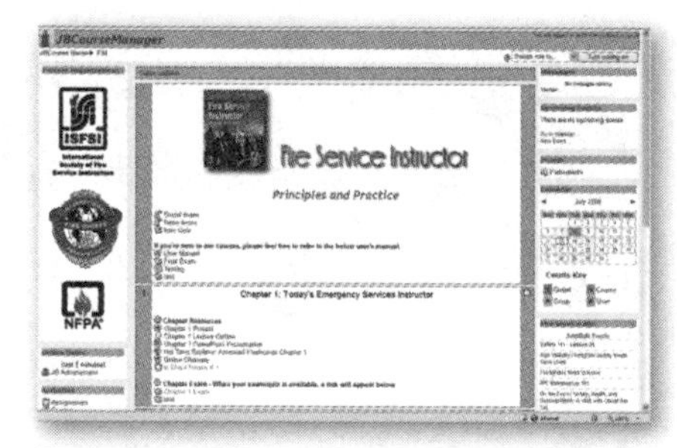


## JBCourseManager

Combining our robust teaching and learning materials with an intuitive and customizable learning platform, JBCourseManager enables instructors to create an online course quickly and easily. The system allows instructors to readily complete the following tasks:

- Customize preloaded content or easily import new content
- Provide online testing
- Offer discussion forums, real-time chat, group projects, and assignments
- Organize course curricula and schedules
- Track student progress, generate reports, and manage training and compliance activities

JBCourseManager is free to adopters of *Fire Service Instructor: Principles and Practice*. Contact your sales specialist at 800-832-0034 for details today.

# Student Resources

To help students retain the most important information and assist them in preparing for exams, Jones and Bartlett Publishers has developed a complete set of student resources.

### Student Workbook

ISBN-13: 978-0-7637-6035-9

This resource is designed to encourage critical thinking and aid comprehension of the course material through use of the following materials:

- Case studies and corresponding questions
- Matching
- Fill-in-the-blank
- Short-answer
- Multiple-choice questions
- Answer key with page references

### www.fire.jbpub.com

This site has been specifically designed to complement *Fire Service Instructor: Principles and Practice* and is regularly updated. Resources available include:

- **Chapter Pretests** that prepare students for training. Each chapter has a pretest and provides instant results, feedback on incorrect answers, and page references for further study.
- **Interactivities** that allow students to reinforce their understanding of the most important concepts in each chapter.
- **Hot Term Explorer**, a virtual dictionary, allowing students to review key terms, test their knowledge of key terms through quizzes and flashcards, and complete crossword puzzles.

# Acknowledgments

## Editors

### Alan E. Joos

Alan E. Joos serves as an Assistant Director of Certification and Accreditation at Louisiana State University Fire and Emergency Training Institute (FETI). Prior to employment at FETI, he worked at the Utah Fire and Rescue Academy, 1 year in the training division and 11 years as the Assistant Director of the Certification System. Mr. Joos received an associate and bachelor's degree from Utah Valley State College in Fire Science and Business Technology Management and a Master's degree from Grand Canyon University in Phoenix, Arizona. He is also a graduate of the Executive Fire Officer's Program at the National Fire Academy. Alan's fire service background began in 1985 as a career firefighter in a combination fire department and then as an on-call firefighter/EMT-I during his time while employed at the state fire academy in Utah for a total of 22 years. In both agencies Alan held the position of Training Officer and was a state tester for EMS. Alan is married to Carla Joos, and has three sons, Nathan, Jordan (deceased), and Dallan.

### Forest F. Reeder

Forest F. Reeder began his fire service career in 1979. He currently serves as Battalion Chief/Director of Training & Safety with the Pleasantview Fire Protection District, LaGrange Highlands, Illinois, and is the Director of Training for Southwest United Fire Districts.

Forest is the author of the Weekly Fire Drills feature at www.firefighterclosecalls.com and a weekly contributor to *Fire Engineering Magazines* online drill feature. Forest is the curriculum coordinator and a staff instructor at Moraine Valley Community College in the fire science program. Additionally, he is the coordinator of the Fire Officer 1 and 2 programs for the Illinois Fire Chiefs Foundation and chairman of the State Fire Marshals Office Fire Officer Ad Hoc committee.

His educational background includes numerous Illinois fire service certifications including Fire Officer 3 and Training Program Manager and a Masters Degree in Public Safety Administration from Lewis University. He was awarded the George D. Post Instructor of the Year Award for the International Society of Fire Service Instructors in 2008.

## Contributing Authors

### Stephen H. DiNolfo

Stephen H. DiNolfo of the law office of Ottosen Britz Kelly Cooper & Gilbert, Ltd. focuses his practice on local government litigation and client counseling. A significant portion of his practice is dedicated to defending fire service employees and EMS providers. Mr. DiNolfo has defended clients in professional misconduct matters, federal civil rights claims, as well as wrongful death claims. In addition, Mr. DiNolfo devotes much of his time to counseling and advising the fire service on liability issues. He has taught numerous classes on legal issues for training officers and fire service personnel. Mr. DiNolfo is a speaker at numerous conferences and for numerous associations where he lectures on liability issues in the fire service.

### Ben Hirst

Dr. Ben A. Hirst is President of Performance Training Systems (PTS). Over the past 19 years, PTS has become the leading provider of valid testing materials for the certification, promotion, and training of fire and emergency medical personnel. These testing materials are used by 77 certification agencies, over 300 fire departments, and over 100 training academies and colleges worldwide.

### Barbara Klingensmith

Dr. Barbara Klingensmith is a fire fighter/paramedic with educational mastery in education, fire and EMS training, management, organizational development, and planning. Barbara teaches at the local, state and national levels with experience in the traditional classroom as well as online, and is an educational leader and curriculum developer for fire, EMS and emergency services. Dr. Klingensmith holds a PhD in Educational Leadership from Walden University, a Masters Degree from Johns Hopkins University, and a BA from Hood College in Frederick, Maryland. Barbara is trained as a fire fighter and paramedic through the Maryland Fire and Rescue Institute and the University of Maryland's Maryland Institute for Emergency Medical Services Systems.

### Bryant Krizik

Bryant Krizik is the Fire Chief of Orland Fire Protection District, Orland Park, Illinois, a career department of 120 fire fighters in the Southwest suburb of Chicago. He has been in the fire service for 25 years. Chief Krizik has presented at FDIC, and teaches for Illinois Fire Chiefs, Illinois Society of Fire Service Instructors, University of Illinois Fire Service Institute, and Moraine Valley Community College. He has a Bachelor's Degree from Southern Illinois University in Fire Service Management. Chief Krizik comes from a family of fire fighters. He teaches extensively on conducting live fire training.

### Kurt Larson

Kurt Larson, co-founder of the Florida Institute of Research & Education, Inc. (www.fire-help.org), is a highly regarded and experienced instructor, executive coach and author. As the recipient of the 2007 Florida Instructor of the Year award, Kurt's focus on helping public safety organizations create a culture of excellence through training and education is well-served.

His experience includes positions with both career and volunteer fire departments, advancing to Fire Chief in both avenues of service. He is a graduate of the National Fire Academy's Executive Fire Officer program and holds the designation of Chief Fire Officer from the Center for Public Safety Excellence. Kurt holds degrees in Safety Engineering, Fire Science, and Communications Engineering.

### David Peterson

David Peterson serves as fire chief for Plainfield Fire Department, Plainfield, Michigan, and is President of the Michigan Association of Fire Chiefs for the term 2008 to 2010. He is a 2002 graduate of the Executive Fire Officer Program at the National Fire Academy. He has been a State Certified Instructor since 1982 and was selected as Fire Service Instructor of the Year for Michigan in 2005. Chief Peterson was instrumental in the development of the Fire Chief 101 curriculum; a class that he frequently delivers around the state for prospective chief officers. He is active as an instructor of Company and Chief Officer Programs throughout Michigan.

### David Purchase

David Purchase has 32 years in the fire service with the Norton Shores Fire Department, Norton Shores, Michigan, and was appointed Fire Chief in 1999. He has been a certified fire instructor for over 17 years and is an Executive Fire Officer program graduate of the National Fire Academy. Chief Purchase currently serves as a Director on the Michigan Fire Chiefs Executive Board and Vice Chair of the Michigan Firefighters Training Council. In 2004, the Michigan Fire Service Instructors Association named him State Fire Instructor of the Year and in 2006 he was named Michigan's Fire Chief of the year by the Michigan Association of Fire Chiefs. He has also authored the chapter on Fire Department Training and Education in Jones and Bartlett's recently published *Chief Fire Officer's Desk Reference*.

### Terry Vavra

Terry Vavra is a thirty-two year veteran of the fire service. He began as a paid on-call fire fighter in 1976 and became a career fire fighter and paramedic for Lisle Woodridge Fire District, Illinois in 1979. He retired from Lisle Woodridge at the rank of Deputy Chief after more than 28 years of full-time service serving as Company Officer, Training Safety Officer and Battalion Chief. Currently he is the Fire Chief for the Buffalo Grove Fire Department in Illinois. He has taught classes in fire fighter safety, management, and operations throughout the country and internationally. He has had several articles published in trade magazines and assisted in the development of the class "Preparing the Fireground Safety Officer" and continues to teach the class. Terry has instructed hundreds of students from recruit fire fighters to Chief Officers across the country. Chief Vavra holds two Associates Degrees, a Bachelor of Arts Degree and a Master of Science degree in Management and Organizational Behavior. He is a charter member of the Fire Department Safety Officers and a member of the Illinois Fire Chiefs and International Association of Fire Chiefs.

### Chris Watson

Chris Watson began his nineteen-year fire service career with the Detroit Fire Department, Detroit, Michigan. Currently, he is the Battalion Chief of the Austin Fire Department Training Division in Austin, Texas. He has earned an Associates Degree in Fire Protection Technology and a Bachelors Degree in Fire Service Administration. He is an instructor for both the Austin Fire Department Training Academy and Austin Community College. Chief Watson has served on the Texas Commission on Fire Protection committees that developed the Fire Instructor and Fire Officer curricula for the State of Texas.

Jones and Bartlett Publishers, the National Fire Protection Association, the International Association of Fire Chiefs, and the International Society of Fire Service Instructors would like to thank all of the editors, contributors, and reviewers of *Fire Service Instructor: Principles and Practice*.

## Editorial Board

Shawn Stokes—International Association of Fire Chiefs, Fairfax, Virginia

Craig Richardson—International Society of Fire Service Instructors, Pleasant View, Tennessee. Nanaimo Fire Rescue, British Columbia, Canada.

Kendall Holland—The National Fire Protection Association, Quincy, Massachusetts

Forest Reeder— Pleasantview Fire Protection District, LaGrange Highlands, Illinois

Alan Joos—Louisiana State University, Fire and Emergency Training Institute, Baton Rouge, Louisiana

David Hall—Springfield Fire Department, Springfield, Missouri

## Contributors and Reviewers

Raul A. Angulo, Seattle Fire Department, Lake Tapps Island, Washington
Tom Beatty, Ohio Society of Fire Service Instructors, Springfield, Ohio
John P. Binaski, College of the Sequoias, Visalia, California
Alan Braun, University of Missouri Fire and Rescue Training Institute, Columbia, Missouri
Dwight Scott Burt, Northland Pioneer College, Snowflake, Arizona
Melvin Byrne, Virginia Department of Fire Programs Division 7, Leesburg, Virginia
Sean Campbell, Humboldt County Office of Education, Eureka, California
Ted Cocco, Providence Fire Department, Providence, Rhode Island
Jeff Dean, Georgia Fire Academy, Forsyth, Georgia
John A. DeArmond, Pier 9 Medical, Safety Education and Consulting, San Francisco, California
M. John Dudte, Fire EMS Department, District of Columbia, Washington, D.C.
David R. Fischer, Nevada Department of Public Safety, Las Vegas, Nevada
Robert T. Foraker II, Delaware Technical and Community College, Newark, Delaware
Neil R. Fulton, Norwich Fire Department, Norwich, Vermont
Chris W. Gibson, East Side Fire Department, Baton Rouge, Louisiana
Todd Gilgren, Arvada Fire Protection District, Arvada, Colorado
Cathleen Connor Goetz, Connecticut Fire Academy, Windsor Locks, Connecticut
Joseph Guarnera, Anna Maria College, Paxton, Massachusetts
Darren Hall, Coronado Fire Department, Coronado, California
Frank H. Hammond, Jr., Maine Fire Training and Education, South Portland, Maine
Jeffry J. Harran, Lake Havasu City Fire Department, Lake Havasu City, Arizona
Joe D. Hanson, Nebraska State Fire Marshal and Training Division, Grand Island, Nebraska
Ronald Hassan, Maryland Fire and Rescue Institute, College Park, Maryland
Richard Hilinski, Community College of Allegheny, Pittsburgh, Pennsylvania
Jeffrey L. Huber, Lansing Community College, Lansing, Michigan
Larry Hughes, North Carolina Office of State Fire Marshal, Raleigh, North Carolina
Brian P. Kazmierzak, Clay Fire Territory, South Bend, Indiana
Barron D. Kennedy III, Tennessee Fire Service and Codes Enforcement Academy, Bell Buckle, Tennessee
James H. Maxon III, Sandoval County Fire Department, Bernalillo, New Mexico
Glen B. Munn, Travis Fire and Emergency Services, Travis AFB, California
Philip J. Oakes, Laramie County Fire District #6, Burns, Wyoming
Marybeth O'Leary, Vision Fire Training, Snohomish, Washington
Gaudenz Panholzer, San Jose Fire Department, San Jose, California
Rick Paulsen, Missoula Rural Fire District Training Division, Missoula, Montana
Louie Robinson, South Charleston Fire Department, South Charleston, West Virginia
Bruce Roed, Minnesota State Colleges and Universities, Mentor, Minnesota
Dennis C. Rosolen, New Hampshire Fire Academy, Concord, New Hampshire
Christopher Rosseau, New Hampshire Division of Fire Standards and Training and Emergency Medical Services, Concord, New Hampshire
Steve Schreck, Alaska Division of Fire and Life Safety, Anchorage, Alaska
Jerry L. Schroeder, Idaho Emergency Services Training, Boise, Idaho
Timothy Sendelbach, International Society of Fire Service Instructors, Pleasant View, Tennessee
W. Douglas Whittaker, Onondaga Community College, Syracuse, New York
Gary D. Young, Copperas Cove Fire Department/EMS, Copperas Cove, Texas

## Photographic Contributors

We would like to extend a huge "thank you" to Glen E. Ellman, who was the photographer for this project. Glen is a commercial photographer and fire fighter based in Fort Worth, Texas. His expertise and professionalism are unmatched!

Thank you Frank Becerra, from the Fort Worth Fire Department, for his technical assistance and coordination during the photo shoots.

We would also like to thank the following for opening up their facility for the photo shoot:

Tommy Abercrombie, Administrative Coordinator
Fire Service Training Center and Homeland Security Training Center
Tarrant County College, Northwest Campus
Fort Worth, Texas

PART

# I

# Introduction

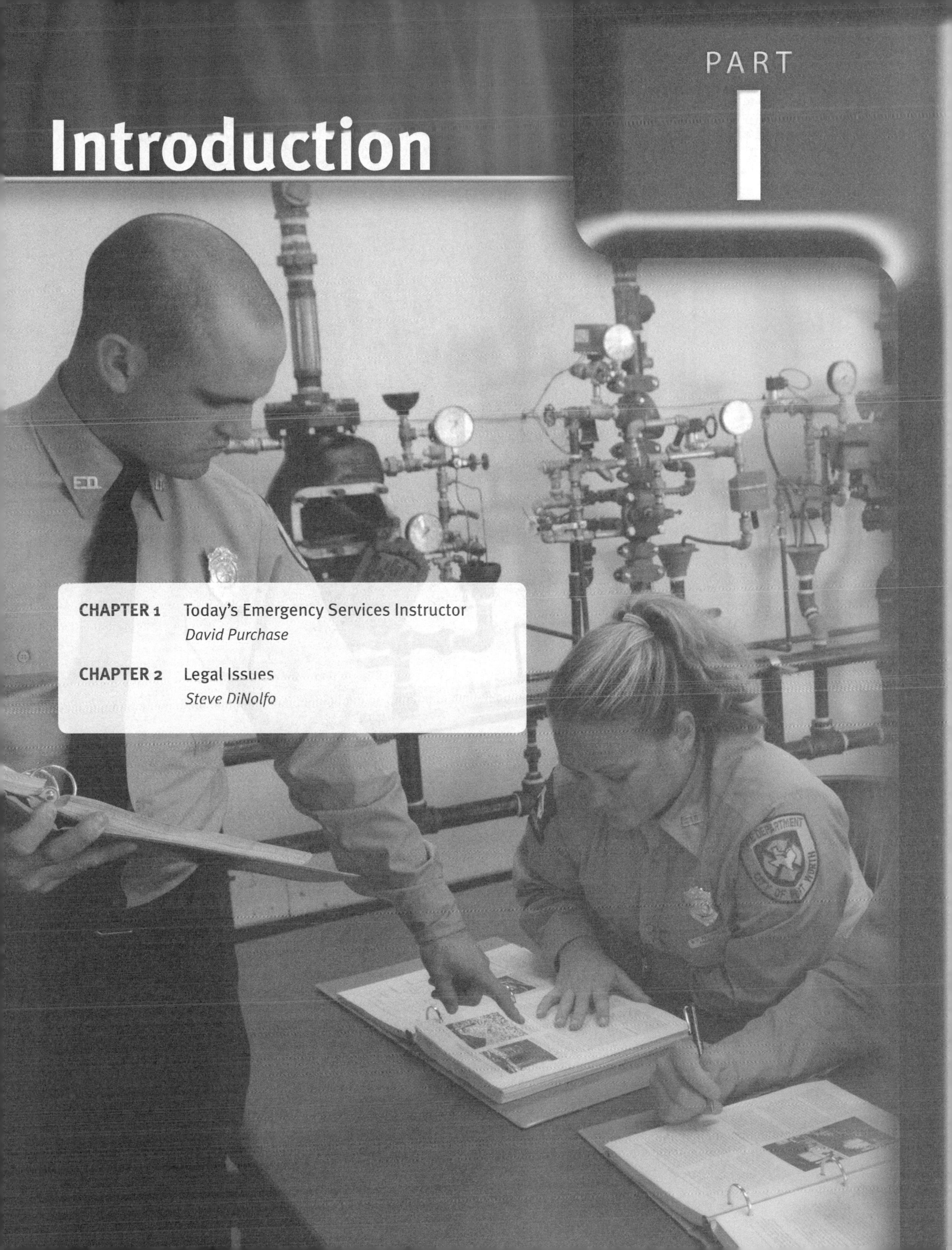

# Today's Emergency Services Instructor

# CHAPTER 1

## NFPA 1041 Standard

**Instructor I**

**4.1** **General.**

**4.1.1** The Fire Service Instructor I shall meet the JPRs defined in Sections 4.2 through 4.5 of this standard. [p. 4–19]

**4.2** **Program Management.**

**4.2.1** **Definition of Duty.** The management of basic resources and the records and reports essential to the instructional process. [p. 4–19]

**4.2.3** Prepare training records and report forms, given policies and procedures and forms, so that required reports are accurately completed and submitted in accordance with the procedures. [p. 16–17]

**(A) Requisite Knowledge.** Types of records and reports required, and policies and procedures for processing records and reports.

**(B) Requisite Skills.** Basic report writing and record completion.

**Instructor II**

**5.1** **General.** The Fire Service Instructor II shall meet the requirements for the Fire Service Instructor I and the JPRs defined in Sections 5.2 through 5.5 of this standard. [p. 4–19]

## Knowledge Objectives

After studying this chapter, you will be able to:

- Define the roles and responsibilities of the Fire Service Instructor I.
- Define the roles and responsibilities of the Fire Service Instructor II.
- Identify and explain five roles of the fire service instructor.
- Identify physical elements of the classroom.
- Cite the importance of visioning and succession planning for the instructor.
- Identify instructor credentials and qualifications.
- Identify four issues of ethics for the fire service instructor.
- Identify three ways to assist the instructor in managing multiple priorities.

## Skills Objectives

After studying this chapter, you will be able to:

- Demonstrate the ability to manage the five roles of the fire service instructor.
- Demonstrate ethical behavior in the classroom.
- Demonstrate the ability to manage multiple priorities as a fire service instructor.

# You Are the Fire Service Instructor

Today marks your first day on the job as the department's training officer. Your position as a fire service instructor is one that you always dreamed of. As you sit behind your desk, you reflect on how you came to this position and think about whether you're prepared to stand in front of a classroom of students just one week from today. You ask yourself three questions:

1. What authority do I have as an instructor, and what responsibility comes with that authority?
2. What direction should my training program take?
3. Now that I'm here, how do I become the best instructor I can be?

## Introduction

Remember the first day you began your career in the fire service? Somewhere in your memory is probably a great fire service instructor. He or she just may be the person who introduced you to the greatest career in America. Fire service instructors are the guardians of knowledge, skills, and ability in the fire service; their knowledge is wrapped up in a long tradition that has served many generations of fire fighters well. As protectors of that tradition, instructors look to and rely upon innovation and creativity while constantly striving to perfect training.

In years past, firefighting training was often left to on-the-job education. New recruits would be handed coats, boots, helmets, and gloves and told to jump on the back of the rig as it rushed to the scene of the emergency. In those days, the fire ground was the recruit's training ground—and the title of "fire service instructor" may have been handed out too generously. Formal instructors and, in some cases, formal instruction were left to the larger departments that had the budgets to hire specialized personnel and build training facilities.

Obviously, times have changed. The fire service is now being pushed to the limit by communities that expect more out of the types of services provided. Retired Phoenix Fire Chief Alan Brunacini often referred to today's recipient of that service as "Mrs. Smith." As a pioneer of customer service, Chief Brunacini knew the value of the instructor in preparing the troops to meet Mrs. Smith's demands. Today, fire departments across the United States rely upon their instructors both to train new recruits and to maintain the skill levels of existing fire fighters.

What does it take to join this elite group of fire fighters on whose shoulders rest the success and safety of emergency operations? While excellent fire service instructors possess many attributes, a few important qualities come to mind and rise to the top of the list. First, there is desire—a desire to be of assistance to those in need. In this case, we are not referring to those persons who require our assistance during times of trouble, but rather to those individuals who would require the knowledge needed to assist those in trouble. These fire fighters must be trained so that they will gain the knowledge, skills, and abilities necessary to serve. The fire service instructor gives freely without hesitation in the pursuit of excellence FIGURE 1.1.

FIGURE 1.1 The fire service instructor gives freely without hesitation in the pursuit of excellence.

The instructor should certainly have experience in the subject matter taught. Students have more confidence in fire service instructors who have demonstrated competence in the subject matter through experience. Experience opens the door for the fire service instructor to reach the minds of the students. Good fire service instructors also understand that they cannot rely on past experience alone: They must remain ever vigilant to their changing environment and gather the newest information, technology, and skills to remain out front and stay current.

Good fire service instructors also demonstrate flexibility: They are able to work in a variety of environments while offering instruction in a variety of topics to a class full of students with a variety of talents. Instructing the adult learner can be

**Theory into Practice**

The relationship between training and education is often confusing. Education is the process of imparting knowledge or skill through systematic instruction. Education programs are conducted through academic institutions and are primarily directed toward an individual's comprehension of the subject matter. Training is directed toward the practical application of education to produce an action, which can be an individual or a group activity. There is an important distinction between these two types of learning.

Within the fire service, training has been considered as essential for many years. The emphasis on fire fighter education is a much more recent development. The First Wingspread Conference on Fire Service Administration, Education and Research was sponsored by the Johnson Foundation and held in Racine, Wisconsin in 1966. This conference brought together a group of leaders from the fire service to identify needs and priorities. They agreed that a broad knowledge base was needed and that an educational program was necessary to deliver that knowledge base. This became the blueprint for the development of community college fire science and fire administration programs as well as the degrees at a distance program. The transition from training to education had begun.

In 1998, the U.S. Fire Administration hosted the first Fire and Emergency Services Higher Education (FESHE) conference. That conference produced a document "The Fire Service and Education: A Blueprint for the 21st Century," that initiated a national effort to address and update the academic needs of the fire service. Participants at the 2000 FESHE conference began work to develop a model fire science curriculum that would go from community college through graduate school. At the 2002 conference, U.S. Fire Administration Education Specialist Edward Kaplan compared the results of the FESHE effort with the Wingspread higher education curriculum. The FESHE work affirmed the soundness of the original Wingspread model, with information technology the only new knowledge item added to the curriculum.

challenging, and the fire service instructor must be willing to alter the approaches to education used to reach all students. In some cases, the fire department's schedule provides the greatest challenge, as instruction time is often interrupted by responses. In other fire departments, it is the fire fighters' personal schedules that present the challenge, as training time competes with family time and full-time work schedules.

Motivation is the key to bringing excitement to the training environment. Nothing breeds fire fighter motivation like a motivated instructor. The right kind of motivation leads to the attitude, "I can't wait to teach," and, just as important, "I can't wait to learn." Motivation is contagious; it can spread from the classroom to the station floor, and ultimately it can drive the quality of the service provided to the community. Motivated fire service instructors bring creativity and ingenuity to the classroom as a means to create excitement in the learning process.

**Safety Tip**

Do not attempt to teach a subject without being competent and having knowledge and experience of the subject. Failure to do so can result in injury not only during the course, but also on the fire ground, where improperly trained fire fighters may become injured.

Are you up for the challenge? Can you accept the responsibility? Can you be a steward of tradition while remaining open to teaching new ideas? Can you strive to maintain desire, experience, flexibility, and motivation as you pursue operational excellence? If your answers to these questions are all "yes," then you are on your way to joining the ranks of the greatest profession. Ideally, your curriculum will serve as the launching pad for fire fighters and serve them well as a critical reference for years to come.

This book provides information to meet the standards outlined in National Fire Protection Association (NFPA) 1041, *Fire Service Instructor Professional Qualifications* at the Instructor I and II Levels. The professional qualifications standards for instructors are documented in NFPA 1041, which defines three levels of fire instructors. These definitions were the result of a task analysis intended to validate these levels and to create specific requirements that would apply to each level. The three levels are identified as instructor qualifications and, in many states, may become certifications at these levels with candidate prerequisites to be completed before a certification is granted.

## Levels of Fire Service Instructors

According to NFPA 1041, the separation between the duties that are performed by specific instructors are broken down into three distinct levels. These three classifications build on one another and progressively give the fire service instructor additional duties and responsibility.

Instructor I is defined as follows:

> A fire service instructor who has demonstrated the knowledge and ability to deliver instruction effectively from a prepared lesson plan, including instructional aids and evaluation instruments; adapt lesson plans to the unique requirements of the students and authority having jurisdiction; organize the learning environment so that learning is maximized; and meet the record-keeping requirements of authority having jurisdiction. (3.3.2.1)

Stated in the most basic terms, the Instructor I delivers instruction from prepared materials at the direction, and often under the supervision, of an Instructor II or higher. Emphasis of this level of instructor is on the ability to communicate effectively and use various methods of instruction, including hands-on training and lecture.

The Instructor II is a fire service instructor who, in addition to meeting Instructor I qualifications, satisfies the following criteria:

> Has demonstrated the knowledge and ability to develop individual lesson plans for a specific topic including learning objectives, instructional aids, and evaluation instruments; schedule training sessions based on overall training plan of authority having jurisdiction; and supervise and coordinate the activities of other instructors. (3.3.2.2)

The Instructor II functions at a higher level of authority and responsibility than the Instructor I; he or she is responsible for all duty areas of the Instructor I, plus is able to create the training materials. In the purest sense, the Instructor II will create the training materials for distribution to the Instructor I to present to the students. In reality, both tasks may be completed by the same person.

An Instructor III is defined as follows:

> A fire service instructor who, in addition to meeting Instructor II qualifications, has demonstrated the knowledge and ability to develop comprehensive training curricula and programs for use by single or multiple organizations; conduct organization needs analysis; and develop training goals and implementation strategies. (3.3.2.3)

Not all fire departments find it necessary to develop a fire service instructor to this level, which is almost equivalent to being a specialist. The Instructor III typically works as an overall training program manager and oversees the entire spectrum of a comprehensive training program. This text does not cover the JPRs required for Instructor III classification because the content of that level relates to a different level of performance and is outside of the scope of this book.

## Roles and Responsibilities of an Instructor

Throughout our lives, we are asked to conform to someone else's idea as to how we should act or behave. Child, adolescent, adult, parent, spouse, employee, supervisor, owner, citizen—all are examples of roles that we fill while negotiating through life. With each of these roles, expectations direct our actions and allow us to evaluate our success or failure at that role.

The fire service mirrors life in that it also contains various roles, each with its own expectations and responsibilities. At every step along the path from recruit to fire chief, we strive to meet these expectations and responsibilities. The fire service instructor is one of those important roles in the fire service that requires dedicated and competent individuals who can positively influence the entire fire department.

The roles and responsibilities of a fire instructor vary greatly by fire department according to the size, make-up, and delivery system used. Understanding the roles and responsibilities for Fire Service Instructors I and II is essential for success in these key positions.

### Roles and Responsibilities of the Fire Service Instructor I

The roles and responsibilities of the Fire Service Instructor I include the following:

Manage the basic resources and the records and reports essential to the instructional process.

- Assemble course materials.
- Prepare training records and report forms.

Review and adapt prepared instructional materials.

- Deliver instructional sessions using prepared course materials.
- Organize the classroom, laboratory, or outdoor learning environments.
- Use instructional media and materials to present prepared lesson plans.
- Adjust presentations to students' different learning styles, abilities, and behaviors.
- Operate and utilize audiovisual equipment and demonstration devices.

Administer and grade student evaluation instruments.

- Deliver oral, written, or performance tests.
- Grade students' oral, written, or performance tests.
- Report test results.
- Provide examination feedback to students.

### Roles and Responsibilities of the Fire Service Instructor II

The Instructor II must meet and perform all of the duties of the Instructor I. In addition, the Fire Service Instructor II is responsible for performing the following tasks:

Manage instructional resources, staff, facilities, and records and reports.

- Schedule instructional sessions.
- Formulate budget needs.
- Acquire training resources.
- Coordinate training recordkeeping.
- Evaluate instructors.

Develop instruction materials for specific topics.

- Create lesson plans.
- Modify existing lesson plans.

Conduct classes using a lesson plan.

- Use multiple teaching methods and techniques to present a lesson plan that the instructor has prepared.
- Supervise other instructors and students during training.

Develop student evaluation instruments to support instruction and evaluation of test results.

- Develop student evaluation instruments.
- Develop a class evaluation instrument.
- Analyze student evaluation instruments.

## Applying the JOB PERFORMANCE REQUIREMENTS (JPRs)

As a fire service instructor, you are charged with a tremendous responsibility: training and educating personnel on the diverse aspects of the job that ultimately can affect their safety and survival. That is a lot to ask of one person. In many cases, an entire training team is charged with carrying out this mission. Instructors function in many capacities and at different levels of professional qualifications. Understanding the relationships between the levels of instructor qualifications is an essential task in the application of job skills at all levels. It requires understanding both the process of instruction and the process of the delivery of instruction.

### Instructor I

The relationship between an Instructor I and an Instructor II is based on an understanding of the role each person plays in the overall delivery process. The Instructor I must demonstrate an ongoing desire to improve his or her instructional skills to enhance the instructor's ability to get the training message across.

### JPRs at Work

It is important to note the importance of staying up-to-date on the current events and issues facing instructors today.

### Instructor II

The Instructor II will often develop the instructional materials by developing lesson plans, class content, and evaluation materials. Staying on top of the latest changes in instructional delivery and current subject matter allow such instructors to maximize their ability to train fire fighters.

### JPRs at Work

It is important to note the importance of staying up-to-date on the current events and issues facing instructors today.

### Bridging the Gap Between Instructor I and Instructor II

Both the Instructor I and the Instructor II must collaborate on challenges they face in the classroom and on the drill ground, making sure that they account for any successes, failures, obstacles, and new techniques that allow for better delivery of training. Consider scheduling frequent planning meetings between training members or creating an e-mail/bulletin board system to facilitate good communication flow between positions.

### Instant Applications

1. Identify the changes in training you have seen or experienced since you began your career in the fire service.
2. Check industry Web sites and reference materials to identify the hot topics facing fire service instructors today.
3. Identify a new piece of equipment purchased by your department and review how the initial and ongoing training for that equipment was accomplished.
4. Ask the newest or youngest member of your organization about his or her training. Then identify the similarities and differences between current fire service training and the traditional education of a fire fighter.

The Instructor I delivers instruction from prepared material, while the Instructor II develops course materials.

## Where Do I Fit in?

Much has been written about the roles and responsibilities of positions within various types of organizations, and the fire service is no different. Fire chiefs have their role, rooted in visions of leadership, as the commander-in-chief of the organization. They have a defined responsibility that not only allows them to assume this role, but also gives them the authority to shape their own destiny.

Within the fire service's remaining command structure lie the middle management positions of supervision, including front-line lieutenants, captains, and various chief officers. These ranks have had their roles defined through a history of developed policies and job descriptions created by necessity, the chief's vision, and, yes, tradition. These positions are often referred to as middle management, as they are created and caught in the middle of the chain of command. They are pushed and tested by fire fighters at the bottom and by fire chiefs at the top.

As tough as these positions are, there is one job that still finds itself in a somewhat more precarious spot: the fire service instructor. All too often, a fire service instructor may ask, "Where do I fit in?" Is the fire service instructor a line officer, a training officer, a fire fighter, or someone with a specific specialty? In some fire departments, line officers are assigned to the training division as part of their duties. In volunteer organizations, the training officer often volunteers for the job; he or she may also be assigned to it. Regardless of how you come to be assigned to training, you may feel challenged if you are not given the proper authority to carry out that job.

In examining most fire departments' **organizational charts**, you will find positions identified for operational activities, clearly outlining the chain of command and the corresponding responsibilities for firefighting personnel and their fire officer counterparts. Fire service instructors, if shown at all, are usually found off to the side on the organizational chart, preferably within a training division. This disconnect can often create a sense of isolation from the fire fighters who fire service instructors are expected to train. It may leave the fire service instructor with a void, feeling unfulfilled in terms of a management role that demonstrates the fire department's lack of respect for the role FIGURE 1.2. Fire fighters and even some fire officers are often seen as resisting or even obstructing the training process, because all too often those within the operations staff do not view training as having a high priority.

As a fire service instructor, you must remain focused, upbeat, positive, proactive, and true to your role FIGURE 1.3. You must remind yourself that you hold the power to shape the future fire service. Every fire officer—from the fire chief to lowest line supervisor—along with all the fire fighters, has had a fire service instructor affect their careers. That fact demonstrates the importance of your position within the fire department.

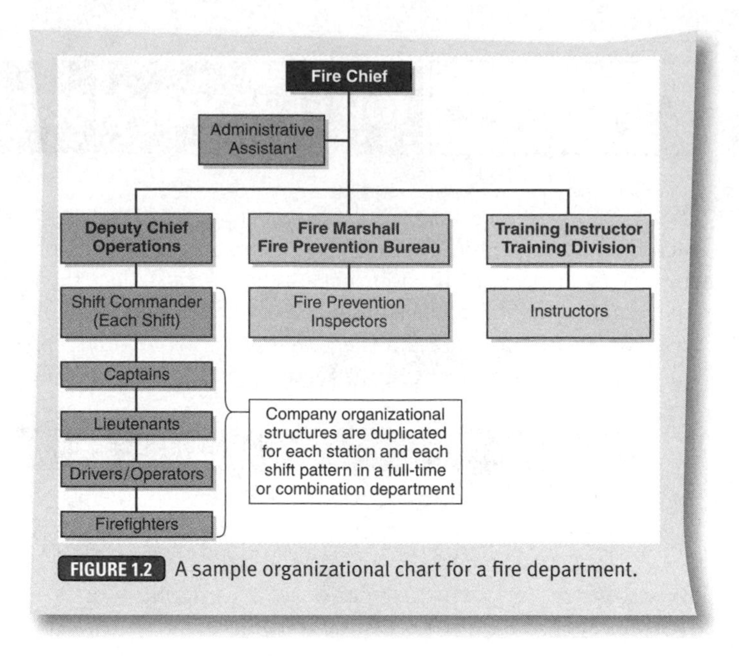


FIGURE 1.2 A sample organizational chart for a fire department.

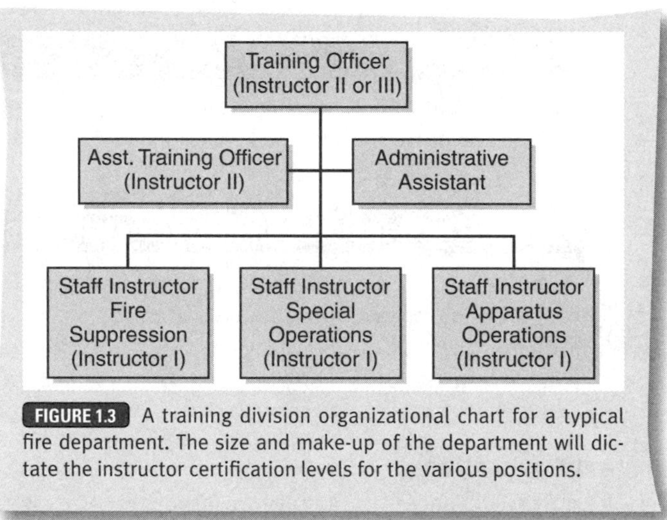


FIGURE 1.3 A training division organizational chart for a typical fire department. The size and make-up of the department will dictate the instructor certification levels for the various positions.

## Managing the Fire Service Instructor's Role

As a fire instructor, you must look beyond charts and titles and instead focus on those important, yet sometimes invisible, roles that produce lasting contributions to overall fire department organizational health and success. Today's fire service instructor is asked to fulfill these roles. These roles are not unique to the fire service; indeed, examples of each can be found in many different professions. Each role is important and will help you to develop the talents found within the organization. Understanding each of these roles can assist you in creating and building your own tradition within the fire department.

Fire instructors must learn to manage the following roles FIGURE 1.4:

- Leader
- Mentor
- Coach
- Evaluator
- Teacher

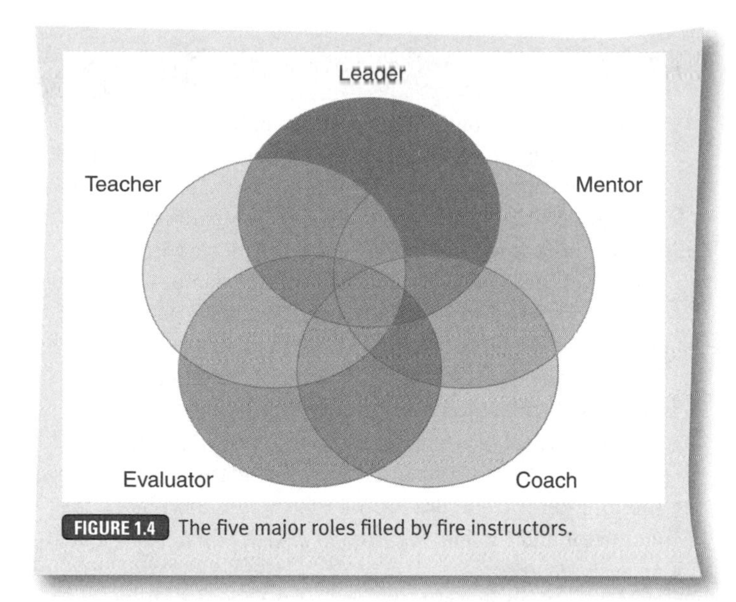


FIGURE 1.4 The five major roles filled by fire instructors.

### The Fire Service Instructor as a Leader

Fire service instructors are often asked to lead fire departments into the future by preparing fire fighters for new missions. Fire service instructors set the example for all fire fighters to follow in terms of performance excellence. As such, they must remain true to the direction set by upper management and supportive of the mission as defined by the fire chief.

Your place in front of the classroom places you in a visible and sometimes vulnerable position where you may have to endure the anxieties of an impatient rank and file. Do not waver from the training mission by letting the instructional environment be turned into a debate on issues of management within the fire department. The most powerful leadership tool is the one you have the most control of: "Lead by example."

#### Teaching Tip

Leading others is a major responsibility. Part of that responsibility is to lead in an ethical manner. Belittling a new recruit over a mistake may cause the recruit to hide future mistakes, which places both the recruit and others at risk.

### The Fire Service Instructor as a Mentor

Good leaders are also good at identifying future talent. Fire service instructors are in the best position to observe firsthand both the raw talent of recruits and the ongoing growth of personnel. They can support and enhance the careers of fire fighters by identifying their talents and mentoring them appropriately. Fire service instructors with mentorship abilities enhance the fire department's survival and success by evaluating and developing the talent pool. For example, established fire service instructors assist future fire service instructors by showing confidence in their abilities and recommending additional training and opportunities. In short, good fire service instructors mentor future good fire service instructors.

#### Theory into Practice

As a fire service instructor, you must have the respect of the trainees. Respect may come from a title or position, but it can also come from the individual, based on his or her expertise and knowledge. If you find yourself charged with the task of teaching those of higher rank, be prepared to work for their respect.

Earning someone's respect can be done in a number of ways. Do not wait to begin building credibility until you are asked to teach—be a professional at your job from the first day you join the fire department. Read books, trade magazines, and other material so that you will become well versed in the trade. Take additional courses and training. Get out and practice what you know: Crawl under the engine to see how the pump works; organize the medical kit. Demonstrate that you are an avid learner and practitioner of your craft. Other fire fighters will notice and respect your thirst for knowledge, which builds your credibility when you are teaching any topic.

Credibility is a direct result of your knowledge of the subject. Rank does not always equate with knowledge—but hard work does. Go over the material so that you can determine the flow of the course. Read the activities carefully so you have the right materials on hand and you understand the intended result. Write notes in the margins of the instructor's guide about comments you want to bring up and examples of the practical application of the material. An ill-prepared instructor is doomed.

In addition to your before-class preparation, be ready to build credibility in the classroom. Respect the knowledge that the students possess. Take advantage of those individuals with specialized knowledge to add value to the course. Don't allow yourself to be chastised. Be confident in your skills—obviously whoever asked you to teach believed in you. Take control of the class and be a professional. In the long run, you will be respected for your teaching skills if you have the credibility to support it.

Building respect and credibility with students is essential for those who do not have a formal position in the fire department or when teaching those with greater rank. Although it can be more challenging, success can be achieved with proper preparation and implementation.

### The Fire Service Instructor as a Coach

In their book *A Passion for Excellence*, Tom Peters and Nancy Austin define coaching as follows:

> Face-to-face leadership that pulls together people with diverse backgrounds, talents, experiences and interest, encourages them to step up to responsibility and continued achievement, and treats them as full-scale partners and contributors.

In the sports world, the coach receives the majority of attention and pressure, because he or she is cast in the role of determining the success or failure of the team. In the fire

FIGURE 1.5 The fire service instructor builds team skills through practice and repetition on the drill ground.

service, we often stress the need for teamwork in all aspects of the job. Teamwork is an important ingredient on the fire ground as the incident commander orchestrates the company through various tactical assignments. Even in the fire station, much of the nonemergency activity is accomplished not by an individual, but rather by a dedicated team of professionals. Whether on the fire ground or in the station, the incident commander acts much like the quarterback of a football team, calling out the formations and plays. The instructor plays a role analogous to that of the football coaching staff: Whereas the coaching staff works on drills to prepare the players for the next game, the instructor works to prepare the fire fighters for the next incident response.

The fire service instructor's job within the fire service corresponds to that of a coach in the sports world in many ways. It is the fire service instructor who must prepare the interior team for battle. The fire service instructor studies the team's opposition (the fire) and prepares the team for what they may face on game day. He or she builds team skills through practice and repetition on the drill ground FIGURE 1.5. The fire service instructor, much like a coach, must enthusiastically encourage fire fighters as they struggle with new ideas and techniques. Finally, the instructor must remember to remain positive even in tough times and to support the direction in which the team is headed as determined by the general manager (fire chief) and the owners (residents). Coaches and instructors lead, train, and drive their teams' performance.

## Ethics Tip

Have you ever gone through a class, only to find out that there is a written exam at the end? The mere thought of this kind of test may make you nervous. Too often, some instructors go over the exam questions just prior to the exam to reduce this apprehension. But is this practice ethical?

Many new fire service instructors believe that they need to be liked to be effective instructors. This mindset leads them to go preview answers to written exams or intentionally overlook mistakes during practical evaluations. A certificate is designed to show that the student has demonstrated a level of knowledge. Giving a meaningless certificate is detrimental to the student, undermines the long-term credibility of the instructor, and places the instructor and the fire department at risk of liability should an accident occur.

Be upfront at the beginning of the course about your expectations for your students, and remain firm in that stance. Examinations encourage students to be active participants rather than just passive observers. Students take great pride in the certificates that they actually earn.

## The Fire Service Instructor as an Evaluator

Who is in the best position to evaluate the capabilities of a fire department? The answer is found standing in the front of the classroom: the fire service instructor. As obvious as that answer is, how many fire chiefs have actually taken advantage of the knowledge fire instructors have assembled through their daily interactions with the troops? Fire service instructors must evaluate students' learning and competencies at many points during the training process. To do so, instructors must sharpen their skills at evaluating the proficiency and knowledge retained by their students. The ability to accurately assess ability is important if the fire service instructor is to be able to determine the pace of the training program. Determining the right time to introduce new techniques or to review previously taught topics cannot be done without the ability to honestly evaluate students. Proper evaluations done in the course of training save the fire department from the heartache of critical evaluations performed by the public at the scene of a mistake.

Fire service instructors are in a unique position to evaluate the capabilities and limitations of fire fighters. In many fire departments, input from the training division and fire service instructors is regularly sought by fire department leadership to help them develop policies and procedures. The fire service instructor can evaluate fire fighters, suggest operational directives, and field-test new standard operating procedures (SOPs).

Fire service instructors should also be evaluators of the classroom. That is, they must always be able to evaluate their effectiveness, recognizing what the students need and what the students are actually learning. They must evaluate themselves as well, looking for ways to improve their own teaching techniques. Classroom and self-evaluations are essential in keeping the training program proactive and on the road to continuous improvement.

### The Fire Service Instructor as a Teacher

As simple as it sounds, the fire service instructor must be a teacher. What comes to mind as you hear the word "teacher"? Perhaps it is someone who you remember from your elementary school days. Why does that name come to mind? Teachers affect students in many ways—many times positively, but unfortunately sometimes negatively. Fire service instructors must teach students new skills and abilities, thereby shaping the abilities of the team. They must remember the potent influence they have with their students. It's not just what the students learn that is important; it's also the shaping of a fire fighter's demeanor, attitudes, and desires that ultimately determines how the instructor will be remembered.

Being an effective teacher should be the goal of each fire service instructor. Mastering the interpersonal side of the classroom goes a long way toward reaching that goal. The best and most remembered teachers shape and mold the total person.

## Setting Up the Learning Environment

Every fire service instructor should understand the effect that the **learning environment** has on the ability of students to grasp the material being presented. You need to take command of that environment and use it to your advantage. The learning environment includes those environmental factors that influence the learning process and can include multiple elements, both physical (i.e., lighting, temperature, setup) and emotional (i.e., attitudes, comments, learning abilities).

Physically, the classroom and drill ground must provide a safe, comfortable, and distraction-free environment. Emotionally, students' minds must be kept focused on the instructional material in front of them and away from the sometimes-contentious issues found in many fire departments. To make matters even more challenging, some of the factors affecting the learning environment are outside of the fire service instructor's control. If you are not careful, you may bring some frustrations into the classroom, thereby impeding the learning process. To avoid these mistakes, familiarize yourself with the elements that affect the learning environment.

### Physical Elements Affecting the Learning Environment

Imagine yourself as you walk into a classroom. What do you first notice? Lighting, cleanliness, temperature, or perhaps how far from the front of the room you can sit. Every student who has ever taken a class has evaluated the learning environment. The simple arrangement of tables and chairs can affect how we interact and ultimately learn. The best arrangements allow for the free exchange of information, both from student to student and from student to instructor.

Traditional school classroom setups can lead to the formation of groups in distinct areas of the room, as students tend to sit next to those with whom they are most comfortable. Both the quiet and the uninterested students vie for the seats farthest from the front, not wanting to become involved in the learning discussion. Arrangements such as squares or U-shaped tables, which allow face-to-face interactions, can sometimes neutralize the hierarchy of traditional settings **FIGURE 1.6**. These arrangements also allow you to move among the students freely, improving the exchange of information and increasing the attentiveness of students.

**FIGURE 1.6** Arrangements such as squares or U-shaped tables, which allow face-to-face interactions, can sometimes neutralize the hierarchy of traditional settings.

In setting up the classroom, you must also take into account both the natural light and the installed lighting. Lighting affects many aspects of the learning environment. If projectors are used in the presentation, lighting becomes critical to the ability of the students to view the projected information. If natural light is a problem, a simple rearrangement of the room may allow the movement of a screen to a better location. If that is not possible, perhaps the addition of window blinds can correct the problem.

You must also evaluate the installed lighting in the room. Improperly installed lights may make it difficult to dim the lights enough to ensure quality projection. Conversely, dimming the lights too much may create reading or note-taking difficulties for students. If installed lighting is causing problems, your supervisor might be able to recommend improvements to the system through the annual department budget process.

Room temperature can also create problems for the learning environment. Take the entire class into account when setting the classroom temperature. Finding a comfortable

**Teaching Tip**

Conduct a review of your classroom, drill-ground facilities, and equipment. Submit your results to department administration in the form of a recommended capital improvement plan so that it can be considered in the budget planning process.

compromise for all may turn you into a negotiator. Of course, some classrooms have environmental controls that cannot be changed by those using the facility. In those cases, you may have to alter lecture and break times if temperatures are uncomfortable. You may also have to become a student advocate and seek improvements to the environmental controls.

### Emotional Elements Affecting the Learning Environment

You must also protect the individual student from the emotional letdown that comes with the inability to learn a new task or subject. The fire service training environment includes both rookies and senior personnel, adult learners and students fresh out of school. It contains students who know what to do and those who think they know what to do. Fast learners and those requiring more individual efforts will challenge even the most seasoned instructors.

The best fire service instructors can present material effectively for all types of learners. Instructors who protect those who struggle by placing them in positions where they can make progress without the embarrassment of failure in front of the greater group will be successful. Good fire service instructors learn how to use the more talented department members to coach and teach those with lesser skills.

Learning to work within the environment challenges you to be flexible, loyal, confident, and fair. Building a bond of trust with your students will enable you to present new ideas and make needed revisions to old traditions. As the fire service instructor, you are the visionary of the fire service.

## Staying Ahead of the Curve: Vision

Have you ever thought about what drives an organization? Why do some survive and some struggle as time forces change?

In examining the structure of a typical fire department, it is easy to point out the formal leader: It is the individual at the top of the pyramid, the chief who provides the formal direction for the troops to follow. The fire chief is responsible for the ultimate success or failure of the fire department. Of course, organizational charts are filled with many other positions as well, each of which has its own responsibility for providing direction in support of organizational objectives. It is within these ranks that you find assistant, deputy, battalion, and division chiefs; line officer positions including captains and lieutenants; fire marshals; and inspectors.

Where does the fire service instructor fall within this scheme? In many fire departments, the fire service instructor is treated as an operational support assignment more than an official rank or position. The fire service instructor position is often viewed from one of two extremes: as vitally important to the overall operation or as unnecessary by the higher-ups. Wise leaders understand that fire service instructors are important members of the team. They use their instructors to maintain their department's state of readiness and prepare their fire fighters for future missions.

It is this role—preparation for future activities—that requires you to keep ahead of the curve. To fulfill it, you must become a visionary force within the fire department. It is **vision** that drives fire service instructors to keep abreast of the ever-changing world of firefighting. Changes and advancements in firefighting tactics must be reviewed and implemented through revisions in training curriculum. If the fire department desires a change in operations, the fire service instructor and training division will be charged with educating and training fire fighters to make that change. New techniques and advancements must be tested and made applicable to each organization, because many changes are not "one size fits all" measures. You will be challenged as you introduce new ideas that seem to conflict with established traditions.

As the mission of the fire service evolves, you must be able to prepare the troops to carry out the new mission. The history of the fire service is full of pertinent examples, as the fire service instructor has had to evolve so as to provide training for medical, hazardous materials, and technical rescue services. Today the threat of terrorism and the use of weapons of mass destruction (WMDs) presents its own unique challenges for the fire service, with fire service instructors once again being called upon to lead the troops into these new areas of service.

Fire service instructors must monitor the ever-changing learning environment as well. Struggles with budgets and staffing present ongoing problems for fire service instructors, who must continually seek to keep training at the forefront of nonemergency operations. You must find new and creative ways to reach students and present training. Today visionary fire service instructors are turning to the cyberworld as online training programs gain a foothold in the instructional world. Technology will continue to advance and, consequently, affect the delivery of training. Today fire service instructors with the vision to see how this new technology can be used for training purposes are establishing virtual classrooms and using satellite and video classrooms as effective learning media.

Given the unique challenges apparent in the modern-day fire service, the fire chief would be wise to select the very best for the position of fire service instructor. A fire service instructor with vision can greatly assist the fire chief in meeting the fire department's future challenges.

## The Fire Service Instructor's Role in Succession Planning

Preparing fire fighters for battle is not the only job of the fire service instructor. For any organization to survive and grow, a continuity plan must be in place. Continuity of the organization provides security for the community that the organization serves. The fire service instructor can assist in that regard by becoming involved in **succession planning**.

As trainers of fire fighters, fire service instructors are often in the best position to recognize potential talent and leadership qualities. The fire service instructor's role is to nurture and challenge these fire fighters through the training program. During the course of training, the instructor may use some fire fighters to assist in training other members. By placing fire fighters in leadership positions within the training environment, you allow them to refine, enhance, and demonstrate leadership qualities.

The fire service instructor may also be in the position of providing input to ranking officers on the performance of fire fighters in training. By recommending high achievers to fire department administrators, you assist in the identification of future fire officers and fire service instructors. You may also be in the best position to identify those fire fighters who do not fit the traditional mold of fire officer that the fire department has established.

**Teaching Tip**

You can increase your visibility within the fire department by volunteering to assist in hiring and promotional processes. Sell your fire chief on the knowledge you have of each individual fire fighter. You can share the training performance of fire fighters vying for promotion with the fire chief.

The fire service instructor walks a fine line between operations and training. In organizations with a weak operations leadership, the instructor may become absorbed with setting the operational direction and standards simply because he or she has expertise in specific areas. In other cases, the fire service instructor has a formal role in establishing policy. Of course, you should be careful to ensure that those policies exist to support the direction of the operations division—not the other way around. While you may not always agree with the standards chosen by the operations chief, you must accept that direction and train fire fighters accordingly.

**Teaching Tip**

A degree is awarded by an institution after a person has completed acquisition of the required knowledge in a particular field. A certificate is given after a person attends a learning event that has no testing requirement. A certification is awarded after someone has required knowledge and experience and has passed an examination process that is based on a set standard.

## Fire Service Instructor Credentials and Qualifications

You've heard it before: "Walk the talk." It's a phrase that can certainly be applied to the fire service instructor. Your best friend is the confidence that your students have in your abilities and knowledge. This is a quality that cannot be learned from a book. Often, fire service instructors are born in the classroom from energetic students. Others vow to develop their instruction skills after attending a highly charged training session or a lecture delivered by an impassioned instructor. As a fire service instructor, you must be aware that your success or failure might rest with the degree of effort and preparation you put forth at the beginning of your career in the fire service. It is here that dividends are paid.

## Laying the Groundwork

Good fire service instructors are born from good students—students who thrive on the knowledge gained through training, and those who actively participate in training activities and are not afraid to learn from their initial failures. These individuals are the fire fighters who understand the value of education and refinement of skills. They continually place themselves in learning situations, looking to upgrade their skills and knowledge.

Think for a moment about the fire service instructors who have influenced you. Students attending any training program immediately focus their attention upon the instructor as they begin looking for clues about the quality of the program. They might ask what experience the instructor has in the subject area. What is the instructor's firefighting experience level? If the training is being held in-house, your students might remember the days when you were a student. Your credibility in the classroom depends on your past behavior. The past always has a way of finding the future, so it is always wise to protect your future by engaging in proper behavior in the present. Laying the groundwork for becoming a good fire service instructor begins the day you join the profession. While some mistakes will be made, you must always guard your credibility and integrity.

**Teaching Tip**

If you are interested in becoming a fire service instructor for your fire department, talk to current instructors and offer to assist them in their classrooms. Better yet, develop your own training outline for a subject that you are familiar with, submit it to the training officer for review, and ask permission to teach it.

## Meeting Standards

**Standards** dictate many things within the fire service. Training programs are not immune from national standards. Often fire service instructors are judged by their attainment of standards. Standards set the bar for fire service instructors and are intended to maintain a high level of proficiency and knowledge. They seek to establish uniformity for fire service instructors across the entire profession and across the country. NFPA 1041

# Voices of Experience

While I was deeply involved in conversation with several fire fighters from the A shift, the fire chief invited me into his office. He shut the door, then said he had a significant concern to discuss with me. I thought I would be looking for new employment within the week.

The chief said he had just received a telephone call from an insurance company that was asking questions about why there was so much damage to a residence that had a small fire when we arrived. We were at this fire for more than one and a half hours. The fire had all but destroyed the structure.

*"I thought I would be looking for new employment within the week."*

The chief asked me what I thought the problem might be. During our discussion, I revealed my analysis: We were using antiquated fire suppression tactics—surround and drown. As a result, we were unable to get to the seat of the fire quickly, which led to the ensuing damage. I told the fire chief that I had attended a fire suppression course where I had heard that several fire departments were experimenting with an aggressive interior fire attack method of extinguishment. In this strategy, fire fighters don SCBA and go into the building to extinguish the fire before it spreads throughout the structure.

The chief was very excited with this idea. He told me to train all of the shifts and work on implementing this process. He wanted everyone on the department to use this technique.

After gathering data from numerous fire reports, I found that our department's average under-control time was currently one hour for structure fires. The under-control time from the first structure fire incident that occurred after this change was less than 15 minutes.

The success from this change led to numerous additional changes for the fire department. These changes consisted of using large-diameter hose for water supply, using 1¾-inch hose for fire attack, using positive-pressure ventilation, implementing alarm assignments, making apparatus assignments, using the Incident Command System, implementing NFPA 1500, and instituting promotional exams for officers, plus numerous other changes to improve fire suppression operations. These changes occurred in one department because one individual shared what he had learned in a fire course. That one person was able to implement numerous changes over a 15-year period as chief of the training division.

As a fire services instructor, you will have the opportunity to initiate change throughout your career. However, you must prepare yourself to embrace those opportunities through training and education. This field provides lifelong learning opportunities. Never stop learning, because you are the future of the fire service.

*Jeffrey L. Huber*
Lansing Community College
Lansing, Michigan

outlines the instructional levels in the fire service and guides fire service instructor trainer programs; this book is written to be in compliance with this standard. Fire fighters seeking local, state, or national instructor certifications may be asked to demonstrate compliance with NFPA 1041 as well, though it may not be the only standard that the fire service instructor must meet.

Fire chiefs and fire departments are free to establish their own set of qualifications for fire service instructors. If the fire service instructor is asked to teach only one class in one department, he or she may not be asked to obtain a formal certification from another agency. Instead, the instructor may simply receive the training that the fire department deems necessary from other in-house instructors. In some cases, fire service instructors may be asked to have a certain number of years of experience prior to taking on an instructional assignment. In other cases, instructor positions might be reserved for those with command authority, holding a line officer's or chief officer's position. Some fire departments attach little or no additional requirements in an effort to find a fire fighter willing to take on the extra responsibilities as the fire service instructor.

Whatever the case, it is wise to meet these challenges head on. Becoming a fire service instructor may open doors for future promotions. Meeting the qualifications of this position will also prepare you for future leadership assignments.

## Continuing Education

Meeting requirements, whether set by a national standard or through fire department policy, is just the beginning for the fire service instructor. Working within the dynamic world of the fire service means you need to continue your professional growth and development. While some individuals dread the idea of **continuing education**, you need to understand the need to improve your knowledge, as it allows you to provide the very best and up-to-date information in the classroom. The idea of requiring continuing education is not a new one. Many professions, including health care, education, and inspections, require their members to participate in continuing education to remain licensed or certified.

For the fire service instructor, who issues the initial certification decides what, if any, continuing education is necessary. Some states may require only proof of continued instructional activity. Do not rely on the requirements of outside entities as a motivational force to professional growth, however. You should always strive to be on the cutting edge of the fire service. Attend outside trainings, seminars, and instructional conferences to remain current with the most current fire-ground tactics, management, practices, and instructional techniques available.

**Teaching Tip**

Look for state and national instructor organizations to join that will enable you to gain access to the most up-to-date training information. These organizations are also helpful in building networks from which to receive and share knowledge and experiences.

## Building Confidence

By keeping abreast of the latest information, you demonstrate the very value of education. For learning to take place, students must believe in your knowledge of the subject—which is not to say that you will always know everything about a particular topic.

There is a saying that in the classroom, "The instructor is always right." Believing in this old adage could be a fatal mistake and result in the eventual loss of your credibility. It doesn't take many times of being proven wrong in the classroom to lose the confidence of your students.

Building and maintaining student confidence in both the instructor and the training program go hand-in-hand. Instructors can build high levels of confidence in several ways. For example, you can demonstrate your own commitment to lifelong learning through continuing education. Be willing to freely admit if you are in doubt about a particular question and look to find real answers for the inquisitive student instead of trying to make something up on the fly in an attempt to impress your pupils. Be open to the suggestions and ideas of the students. Because fire service instructors often serve in operational roles as well, be very careful about following all teachings when working the emergency scene or when in the supervision role. One sure-fire way to lose important credibility is to project a "Do as I say, not do as I do" attitude.

**Teaching Tip**

Keep current by subscribing to magazines and Web sites, participating in e-mail lists, and tracking blogs that relate to the topics you teach.

## The Fire Service Instructor and Department Rank

The relationship between the fire service instructor and the fire officer has been the subject of much debate. The question usually asked is this: Does the fire service instructor have to hold a supervisory position to be effective? To answer this question, it is necessary to examine the organization and culture of the fire department. The answer also depends on the chief's commitment to the training program. Rank in and of itself is not a requirement for becoming an effective fire service instructor. Instead, the creation of a culture in which the training program can flourish is perhaps a more important component for success.

Holding a supervisory rank does give the fire service instructor additional credibility and authority to complete the assigned mission. Such a rank may enable the fire service instructor to have greater input into training schedules and priorities. In larger fire departments, where training is conducted by more formal training divisions, the creation of division command structures is more common, as training division chiefs, captains, and lieutenants manage the training program. In smaller fire departments, fire service instructor positions are sometimes left to those fire fighters who show the most interest in fire department training. It is these fire service instructors who sometimes are left to rely on the fire chief's mandates to ensure participation in the training program. Many fire fighters have found that efforts put forth in becoming successful fire service instructors place them in better positions when fire departments go looking for future fire officers.

## Issues of Ethics in the Training Environment

Firefighting is perhaps one of the most respected professions in today's society. The task of maintaining the public's trust in its fire service is held in high regard by all those who wear the uniform.

**Ethics** reaches beyond laws and standards to define behavior. Codes of ethics spell out what is acceptable and what is unacceptable within a fire department. For many individuals, ethics are rooted in their own personal assessment of right versus wrong, which begins during their early upbringing. Ethics can reflect the attributes of the people we respect and interact with in our own profession FIGURE 1.7. In the end, it is perhaps that feeling in the pit of your stomach that gives you the best clue about whether a certain behavior is ethical.

The issue of ethical behavior also affects the training program. Fire service instructors are oftentimes the ones who are asked to judge which candidates are ready to provide public service. Failing to hold students accountable for their learning objectives or simply passing fire fighters through the training program in the interest of moving on is unacceptable by any standard.

### Leading by Example: Do as I Say and as I Do

Ethics in training must begin with the fire service instructor. Instruction must extend beyond the classroom and into everyday operations. Fire service instructors demonstrate ethical behavior through their "Do as I say and as I do" lifestyles. Leading by good example lays the cornerstone for training ethics. If you are responsible for teaching the confidentiality of disciplinary measures in a leadership class but are later seen discussing an employee's behavior around the coffee table, you will lose credibility.

FIGURE 1.7 Ethical behavior is often rooted in one's own personal assessment of right versus wrong.

### Accountability in Training

Another ethical issue related to the training program is the need to maintain student accountability for training. As an example, if 10 fire fighters attend training on pump operations but only 4 fire fighters actually operate the pump while the remaining 6 chat away, who should receive credit for pump training?

Never put yourself in a position of simply passing a student by rote. It is important to document students who fail to meet training objectives. Because the training program is tasked with preparing fire fighters for critical life-safety operations, you must be on guard for protecting the program's integrity. The classroom or drill ground is no place for favors and friendship to affect a student's achievement record. At the end of the day, the fire fighter's performance ultimately judges your ability as an instructor.

### Recordkeeping

Just as you must maintain accountability in individual training accomplishments, so you must also maintain accurate records of program achievements. Training that is not fully completed owing to interruptions should never be recorded as accomplished. Good records are important to any training program, but only if those records are accurate.

Less than honest fire service instructors have been known to falsify records by simply recording trainings or skill attainments that did not occur for one reason or another. This act of falsification, also known as "pencil whipping an exercise," may look good on paper, but can lead to disastrous consequences once fire fighters actually face the prospect of having to perform the task under the stress of a true emergency.

All training sessions must be accurately recorded, along with a factual listing of objectives accomplished. Whenever training records are found to be inaccurate, whether due to negligence or just because of an error, your credibility will be called into question. Accurate recordkeeping builds confidence in the training program and sets ethical expectations for all to follow.

The accuracy of training records is not only an ethical issue, but also a legal issue. In recent court cases dealing with fire fighter fatalities, training records have been used both for and against the Authority Having Jurisdiction (AHJ). In one recent case, a homeowner sued a fire department for the loss of a home, and the training and qualifications of the fire fighters were the main focus of the lawsuit. Clearly, accurate and correct training records will have a tremendous influence in the outcome of such lawsuits. Recordkeeping as it pertains to the law is discussed in Chapter 2.

#### Teaching Tip

When developing a course, be sure to create an objective measure of performance. For example, create a checklist of each step that should be performed and notate those steps that are essential for passing the station.

## Sharing the Knowledge Power Base

Much has been written about the use and misuse of **power** within an organization. The unethical use of power that a supervisor holds over an employee has been the catalyst for many lawsuits. Expertise in a subject is considered a power, and you should be aware of its proper use. Individuals increase their expert power by increasing their own knowledge.

## Trust and Confidentiality

Trust and confidentiality go hand-in-hand: Lose one and you risk losing both. Both of these attributes have tremendous influences in terms of how well you are able to lead in the classroom and maintain an effective teaching ability.

Fire service instructors are often one of the first points of contact fire fighters make when joining the fire department. As such, you are placed in a position where your trust and confidentiality are critical to students. Struggling students may relay personal information referencing problems in their personal lives to explain why they are having difficulty in completing a task. A student might also share information about learning disabilities that require special considerations in the classroom. Conversations such as these require that you make every effort to maintain confidentiality. If circumstances require that fire department management become involved in the issue, then you should make that fact known first to the student. You may also provide guidance to students about where to go for assistance within the fire department's command structure.

Another area of concern is the need to maintain the confidentiality of student performance in the program. Information dealing with test scores, student evaluations, attendance, and behavioral issues should always be protected. Guidelines for the recording and sharing of this type of information should be written into policy so as to protect students' rights. Fire service instructors are often required to handle sensitive student information and must consistently demonstrate the ability to do so with professionalism.

## Managing Multiple Priorities as a Fire Service Instructor

In today's fast-paced world, we are bombarded with many priorities, each demanding our attention and time. As a consequence, success or failure is often determined by the ability to understand and prioritize tasks. Recognition, planning, and delegation are skills that can assist those facing multiple tasks while maneuvering through sometimes hectic and complicated assignments.

## Training Priorities

Managing today's modern fire service creates multiple priorities for the training program. Deciding what and when to teach is just one of the challenges. The direction that the program takes may be decided at different levels depending on the fire department's organization. In some cases, upper management may choose to lay out the training schedule for the period and then leave it to the fire service instructor to decide how to accomplish it. If the fire chief takes a hands-off approach to training, then the fire service instructor may be asked to develop both the schedule and the training program's objectives.

Training priorities may also reflect specific community characteristics. For example, a community with a heavy chemical industrial presence may place additional emphasis on hazardous materials training. Large cities with high-rise construction may require additional training on high-angle rescue and high-rise fire operations. Rural communities may require training on water supply shuttle operations and farm rescue. Training priorities are driven by mandatory training requirements to maintain current certifications or Occupational Safety and Health Administration (OSHA) regulations. Clearly, the task of setting training priorities is not a simple one.

Decisions about how precious training time should be spent must include input from all levels of the fire department. Fire company officers may be able to provide valuable insight on fire-ground performance issues affecting operations. Perhaps some problematic issues could be corrected through additional training. Fire chiefs may have knowledge of changes in future missions that will require fire fighters to learn new skills. Training committees may need to be established to provide a broad-based perspective as to the fire department's training needs, tackling the question, "Should we train more on our most serious hazards or on those hazards we respond to most often?" Both elements deserve consideration in the planning process.

## Planning a Program

The level of success of any training program is directly proportional to the planning efforts. In career settings, the schedules of on-duty personnel have to be considered when setting up training programs. By contrast, fire service instructors who are dealing with volunteer and part-time fire fighters must take into account the availability of personnel to leave their full-time jobs and to balance family time with the many department requirements. Holidays, vacations, multiple shifts, injuries, and illnesses all affect the need to reschedule or make up lost training opportunities.

The program must also plan for the use of training facilities and equipment. In some cases, outside instructors with special expertise may need to be scheduled.

No employee likes surprises, so consideration of the fire fighter's efforts to be put forth during training and other personal commitments require that training program schedules and completion requirements be established well in advance and communicated clearly to all. Once established, training schedules need to be followed as closely as possible. While changes will inevitably need to be made, any changes should be implemented with as much advance notice as possible and with sufficient time planned for rescheduling of the training so as to allow fire fighters to readjust their own schedules.

In some locations, a classroom may be shared by the city government, community access groups, and fire department training programs. Nothing is more frustrating to both fire fighters and fire service instructors than to have an overlap of scheduling or other unforeseen event force the cancellation or location change of a training session at the last minute. Given the very nature of emergency services, it is always difficult to schedule training and actually execute it according to schedules without interruption. In a combination or volunteer organization, conflicting priorities of personal and professional life outside the fire station make the program management portion of fire service training all the more important.

Program management aspects of running a training division and training program include many basic management skills learned in fire officer training, albeit focused in this case on the goals and objectives of the training division. Administration of training policy and procedures, training record-keeping systems, selection of instructional staff, review of curricula, and design of programs and courses that meet the current and future needs of the organization require a training team to accomplish many tasks. Chapter 12 of this text provides more information and resources to help plan and organize the training division's responsibilities.

Fire service instructors also need to plan the content of individual training sessions. Without planning, conflicts between competing groups can arise in the use of facilities and lead to a delay or cancellation of a scheduled training session, further eroding confidence in the fire service instructor's abilities. Contingencies for the use of specialized equipment needed in training evolutions must be built in if dedicated training equipment is not available. The fire service instructor should also plan how a missed training session will be made up and assure that make-up sessions accomplish the same learning objectives as the original training. Training time should be planned so as to eliminate distractions and interruptions. This may not always be possible when students attend training while on duty and as part of in-service training, as the need to respond to an emergency can interrupt even the best-laid plans.

## Ethics Tip

Have you ever been to a class where a student has been late to the course? Maybe it was only 5 minutes. Maybe it was 15 minutes. But what if it is 30 minutes or even an hour and 30 minutes? Would you let it go, require the student to make up the time, or fail the student?

You will inevitably face this issue as a fire service instructor. Unfortunately, there is no clear-cut rule for handling the issue, although some principles exist to help guide your decision about how to deal with this problem. The best method is to help avoid the situation by setting clear expectations at the beginning of class. Even with that step, however, at some point student tardiness will occur.

Use the following steps to aid your decision-making process. First, determine if there is a departmental policy on the issue. Next, determine whether the student missed any essential information or whether the missed content would place him or her at risk later. Consider the impact on the other students and their perceptions of the student's non-attendance. If you would be uncomfortable explaining your decision to the fire chief and other students, it's probably not the right decision.

## Training Through Delegation

Fire service instructors are faced with many tasks in managing the fire department's training program. Planning and scheduling for a multitude of priorities, developing and reviewing curricula, evaluating program effectiveness, coaching and mentoring students, and instructing all place enormous demands on the fire service instructor's time. Without assistance, it is easy to burn out, even for the most dedicated fire service instructor.

One way to minimize the potential for burnout is through **delegation**. Be alert to those fire fighters who have shown through recognized efforts the ability to handle additional training assignments. Be observant in the classroom for students who show an interest in instructional activities. It is often said that there is no better way to learn a subject than to have to teach it. Learning can be achieved through delegation, and you can obtain valuable assistance through the assignment of certain instructional tasks to competent fire fighters.

**Safety Tip**

Only delegate those tasks where a student has shown competency and where such a handoff of responsibilities will not jeopardize the safety of other students.

# Wrap-Up

## Chief Concepts

- NFPA 1041 categorizes instructor duties into three classifications.
  - The Fire Service Instructor I is able to deliver presentations from a prepared lesson plan, adapt lesson plans to meet the needs of a jurisdiction, organize the learning environment, and provide appropriate record-keeping.
  - The Fire Service Instructor II meets the requirements of Instructor I and is able to develop lesson plans, schedule training sessions, and supervise other instructors.
  - The Fire Service Instructor III meets the requirements of Instructor II and is able to develop training curricula for single or multiple organizations, conduct needs analysis, and develop training goals and implement strategies.
- Fire service instructors can have a wide range of job titles, ranging from fire fighter to chief of department.
- Fire service instructors will act in a variety of roles, including leader, mentor, coach, evaluator, and teacher.
- Both physical considerations and emotional issues influence the learning environment.
- Fire service instructors have a responsibility to stay current on topics and aid in succession planning.
- Fire service instructors regularly face ethical dilemmas that require them to make value-based judgments.

## Hot Terms

**Certificate** Document given for the completion of a training course or event.

**Certification** Document awarded for the successful completion of a testing process based on a standard.

**Continuing education** Education or training obtained to maintain skills, proficiency, or certification in a specific position.

**Degree** Document awarded by an institution for the completion of required coursework.

**Delegation** Transfer of authority and responsibility to another person for the purpose of teaching new job skills or as a means of time management. You can delegate authority but never responsibility.

**Ethics** Principles used to define behavior that is not specifically governed by rules of law but rather in many cases by public perceptions of right and wrong. Ethics are often defined on a regional or local level within the community.

**Learning environment** A combination of the classroom's physical and emotional elements.

**Organizational chart** A graphic display of the fire department's chain of command and operational functions.

**Power** The ability to influence the actions of others through organizational position, expertise, the ability to reward or punish, or a role modeling of oneself to a subordinate.

**Standards** A set of guidelines outlining behaviors or qualifications of positions or specifications for equipment or processes. Often developed by individuals within the regulated profession, they may be applied voluntarily or referenced within a rule or law.

**Succession planning** The act of ensuring the continuity of the organization by preparing its future leaders.

**Vision** Having an alertness to the future, recognition of potential, and expectations of improvement.

## References

Peters, T., & Austin, N. (1985). A Passion for Excellence: The Leadership Difference. Grand Central Publishing.

## Fire Service Instructor *in Action*

It's your first day as a fire service instructor. You begin by reflecting about how you got to this point in your career. You realize that this was not a position of chance. You have prepared to be a member of this noble profession by your actions in training, in daily station duties, and on the scene. You have done your homework before accepting this new responsibility.

Today you have scheduled a meeting with the fire chief to discuss your new position. You will seek clarity about your position and jointly either review the instructor's job description or seek to develop a job description in its absence. You will ask the chief to provide input on the direction of the training program and seek permission to meet with the department's officer staff to discuss training concerns. Upon conclusion of the meeting, you hope to have a clear understanding of the expectations for your position, including your level of authority to manage the training program, set the schedule, and control the training ground. If one does not already exist, you will draft a mission statement for the training program to establish its goals and objectives, which will be clearly stated for all to embrace.

You will also work to maintain the high level of confidence shown in you through your appointment by remaining humble and showing that your program's success depends on everyone's buy-in to the program. You hope to garner the respect of your fellow fire fighters by maintaining a positive enthusiastic attitude, fulfilling your commitments, controlling the learning environment, and engaging in continued enhancement of your own knowledge, skills, and abilities. After all, the best sales pitch to others for participation in training is the instructor's own continued willingness to learn.

**1.** Why is an instructor job description necessary?
- **A.** To guarantee a higher pay scale
- **B.** To clarify the instructor's responsibilities
- **C.** To set student learning objectives
- **D.** To place the instructor above the middle management ranks

**2.** Who is primarily responsible for controlling the learning environment?
- **A.** The instructor
- **B.** The fire chief
- **C.** The student
- **D.** Both A and C

**3.** Which of the following is *not* a way of earning respect as the instructor?
- **A.** Maintaining a positive and enthusiastic attitude
- **B.** Improving your own knowledge, skills, and abilities
- **C.** Keeping training sessions brief
- **D.** Controlling the learning environment

**4.** What are common attributes of today's fire service instructor?
- **A.** Desire to teach
- **B.** Experience in subject matter
- **C.** Motivational skills
- **D.** All of the above

**5.** Which of the following are *not* responsibilities or duties of the Fire Instructor II?
- **A.** Develop lesson plan materials
- **B.** Develop evaluation instruments
- **C.** Provide overall management of the training program
- **D.** Evaluate other instructors

**6.** You may determine that the requirements to accomplish the training will be too time-consuming for you. Which technique might you use to get the work done?
- **A.** Coaching
- **B.** Mentoring
- **C.** Delegating
- **D.** Testing

# Legal Issues

CHAPTER

# 2

## NFPA 1041 Standard

**Instructor I**

4.2.3 Prepare training records and report forms, given policies and procedures and forms, so that required reports are accurately completed and submitted in accordance with the procedures.

**(A) Requisite Knowledge.** Types of records and reports required, and policies and procedures for processing records and reports. [p. 24–35]

**(B) Requisite Skills.** Basic report writing and record completion. [p. 34]

**Instructor II**

5.2.5 Coordinate training record-keeping, given training forms, department policy, and training activity, so that all agency and legal requirements are met. [p. 24–35]

**(A) Requisite Knowledge.** Record-keeping processes, departmental policies, laws affecting records and disclosure of training information, professional standards applicable to training records, and databases used for record-keeping. [p. 24–35]

**(B) Requisite Skills.** Record auditing procedures. [p. 34]

## Knowledge Objectives

After studying this chapter, you will be able to:

- Describe how to prepare training records.
- Describe how to ensure that reports are kept accurately according to fire department procedures.
- Describe the records and reports required by the fire department.
- Describe the law as it applies to Fire Service Instructors.
- Describe whether training materials are protected by copyright or if they are public domain.
- Describe examples of ethical considerations.
- Describe how to minimize risk.

## Skills Objectives

After studying this chapter, you will be able to:

- Demonstrate how to complete records and reports.
- Demonstrate ethical conduct.
- Demonstrate how to minimize risk.

## You Are the Fire Service Instructor

Your chief has asked you to coordinate a physical combat challenge for multiple fire departments, which will be held at an off-site facility. The chief requests that you prepare an outline of the challenge and a report indicating how the fire department will be protected from liability should something go wrong.

1. As the fire service instructor, which factors must you consider in arranging this challenge?
2. Which documents do you need to minimize the potential liability for your fire department?
3. Which reference sources will you use to ensure that the challenge meets acceptable standards?

## Introduction

In the fire service, training is designed to keep existing skills sharp and to introduce new skills. Every training session exposes the fire service instructor and the fire department to potential liability. For example, inappropriate comments made in a training session based on religion, nationality, gender, or sex—even if done in jest—can subject the fire service instructor and the fire department to liability. A good working knowledge of the law, standard operating guidelines, and the department's own rules and regulations will reduce the potential exposure to liability.

A thorough understanding of the legal issues pertaining to training is critical for fire service instructors. This chapter will explore legal issues inherent in instruction and provide you with a working knowledge of the laws you will most likely encounter. The chapter also describes recordkeeping strategies and offers practical advice on how to deal with commonly encountered difficult situations.

## Types of Laws

Laws are generally established by one of two possible methods. The first method is through the legislative process (federal or state), which usually results in the establishment of **statutes** (TABLE 2.1). The second method usually is performed at the administrative or local level and results in **regulations** or **codes**.

**Table 2.1 Types of Laws**

| Type of Law | Definition | Example |
|---|---|---|
| Statutes | Created by legislative action and embody the law of the land at both the federal and state levels. | Americans with Disabilities Act<br>Motor vehicle codes |
| Codes/regulations | Can be established by legislative action, but are most commonly created by an administrative agency or a local entity with the authority to do so. Codes and regulations usually deal with narrower issues and enhance existing laws. | National Fire Protection Association (NFPA) codes<br>Building codes |
| Standards | Any rule, guideline, or practice recognized as being established by authority or common usage. Standards take on the effect of law if they are adopted by a local entity. | NFPA standards<br>Occupational Safety and Health Administration (OSHA) standards |

## The Law as It Applies to Fire Service Instructors

As a fire service instructor, your conduct is governed by three sources of law:

- Federal law, which is made up of statutory laws
- State law, which is established by each state's legislature
- Policies and procedures crafted by each individual fire department

All three sources set expectations for and restrictions on the conduct of fire service instructors (FIGURE 2.1). Nevertheless, there is an important difference between federal and state law and a fire department's policies and procedures. Federal and state laws apply evenly to all citizens and are the standard by which all fire service instructors are measured across the board and across the United States. The Americans with Disabilities Act (ADA) and the Age Discrimination in Employment Act (ADEA) are two examples of federal laws that govern all citizens across the nation. By contrast, policies and procedures are specific to a particular fire department and vary from region to region within a state, because each fire department must take into consideration numerous variables, such as the size of the community it serves, its locale, and its resources.

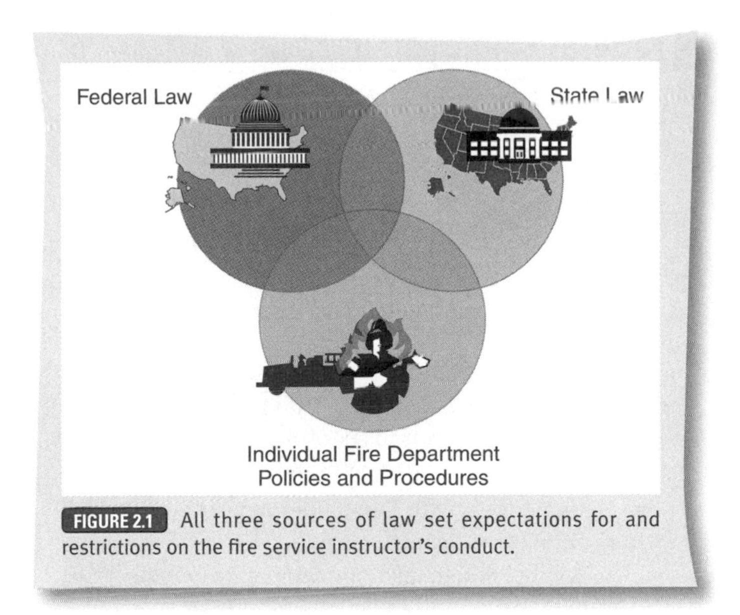


**FIGURE 2.1** All three sources of law set expectations for and restrictions on the fire service instructor's conduct.

A fire department's policies and procedures must not violate any state or federal laws. Indeed, they do not have the same status in the law as state and federal statutes. A fire department's policies and procedures represent the minimum expectations of the fire department and do not necessarily impose a legal duty as does a state or federal law. The only true benchmark against which the departmental policies and procedures are measured is the national standards or guidelines set within the fire service. Departmental policies and procedures can place obligations upon a fire service instructor that are not otherwise required by federal and state laws. For instance, a fire department's policies and procedures may require the department's fire service instructor to keep all department licenses and permits up-to-date, even though there is no legal requirement for the fire service instructor to do so. The policies and procedures may also require additional recordkeeping requirements not set forth by state or federal laws.

This chapter examines the laws that apply to fire service instructors, from those with the broadest coverage to those with the narrowest scope. Federal law is the broadest, in that it applies to all persons and requires the same expectations across the board regardless of your chosen profession or the state in which you live or work. For example, the ADA and the ADEA govern not only fire departments, but private employers as well. State laws are somewhat narrower, in that they apply only to the state in which they are enacted. State laws set the expectations for that state, but are not necessarily specific to a profession. The policies and procedures of a fire department have the narrowest scope, being designed specifically for an individual department and the professionals who work in it.

While the law is designed to set certain standards and expectations, it also can be used to protect fire service instructors from lawsuits. Fire service instructors must have a working knowledge of federal and state laws. They must also be well versed in the policies and procedures of their particular fire department, which also requires that they possess a strong working knowledge of national guidelines. While national guidelines are not legally binding upon a fire department, they clearly set out the consensus of the fire service across the United States. The NFPA standards, for example, are the national standards upon which most fire departments model some of their own standards.

Should a fire service instructor's conduct be called into question, the national guidelines are used as the backdrop against which to judge that conduct. For example, if a controlled burn gets out of control, the fire service instructor's methods will be compared with NFPA 1403, *Standard on Live Fire Training Evolutions*, to see if the instructor's conduct was reasonable. NFPA 1403 specifies a checklist of safety activities when performing a controlled burn, which includes activities such as ensuring that all utilities are disconnected and that proper permits have been obtained. If a fire service instructor does not follow this checklist, his or her conduct may be considered unreasonable and result in liability.

**Safety Tip**

Every fire service instructor who conducts live fire training must be thoroughly familiar with NFPA 1403 to increase the level of safety for the participants.

You will likely encounter situations during training evolutions, classroom participation, and your daily duties when issues pertaining to federal law will arise. You may face questions about compliance with the Americans with Disabilities Act, hostile work environment claims, and complaints of sexual harassment. A strong familiarity with these federal laws will prepare you to properly fulfill your obligations. State law could come into play on issues of privacy, indemnification, injuries to training participants, and equipment failures.

It is only through familiarity with the appropriate laws that you can insulate yourself and the fire department from potential liability. Directly associated with familiarity is proper recordkeeping. Maintaining an appropriate recordkeeping system affords you some level of protection, whereas inadequate or incomplete records may expose you to increased liability.

**Teaching Tip**

Attend local and national seminars to remain current with trends in the fire service. You should also join the appropriate associations that provide periodic updates throughout the year.

## Legal Protection

Once you understand the laws that apply to you as a fire service instructor, the next logical question is this: How do you perform your job so that you abide by the law and have protection should a lawsuit be filed? The first line of defense is always proper and thorough recordkeeping. Creating and maintaining records, reports, memoranda, e-mails, and other records in the course of your duties places you in a good position. The records should be organized appropriately and maintained for a reasonable period of time and in accordance with any local law.

The recordkeeping process starts at the outset of each activity. For example, when organizing a training evolution, the materials prepared should set forth the purpose of the training, the activities to be performed, the equipment to be used, and the goals of the training. You should have available any sources that were relied upon in creating the evolution and should confirm that the training complies with all policies and procedures of your fire department.

If the training will involve participants from other fire departments, additional documentation should be obtained for those fire fighters. In addition, your fire department should enter into a hold harmless and indemnification agreement with any other department or agency involved in the program. Such an agreement provides that your fire department will not be held monetarily or legally liable for injuries or conduct of the other department or agency's employees. This document is necessary to protect your fire department from individuals over whom you have no control. As a representative of the hosting fire department, you should also verify that all outside participants are medically cleared to participate. While this approach seems somewhat burdensome, the protection it affords is priceless.

Off-site training is filled with additional risks not encountered with on-site training. You should perform a site inspection well in advance of the training and prior to development of any handouts to identify any unusual or unsafe conditions. In addition, contact should be made with all other entities (town, city, utility companies) to verify that the proposed training does not violate any law or create any unreasonable risks. For example, you should confirm that no ordinances or local laws prohibit the proposed activity. Document each conversation that you have with other entities as well as the steps you take after each conversation.

**Safety Tip**

When using facilities or equipment you have not used before, be sure to get proper instruction from a qualified person.

Recordkeeping is vital to the smooth operation of the instructional process. In addition, recordkeeping pertaining to certifications, permits, and licenses is needed to assure the continued operation of the fire department. Failure of the fire service instructor to keep track of EMS recertification or continuing education, for example, could impair the staffing of the fire department.

## Federal Employment Laws

The U.S. federal government has created an expansive body of laws designed to provide equality in employment. The scope of this chapter is too limited for a complete discussion of all federal employment laws, but you should always ensure equality of treatment for all participants.

As a fire service instructor, you are not expected to know all the laws of the land, but to perform your job correctly you do need to have an understanding of three laws: the Americans with Disabilities Act, Title VII, and Section 1983 of the Civil Rights Act of 1964. In all likelihood, you will be called upon to teach or arrange instruction on each. You will also encounter situations that will require you to act or react to situations involving these three laws. Here is a brief summary of their content:

- The Americans with Disabilities Act (ADA): Prohibits discrimination against a qualified person because of a disability, where discrimination may include hiring, firing, promotions, and compensation.
- Section 1983 of the Civil Rights Act of 1964: Allows a complainant to not only sue a department or employer, but also to sue the alleged aggressor personally. Applies only to public officials.
- Title VII of the Civil Rights Act of 1964: Prohibits discrimination based on race, color, religion, sex, or national origin. Applies to all employers, public and private.

In addition to those laws, you should be familiar with the following legislation:

- Age Discrimination in Employment Act (ADEA): Prohibits discrimination against persons older than the age of 40 unless there exists a recognized qualification requirement.
- Equal Pay Act: Requires that all persons, regardless of gender, be paid at the same rate when performing the same job.

## Americans with Disabilities Act

The **Americans with Disabilities Act of 1990 (ADA)** is designed to protect individuals with disabilities from discrimination in the workplace. A **disability** is a physical or mental impairment that substantially limits one or more major life activities. **Major life activities** include functions such as caring for oneself, performing manual tasks, walking, seeing, hearing, speaking, breathing, learning, and working. To be limited in the major life activity of working, an employee must be restricted in the ability to perform either a class of jobs or a broad range of jobs; it is not enough to say that an employee is unable to perform one particular job. Conditions such as compulsive gambling, kleptomania, pyromania, and current illegal use of drugs are not considered impairments under the ADA.

In the fire service, it is unlikely that you would encounter physical disabilities such as blindness or the inability to walk. Nevertheless, there are a number of disabilities that you may encounter, including psychological disorders, neurological

## Applying the JOB PERFORMANCE REQUIREMENTS (JPRs)

Legal exposures exist in many training sessions that you will teach or participate in. In the event of almost any accident or injury, the training records and training history of those involved may be examined by attorneys and investigators in an attempt to identify any potential errors or omissions in the training and educational process. Many fire service instructors fail to realize the importance of the documentation part of the training process. When shoddy records are reviewed during investigations, the professional reputation of the fire service instructor may suffer—not to mention the potential for litigation under the heading of "failure to train" or "inadequate training." Training development from established reference sources based on up-to-date standard operating procedures (SOPs) is key to ensuring legally defensible training development.

### Instructor I

The Instructor I must have an understanding of the legal system and process used to validate training by referencing established standards and procedures. Understanding potential legal exposures in the delivery, recordkeeping, and evaluation of training are tasks involved in every training session.

### JPRs at Work

Know the components of the lesson plans you use as part of your instruction so that you can identify references and evaluation criteria. Be able to correctly complete training record report forms documenting each training event.

### Instructor II

The Instructor II serves in many cases as the "author" of training materials that are often used by others within their department. Knowledge of areas of exposures, identification of reference sources, and sound objective and lesson plan development reduce potential exposure to the instructor.

### JPRs at Work

Understand the laws and standards that apply to the development and delivery of training. Prepare lesson plans, evaluate materials, and collect records in accordance with local training policies and industry standards.

### Bridging the Gap Between Instructor I and Instructor II

When an Instructor II develops a lesson plan, he or she should always keep in mind these laws and standards as they apply to the audience to whom the lesson plan will be taught. Make sure the Instructor I understands the requirements of proper recordkeeping, including how a completed training record report should look and how to properly evaluate students after training. The Instructor I should make sure that any inconsistencies in the delivery of training or recordkeeping irregularities are brought forward to the instructor in charge immediately. Any potential areas of legal exposure need to be immediately documented and passed through the appropriate chain of command.

### Instant Applications

1. Review your training policy for requirements in documentation of training sessions and recordkeeping.
2. Identify local ordinances, laws, and standards that apply to the development, delivery, and recordkeeping of training.

FIGURE 2.2 A reasonable accommodation may include allowing for periodic blood sugar testing during the day or during training.

disorders, cardiac issues, respiratory issues, diabetes, epilepsy, dyslexia, and drug and alcohol addiction.

If you are approached by an individual alleging a disability, you should undertake a two-step analysis. First, you must determine whether an actual disability exists. If the answer to that question is yes, then you must determine whether the disability can be reasonably accommodated. The ADA defines **reasonable accommodation** as a modification or adjustment to a job or work environment that will enable a qualified applicant or employee with a disability to perform the essential functions of the job or enjoy the benefits and privileges of employment and that does not create an undue hardship for the employer. For example, if a fire fighter has diabetes, a reasonable accommodation may be to allow for periodic blood sugar testing during the day or during training FIGURE 2.2.

Examples of reasonable accommodations found to be in compliance with the ADA include modifications to facilities to make them accessible to employees with disabilities, restructuring of a job, modifying the work schedule, acquiring or modifying equipment, or reassigning a disabled employee to a vacant position for which the individual is qualified. Accommodation is not automatic—that is, you are not required to make an accommodation if it would impose an undue hardship on the operation. An **undue hardship** exists if the accommodation would require significant difficulty or expense relative to the size, resources, and nature of the employer.

Further, a reasonable accommodation does not have to be provided if doing so would present a **direct threat** to the health or security of others. Four factors are considered when assessing whether a direct threat exists: the duration of the risk, the nature and severity of the potential harm, the likelihood that potential harm will occur, and the imminence of the potential harm.

Like many legal situations, ADA issues are fact driven. There is no bright line test; each case must be analyzed based on the particular facts at hand. Courts have ruled on a number of cases involving fire fighters and the ADA, and those rulings can provide guidance for you in similar situations. For example, the U.S. Court of Appeals for the Fifth Circuit has held that a fire department is not required to accommodate a fire fighter who is unable to perform the essential duties of firefighting owing to a back injury when no light-duty position exists in a small fire department. In Pennsylvania, a fire department was not required to provide a fire fighter with a second chance at the academy course after he failed due to depression. In Maryland, the court held that it was proper for a fire department to prohibit a fire fighter from using an inhaler when the brand used was highly flammable.

It is important for the fire service instructor and managers up the chain of command to be aware of both the requirements of the ADA and those personnel under their command who qualify for reasonable accommodation under the ADA. In situations when an accommodation is requested during a training evolution, you should take great care to evaluate the request and, if needed, contact a higher authority for advice. Once a decision is made, the decision should be documented immediately to protect both you and the fire department.

**Teaching Tip**

Not all medical or psychological issues qualify as disabilities under the ADA. Each issue must be examined separately. Also, any accommodation request must be reasonable.

## Title VII

**Title VII** prohibits employment discrimination based on race, color, religion, sex, or national origin. Discrimination does not have to be intentional to violate Title VII. Any practice that has the effect of discriminating against a person because of his or her race, color, religion, sex, or national origin also violates Title VII. Title VII applies to state and local units of government with 15 or more employees. Although issues such as hiring, firing, promotion, and compensation do not generally arise in the context of training, harassment on the basis of race, color, religion, sex, or national origin may occur in training. You must be on the lookout for harassing behavior and know how to address it if it occurs.

In addition to your primary role of educating employees, you have a responsibility to address harassment and discrimination that you observe or that is reported to you. If you witness the conduct, you must take corrective steps immediately. Be sure to follow any grievance procedure set by policy or agreement in such a case. You should document the event, including statements from any witnesses of the event. All doc-

**FIGURE 2.3** Properly document all information provided by the complainant when an allegation of discrimination is made.

umentation should be completed as soon after the incident as possible.

If a report is made to you about harassing conduct but you did not witness that conduct firsthand, you must follow the steps set out in the fire department's harassment policy. If you are the person responsible for investigating claims, begin the investigation and document all relevant information regarding the allegation. If someone other than you is responsible for investigating the incident, prepare a report to that person documenting the information provided by the complainant and forward it to the appropriate parties **FIGURE 2.3**. A prompt and thorough response to all claims should be made. Finally, follow up with the complainant when and if appropriate.

## Race and Color Discrimination

Race and color harassment includes ethnic slurs, jokes, and derogatory or offensive comments based on a person's race or color if the behavior creates an intimidating, hostile, or offensive work environment or interferes with a person's work performance.

## Religious Discrimination

Religious discrimination can also take the form of harassing slurs, jokes, and derogatory or offensive comments. In addition to harassment, a religious discrimination issue may arise based on the scheduling of training classes. If training classes are scheduled only on days that conflict with an employee's religious needs, classes may need to be rescheduled or repeated on a different day unless doing so would cause an undue hardship to the employer.

## Sexual Discrimination

The most common form of sexual discrimination in the workplace is sexual harassment. **Sexual harassment** consists of sexual advances, requests for sexual favors, and other verbal and physical conduct of a sexual nature that affects an individual's employment either explicitly or implicitly, unreasonably interferes with a person's job performance, or creates an intimidating, hostile, or offensive work environment. Sexual discrimination also includes discrimination based on pregnancy.

Every fire department must have in place a policy prohibiting sexual harassment in the workplace. Generally, two categories of sexual harassment are distinguished: *quid pro quo* harassment and creation of a hostile work environment.

***Quid pro quo* sexual harassment** usually occurs in a supervisor–subordinate relationship where one party controls the other party's future in the organization. It ties employment decisions to the submission to or rejection of some form of sexual conduct. An employment decision might include a promotion, demotion, raise, transfer, termination, or other benefit of employment. For example, a supervisor would be guilty of *quid pro quo* harassment if he or she said to a subordinate, "If you have sex with me, I will give you a passing grade on the training evolution." Whether the employment decision is positive for an individual or has adverse consequences for the person is irrelevant if the decision is based on submission to or rejection of some form of sexual conduct.

The second form of sexual harassment is sexual conduct that creates a **hostile work environment** for an individual. Often, but not always, the inappropriate conduct occurs between co-workers who do not necessarily have the power to affect a person's job. The conduct in this scenario consists of a pattern of behavior that is sexual in nature and that interferes with an employee's work performance. Conduct that can rise to the level of sexual harassment includes inappropriate touching, sexually explicit language, sexually oriented jokes, comments about a person's appearance, and repeated requests for a date.

Some misconceptions exist concerning sexual harassment. For example, one misconception is that a victim has to be of the opposite sex from the alleged harasser. In reality, the sex of the victim or the harasser has nothing to do with the application of Title VII.

Another misconception is that the harasser must be an employee of the fire department and the victim's supervisor. In fact, individuals who are not employed by the fire department can create a hostile environment, which can subsequently create a cause of action against your fire department. For example, an outside instructor who is brought in to provide training can create a hostile environment for department employees. Likewise, fire fighters or instructors from other fire departments involved in joint training can create a hostile work environment.

A third misconception is that only the person who is the object of the behavior can be harassed. In actuality, anyone affected by the harassment—including bystanders who observe the conduct—can make a claim under Title VII.

To constitute harassment under Title VII, the conduct must be unwelcome. Conduct that might seem appropriate to you might constitute harassment to others. Off-color jokes can be harassment if they are unwelcome and pervasive. Horseplay and comments about a person's appearance can constitute harassment. Inappropriate pictures, calendars, and posture can create a hostile work environment.

During training evaluations, especially those involving physical agility, you must make sure that comments and ribbing about each participant's performance are not sexual in nature. While good-natured ribbing is acceptable, comments

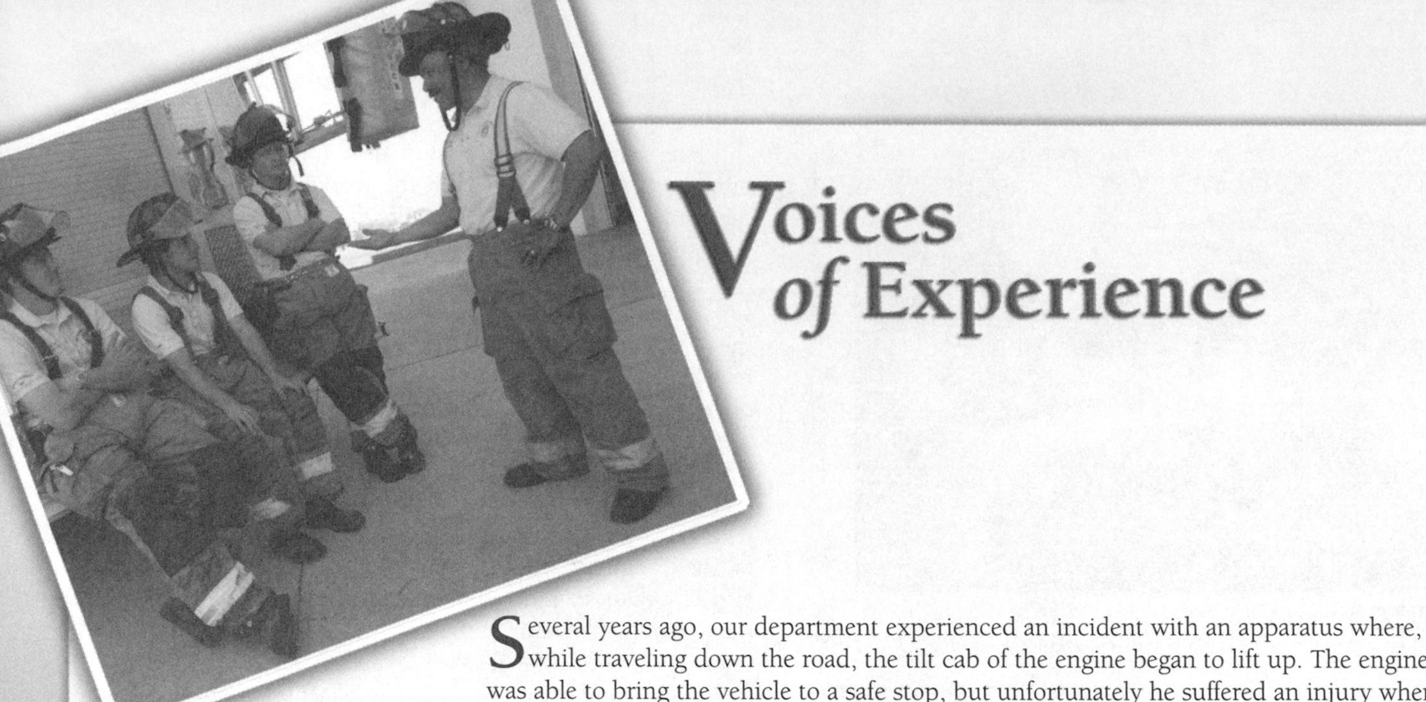

# Voices of Experience

Several years ago, our department experienced an incident with an apparatus where, while traveling down the road, the tilt cab of the engine began to lift up. The engineer was able to bring the vehicle to a safe stop, but unfortunately he suffered an injury when attempting to exit the apparatus.

A couple of years later, when the resulting court cases were being prepared, I was our department's new training and safety officer. The legal issues involved multiple parties. My involvement began when our department was subpoenaed to produce training records that could have a bearing on the issue. This material included the initial and ongoing training records for the driver as well as the training records for our department's mechanic personnel.

*"Because of our attention to maintaining accurate training records, we were able to produce the needed documents."*

Fortunately, the year before we had done an audit of our training records in conjunction with our state fire marshal's office. Because of our attention to maintaining accurate training records, we were able to produce the needed documents. This incident really emphasized for me the importance of tracking both initial training and ongoing training and documenting any certificates of completion and state certification.

Maintaining your training records is not as exciting as conducting a multiple-company drill, but it is just as important to your organization and your members. Many times training officers or instructors get some "grief" when we remind our personnel to "get the training reports done," but this incident was a good example of its importance. Remember—if it isn't properly documented, it didn't happen!

*Rudy Horist*
Elgin Fire Department
Elgin, Illinois

based on sex, gender, or stereotypes can create a hostile work environment in violation of Title VII.

If you, as a fire service instructor, witness conduct that may be a violation of Title VII, or if you receive a complaint from someone that a hostile environment is occurring, you have an obligation to take steps to rectify the situation. Those steps could include taking immediate corrective action if you witness the event or conducting an investigation if a complaint is brought to you by an employee. Your obligations should be set out in departmental policies and procedures. Taking no action or ignoring a complaint could expose the fire department to liability. As always, document all events associated with the matter, including summaries of statements made and steps taken in response to any potential violation.

### National Origin Discrimination

National origin harassment can take the form of ethnic slurs or other verbal or physical conduct based on a person's nationality. "English only" rules can also constitute national origin discrimination unless they are necessary for business purposes and employees are notified in advance of the rule's enforcement.

### Antidiscrimination Policies

As a fire service instructor, you must take the necessary steps to provide an environment that is free of discrimination. The best way to achieve this goal is to ensure that your department has in place a proper policy prohibiting discrimination in the workplace. Once such a policy is in place, all members of the department should be given a copy of the policy and sign an acknowledgment of its receipt and a commitment to abide by the policy. In addition, annual training on the policy should be provided for all employees, and documentation should be maintained indicating which employees attend the training.

You should also take the steps needed to keep the antidiscrimination policy current. As the law changes, the policy should evolve to reflect these changes. Whenever the policy is modified, all employees should be given a copy of the new policy and execute an acknowledgment as discussed previously.

### Retaliation

Title VII also prohibits retaliation against a person for filing a charge of discrimination, participating in a discrimination investigation, or otherwise opposing discrimination. Retaliation occurs when an employer takes an adverse action against a covered individual because that person engaged in a protected activity.

- An *adverse action* is any substantive action intended to discourage a person from exercising his or her rights under Title VII. Minor slights and comments or actions that are justified by an employee's poor work performance do not necessarily rise to the level of a retaliatory adverse action.
- A *covered individual* is anyone who has opposed unlawful discriminatory practices, and those who have a close association with such people. For example, an employer cannot take adverse action against a person because his or her spouse filed a discrimination complaint.
- *Protected activities* include opposing a practice that a person believes to be unlawful discrimination and participating in a discrimination proceeding.

Simply put, the complainant and any witnesses must not be subjected to activities that are designed to punish them for reporting the misconduct. Activities such as not including them at dinner, threatening not to back them up during a call, or refusing to speak to the complainant or witness during work hours can constitute retaliation.

**Ethics Tip**

At some point, you will likely find yourself in a situation where students are harassing other students, yet it is not an illegal form of harassment. For example, a group of fire fighters from a career department may tease fire fighters from a volunteer department when they are slower at the evolutions. Would you prohibit such actions? Would you take the same steps in the reverse situation—that is, if the volunteers were teasing the career fire fighters about their behavior?

These situations arise on a regular basis. As a fire service instructor, it is your responsibility to provide an environment that is conducive to learning for *all* students. This requires a subjective evaluation of the effects of the teasing. It doesn't matter whether the targets "deserve it" or whether the teasing is meant in good nature. If there is any doubt as to whether the behavior interferes with learning, it must be prohibited. Obviously, such behavior must always be prohibited if it is illegal, whether it interferes with learning or not.

## Section 1983 of the Civil Rights Act of 1964

Title VII applies to all employers, whether public or private. By comparison, Section 1983 of the Civil Rights Act of 1964 applies only to public officials, and it has a much broader reach than Title VII. Whereas a Title VII suit is directed only at the employer, a claim under Section 1983 allows a complainant not only to sue the department, but also to sue the alleged aggressor personally. Personal liability is claimed where the alleged conduct falls outside the scope of employment. Inappropriate touching of a co-worker, for example, is not part of one's job description and, therefore, is a personal act. To be sued personally means that a complainant can seek damages directly against the alleged harasser and the assets of the alleged harasser. The complainant can also seek punitive damages against an individual that are barred against units of local government.

### Safety Tip

As a fire service instructor, you have a responsibility to provide for the safety of your students. Generally, the hazards arise from the activities being performed. Sometimes, however, the hazards may come from other students. As a fire service instructor, if you observe or are notified of threats by a student, you have an obligation to take action to provide for the protection of the class.

## Equal Employment Opportunity Commission

The Equal Employment Opportunity Commission (EEOC) is a federal agency charged with investigating and enforcing federal laws pertaining to discrimination in the workplace. It requires that employers take all steps necessary to prevent workplace discrimination from occurring. Actions such as raising the subject of preventing discrimination, expressing strong disapproval for discriminatory conduct, and creating a policy that informs employees of their rights and states that investigations will be performed for all complaints should be discussed and reviewed. Most of the EEOC goals can be accomplished through education of fire department employees by fire service instructors from within the department or by instructors from outside the department who are familiar with that area of the law.

In addition to the federal laws enforced by EEOC, many state and local governments have statutes prohibiting discrimination and agencies that enforce these statutes. These agencies generally cooperate with the federal EEOC in work-sharing arrangements to investigate both federal and state claims of discrimination. Discrimination complaints are generally investigated by the agency with which the complaint is filed.

### Theory into Practice

Many new fire service instructors have never even considered the prospect that they might face a complaint dealing with an area governed by a federal law. In reality, every fire service instructor who teaches long enough must deal with this possibility. For that reason, it is essential for all fire service instructors to understand their responsibility in such cases. Failure to respond appropriately can place you at risk.

To prepare for this possibility, it is important to keep current on federal and state laws. Understand to whom they apply and how they fit into the educational environment. Find out your role and responsibility within the organization and then follow it precisely if a complaint arises. As a fire service instructor, you have both moral and legal obligations to the students. You will be held accountable for your actions.

## Negligence, Misfeasance, and Malfeasance

Negligence, misfeasance, and malfeasance are terms you are likely to encounter in cases where a civil wrong is alleged. Generally, two types of wrongs can be alleged in civil cases when an injury occurs:

- Unintentional conduct, which is best equated to the term accident (**negligence**, **misfeasance**)
- Conduct that include an element of intent or knowledge that a wrong is being committed (**malfeasance**)

Which conduct constitutes negligence/misfeasance and which conduct constitutes malfeasance depends on the facts of each situation. The determination is based on the conduct involved, the fire department's rules and regulations, prior training, and national guidelines.

### Negligence/Misfeasance

Negligence and misfeasance are kindred spirits: They are void of the element of intent, and are based on a duty to prevent harm or injury to others by exercising a degree of care commensurate with the task being performed. Negligence exists when there is a breach of the duty owed to another, there is an injury, and the breach is the proximate cause of the injury.

Once these three elements are established, you can become legally liable for your conduct absent some type of immunity. **Liability** in its simplest form means responsibility. If a person or entity is found liable, the person or entity is responsible for paying for the damages caused by its actions or inactions.

A fire service instructor's conduct is judged according to the reasonable person standard, which is an objective (not subjective) standard. In other words, the question is asked: What would a reasonable person have done in the same or similar situation? In the case of fire service instructors, what constitutes "reasonableness" is gleaned from the policies and procedures, national standards, and federal and state law. For this reason, it is important that you remain apprised of changes in the laws that affect the fire service. It is also important that any changes in the law, technology, and the national guidelines be considered and incorporated into the policies and procedures of the fire department when appropriate.

Misfeasance is the most commonly encountered in the performance of otherwise lawful acts. Simply defined, misfeasance consists of improper or wrongful performance of a lawful act without intent through mistake or carelessness. An example would be the hiring of a family member—a practice that, unbeknownst to you, is a violation of the department's nepotism policy.

Most states offer some level of protection to public employees for suits premised upon negligence or misfeasance. These protections often take the form of immunity laws that bar certain causes of action outright or require a complainant to allege and substantiate a higher level of proof. The higher standard of proof usually requires that an element similar to intent be proved to prevail on a claim. Quite often, the highest standard requires proof of willful and wanton conduct or establishment of gross negligence (discussed next in this section). Gross negligence could be found, for example, if a fire service instructor fails to follow a written protocol or a standing order. For example, a standing order might require personnel to wear protective equipment when dealing with a live fire; failure to do so could constitute gross negligence. The purpose of the higher standard of proof is to allow public employees to perform their jobs without fear that every little mistake will expose them to potential liability.

**Willful and wanton conduct** is defined by some courts as an act that, if not intentional, shows an utter indifference or conscious disregard for the safety of others. For example, it might include the failure to follow a written protocol or a standing order such that this omission creates a substantial risk to another person. That standard can be established if it is shown that, after having knowledge of an impending danger, a person failed to exercise ordinary care to prevent the injury. In addition, if, through recklessness or carelessness, a danger is not discovered by the use of ordinary care, the willful and wanton conduct standard may be satisfied.

Similarly, **gross negligence** is defined as an act (or a failure to act) that is so reckless that it shows a conscious, voluntary disregard for the safety of others. Gross negligence involves extreme conduct but not intentional conduct. It falls somewhere between a mere inadvertent act that causes harm and acting with an intent to harm.

Whether a particular act constitutes willful and wanton conduct or gross negligence depends on the facts of the situation. The most common evidence considered when analyzing whether an act is willful and wanton are your departmental policies and procedures, surrounding departments' policies and procedures, training records, and national guidelines or standards. It is vital that your policies and procedures are reviewed annually to assure that they are current and up-to-date with national standards set for the fire service. The same is true for the training evaluations used. Prior to revising an evaluation, it is important to verify that it still satisfies the standards of the fire service.

## Malfeasance

Malfeasance, unlike negligence and misfeasance, is more than a simple mistake or accident. Most often malfeasance is associated with public officials who partake in unlawful conduct even though they know that the conduct is illegal and contrary to their duties as public employees. Such conduct usually results in criminal charges, but could be made part of a civil suit. An example of malfeasance would be the acceptance of a bribe by a fire service instructor in return for a passing grade.

### Teaching Tip

Make sure that you have all standards, policies, or rules relied upon in creating a training session readily available. The resources used to develop a training program should always be current and consistent with all other fire departments in the area. Provide copies of any applicable policies to the participants during training.

## Copyright and Public Domain

Many materials you may want to use in training are protected by copyright. In 1976, Congress passed the Copyright Act of 1976, which governs copyright issues. The Act divides work into two general categories: those protected by the law and those considered to be part of the public domain. Works include written words, photographs, and some art work. The Copyright Act sets out some specific time frames pertaining to copyrighted material. Any work published before 1923 is considered to be in the public domain. Any works published between 1923 and 1978 have a 95-year copyright protection from the date of actual publication. For any works published after 1977, the author, artist, or photographer has a copyright for life plus 70 years. Violation of the copyright law can result in liability, with the creator of the work being eligible to recover any damages suffered or lost profits.

When providing training materials, many of the handouts used are copied from copyrighted materials. The Copyright Act does not prohibit their use outright or require that permission be sought before they are used. Indeed, the fact that some material is protected by a copyright does not necessarily prohibit you from using it in your training, because a "fair use" exception exists relative to the protection afforded by copyrights. Examples of fair use include copying materials for the purpose of criticizing, commenting, news reporting, teaching, and research. Copies for the class in a training session generally fall into this category. Ultimately, fair use depends on the nature of the copyrighted work and how much of the work is copied. For example, it would not be fair use to purchase one copy of a textbook and make copies of the entire book for the class. By contrast, if you want to use an article from a magazine or newspaper for classroom purposes, that would likely fall within the fair use exception.

If a particular use does not constitute fair use, you may still be able to use the material in a classroom setting by contacting the copyright holder and seeking permission in advance to use the copyrighted material. You should obtain permission in writing prior to using the materials.

Whether a work is copied under the fair use exception or with permission, it should include a notice indicating who holds the copyright to the work.

## Ethical Considerations

Ethical conduct should be the goal for all those involved in the fire service. Society has established certain socially acceptable conduct that matches the morals and values held by its citizens. Simply put, ethics is the difference between right and wrong. Ethical issues must be considered in all aspects of the fire service instructor's job. Ethical considerations include those held by the fire department as well as your own personal ethics.

Unethical conduct by any member of the fire service has a chilling effect on how society views fire fighters. During training, it is important that you reinforce the ethical standards that members of the fire service must abide by. To do so, first ensure that your department has a clear set of standards that sets forth expectations for conduct by fire fighters not only in terms of their job performance but also in terms of their learning environments. Set the tone for this ethical behavior by acting as a role model for those you teach. By showing respect to those who are different because of gender, race, or religious belief, for example, you will foster an environment that is both ethical and positive.

As part of your ethical conduct, place the education of your students above all else, except safety. This is done by creating an environment that promotes growth through challenging evolutions. It is also done by following the law as it pertains to discrimination, copyright issues, and criminal conduct such as taking bribes.

### Ethics Tip

The electronic age has greatly changed the world we live in. Cameras are everywhere, recording everything we do—from walking through a checkout line to driving through an intersection. Many of these videos are only a mouse click away. When you are creating a presentation, these images provide powerful lessons for students. Images such as a fire engine demolishing a car as it runs through a red light will have a lasting impact on students who will soon be making the decisions behind the apparatus wheel, for example. It is often impossible to know whether these videos and images are protected by copyright, although sometimes their copyright status is readily apparent. If you wanted to use video from a TV news program for your class, would you contact the TV station to get permission for its use?

## Records, Reports, and Confidentiality

Lawsuits evolve over long periods of time, which means that the thoroughness and accuracy of your records is vital. Proper recordkeeping is as important as the training you provide to your department. In cases where an injury occurs to a participant or observer of a training session, it is crucial to document what occurred, who was present, and what each person observed. The same holds true for claims of sexual harassment: It is absolutely necessary for the person who receives the report to document what was said and what actions were taken once the complaint was received.

Proper documentation is essential to the proper performance of your job, as is storage of the records in an organized and secure setting. Both state and federal laws (Privacy Act of 1979) prohibit the disclosure of certain information. Most likely your department has a policy or procedure that requires the protection of personal information that is kept in the control of the fire service instructor. Failure to abide by the **confidentiality** provisions required by law or your department could lead to liability or adverse action against you.

Records that should be kept secure and confidential include the following:

- Personnel files: Usually include information such as date of birth, Social Security number, dependent information, and medical information.
- Hiring files: Include test scores, pre-employment physical reports, psychological reports, and personal opinions about the candidate.
- Disciplinary files: Any report or document about an individual's disciplinary history and related reports.
- An audit of your record-keeping system should occur on a yearly basis. This will identify inadequacies and maximize efficiency.

## Risk Management

A vital function of the fire service instructor is to minimize risk. Instructors should incorporate risk management into all aspects of their job. In particular, they need to focus on minimizing risk as it pertains to fire fighters performing on fire or EMS scenes FIGURE 2.4. This is done by ensuring training is completed on all pieces of apparatus and equipment used by the fire department, as well as providing adequate training related to fire suppression tactics and EMS standards. Risk management techniques in fire suppression tactics and EMS standards include repetitive training in all areas to lower the

FIGURE 2.4 Focus on minimizing risk as it pertains to fire fighters performing at fire incidents.

risk of liability. The goal is to make the task so routine that when an actual event occurs, those involved in responding to the incident are thoroughly familiar with their responsibilities and the equipment being used. The goals of all training should be to educate, eliminate mistakes, and refine existing skills.

As important as practical training is in the fire service, equally—if not more—important is the non-operational training and duties of the fire service instructor. First and foremost, you must take steps to educate all members of the fire department about certain aspects of the law—specifically, hostile work environment (Title VII) claims and sexual harassment claims. Annual training should be given on these topics and you should ensure that up-to-date policies on both issues are maintained. In addition, you should provide training on all policies and procedures in place at the fire department—if not annually, at least every two years. Lastly, you should review all areas of the department to eliminate any potential risks to the fire department and its members, including lifting procedures, apparatus storage, and daily duty requirements.

Regardless of the type of training being provided, recordkeeping should be a top priority as part of risk management. Records with sign-in sheets and the topic presented should be maintained for all training sessions for a period of at least five years, or longer as required by law. You should also keep copies of the handouts with the sign-in sheet. Records, reports, or memoranda related to any unusual occurrences that happen during training or that are reported to you should be kept for at least five years. Because many lawsuits are filed long after the event occurred, the only real protection from this risk for the fire department may be the records kept. Incomplete or missing records often make defending suits more difficult.

On an ongoing basis, you should take the steps needed to keep the fire department and its training up-to-date. Changes are constantly occurring in the fire service pertaining to operational activities, as are changes in laws and national guidelines such as those formulated by the NFPA. You may need to modify the training presentation or policies in place to ensure that they continue to reflect the latest laws and standards.

## Teaching Tip

Good recordkeeping is an essential part of risk management. Keep all reports and records in a safe and secure area. All records should be kept for a minimum of five years, or longer as required by law.

# Wrap-Up

## Chief Concepts

- Every training session exposes the fire service instructor and the fire department to potential liability.
- Laws are either established through the legislative process (statutes) or at the administrative or local level (regulations or codes).
- As a fire service instructor, your conduct is governed by three sources of law: federal law, state law, and your fire department's procedures.
- Documentation and recordkeeping is essential in protecting yourself and the fire department in case of a lawsuit.
- The three federal employment laws you must be familiar with are: the Americans with Disabilities Act, the Civil Rights Act of 1964, and Title VII of the Civil Rights Act.
- The Equal Employment Opportunity Commission is a federal agency charged with investigating and enforcing federal laws pertaining to discrimination in the workplace.
- Negligence/misfeasance is unintentional conduct. Malfeasance is conduct that includes an element of intent or knowledge that a wrong is being committed.
- Materials that you want to use in training may be protected by copyright laws.
- Ethical conduct should be the goal for all those involved in the fire service.
- Lawsuits evolve over long periods of time, which means that thoroughness and accuracy of your records is vital.
- A vital function of the fire service instructor is to minimize risk.

## Hot Terms

**Americans with Disabilities Act of 1990 (ADA)** A federal civil rights law that prohibits discrimination on the basis of disability.

**Confidentiality** The requirement that, with very limited exceptions, employers must keep medical and other personal information about employees and applicants private.

**Direct threat** A situation in which an individual's disability presents a serious risk to his or her own safety or the safety of his or her co-workers.

**Disability** A physical or mental condition that interferes with a major life activity.

**Gross negligence** An act, or a failure to act, that is so reckless that it shows a conscious, voluntary disregard for the safety of others.

**Hostile work environment** A general work environment characterized by unwelcome physical or verbal sexual conduct that interferes with an employee's performance.

**Liability** Responsibility; the assignment of blame. It often occurs after a breach of duty.

**Major life activity** Basic functions of an individual's daily life, including, but not limited to, caring for oneself, performing manual tasks, breathing, walking, learning, seeing, working, and hearing.

**Malfeasance** Dishonest, intentionally illegal or immoral actions.

**Misfeasance** Mistaken, careless, or inadvertent actions that result in a violation of law.

**Negligence** An unintentional breach of duty that is the proximate cause of harm.

***Quid pro quo* sexual harassment** A situation in which an employee is forced to tolerate sexual harassment so as to keep a job, benefit, raise, or promotion.

**Reasonable accommodation** An employer's attempt to make its facilities, programs, policies, and other aspects of the work environment more accessible and usable for a person with a disability.

**Regulation (code)** A law that can be established by legislative action, but is most commonly created by an administrative agency or a local entity.

**Sexual harassment** Unwelcome physical or verbal sexual conduct in the workplace that violates federal law.

**Statute** A law created by legislative action that embodies the law of the land at both the federal and state levels.

**Title VII** The section of the Civil Rights Act of 1964 that prohibits employment discrimination based on personal characteristics such as race, color, religion, sex, and national origin.

**Undue hardship** A situation in which accommodating an individual's disability would be too expensive or too difficult for the employer, given its size, resources, and the nature of its business.

**Willful and wanton conduct** An act that shows utter indifference or conscious disregard for the safety of others.

Fire.jbpub.com

## Fire Service Instructor *in Action*

You are in charge of a department session involving hose training. The participants, both male and female, must attach lines to a hydrant and pull charged lines as part of this training evaluation. During the training, one of the participants asks to be excused to check his blood sugar just as his turn to perform the evaluation approaches. Some of the other fire fighters hear his request and begin to call him derogatory names and question his sexuality. One of the female fire fighters complains to you about the language of the jeering fire fighters.

**1.** Prior to beginning the training evaluation, what preliminary steps must you take to properly prepare for the training?
- **A.** Prepare an outline with goals listed
- **B.** Ensure that all participants have undergone physical examinations
- **C.** Have each participant sign a waiver of liability
- **D.** Provide no input prior to evaluation

**2.** To deal with the participant with the blood sugar issue, what should you do?
- **A.** Excuse the participant only if presented with a doctor's note
- **B.** Allow the participant reasonable time to check his blood sugar prior to performing the evaluation
- **C.** Allow the participant reasonable time to check his blood sugar after the evaluation
- **D.** Ignore the request

**3.** How should you deal with the inappropriate language being used by some of the participants?
- **A.** Tell the participants to stop the conduct
- **B.** Report the conduct
- **C.** Ignore the conduct
- **D.** Join in the conduct

**4.** Once you have been approached by a fire fighter indicating that she found the language to be offensive and hostile, what should you do?
- **A.** Follow the steps set forth in the department's policy
- **B.** Meet with the complainant to document her complaint
- **C.** Notify higher members of the chain of command about the complaint
- **D.** All of the above

**5.** After completion of the evaluation, a student asks for the score of his friend because he had to leave early. What is the most compelling reason why you should not provide the information?
- **A.** It isn't any of his business.
- **B.** It may be illegal.
- **C.** You don't have time to do that for everyone.
- **D.** The two may not be friends.

**6.** During the course, you distributed a copy of an article from *Fire Engineering* magazine. Is this legal?
- **A.** Probably not, because it is copyrighted material
- **B.** Probably not, because it is patented
- **C.** Probably yes, because it is a limited amount of material and is being used for educational purposes
- **D.** Probably yes, because it is in the public domain

**7.** When documenting the training, you include the title of the course and indicate who attended but nothing else. Which statement is most correct?
- **A.** This is adequate information should a legal question later arise.
- **B.** This is inadequate information should a legal question later arise.
- **C.** It is a departmental decision on which information is important to retain.
- **D.** There is no need to document training.

**8.** If you accepted a student's offer of free concert tickets to let him pass the exam, you might be guilty of which offense?
- **A.** Malfeasance
- **B.** Misfeasance
- **C.** Negligence
- **D.** Creating a hostile work environment

# Instructional Delivery

PART

# Methods of Instruction

CHAPTER

# 3

## NFPA 1041 Standard

**Instructor I**

**4.4** **Instructional Delivery.** [p. 42–53]

**4.4.1** **Definition of Duty.** The delivery of instructional sessions utilizing prepared course materials. [p. 42–53]

**4.4.3** Present prepared lessons, given a prepared lesson plan that specifies the presentation method(s), so that the method(s) indicated in the plan are used and the stated objectives or learning outcomes are achieved. [p. 49–53]

- **(A) Requisite Knowledge.** The laws and principles of learning, teaching methods and techniques, lesson plan components and elements of the communication process, and lesson plan terminology and definitions. [p. 49–53]
- **(B) Requisite Skills.** Oral communication techniques, teaching methods and techniques, and utilization of lesson plans in the instructional setting. [p. 42–53]

**4.4.5** Adjust to differences in learning styles, abilities, and behaviors given the instructional environment, so that lesson objectives are accomplished, disruptive behavior is addressed, and a safe learning environment is maintained. [p. 51–53]

- **(A)* Requisite Knowledge.** Motivation techniques, learning styles, types of learning disabilities and methods for dealing with them, and methods of dealing with disruptive and unsafe behavior. [p. 42–43, 51–53]
- **(B) Requisite Skills.** Basic coaching and motivational techniques, and adaptation of lesson plans or materials to specific instructional situations. [p. 42–51]

**Instructor II**

**5.4** **Instructional Delivery.** [p. 42–53]

**5.4.1** **Definition of Duty.** Conducting classes using a lesson plan. [p. 42–53]

**5.4.2** Conduct a class using a lesson plan that the instructor has prepared and that involves the utilization of multiple teaching methods and techniques, given a topic and a target audience, so that the lesson objectives are achieved. [p. 42–53]

- **(A) Requisite Knowledge.** Use and limitations of teaching methods and techniques. [p. 42–53]
- **(B)* Requisite Skills.** Transition between different teaching methods. [p. 50]

## Knowledge Objectives

After studying this chapter you will be able to:

- Describe motivational techniques.
- Describe how to adjust the classroom presentation and still meet the objectives of the lesson plan.
- Describe the laws and principles of adult learning.
- Describe characteristics of generations X, Y, and the Baby Boomers.
- Describe communication techniques that will improve your presentation.
- Describe how to deal with disruptive and unsafe behaviors in the classroom.

## Skills Objectives

After studying this chapter you will be able to:

- Demonstrate basic coaching and motivational techniques.
- Demonstrate professionalism during the learning process.
- Demonstrate how to manage disruptive and unsafe behaviors in the classroom.

## You Are the Fire Service Instructor

During your introductory discussion with a group of new students, you inform the class that if any students have special needs, they should see you in private so that you can discuss together how to assist and provide the best possible learning experience. During a break in the class, one of the students comes to you and says that he is dyslexic; he can provide documentation stating that he is to receive special accommodations for testing. As you talk with this student, you also discover that he has a difficult time taking notes because he must concentrate to write and spell.

You and the student discuss how you can make him comfortable in the classroom and what you can do to help him be successful. One suggestion from the student is that you grant him permission to tape-record the lecture portion of the class. To accommodate the student's testing needs, you offer to print quizzes and tests with one question per page, and allow the student to take the test in an isolated room to help him concentrate and reduce distractions.

1. As a fire service instructor, how can you determine which students need extra attention?
2. How can you help all of your students to learn?

## Introduction

As a fire service instructor, you must constantly ask, "How can I best help my students to learn?" It is your responsibility to present information and have a clear understanding of the course objectives. Instruction is the number one factor in students' success: Teaching has a greater impact on achievement than all other factors combined.

Whether the class is a large introductory course for recruits or an advanced course for fire officers, it makes good sense to start off well. Students decide very early—maybe even within the first class—whether they will like the course, its contents, the instructor, and their fellow students.

Fire fighters are adults. This point may seem obvious, but sometimes we can lose sight of that fact. Some of the strategies that work with second graders also work well with adults, but there are important differences in the way adults and children learn. Never treat your fire fighter students like kindergarteners—you are teaching adults.

## Motivation and Learning

**Motivational factors** make the students feel enthusiastic and eager to learn. Motivation affects how students practice, what they observe, and what they do—all of which are critical to learning. As a fire service instructor, you need to understand how learning and motivation work together to make a great fire fighter.

Learning is the potential behavior, whereas **motivation** is the behavior activator. Learning and motivation combine to determine performance in training. Without a motivated student and an expert instructor, learning will not take place. If the student's motivation is poor, then his or her performance in training will be sloppy and inaccurate. If the student's performance in training is sloppy and incorrect, then the fire fighter's response at an incident will be sloppy and inaccurate. This is not acceptable on the fire ground.

## Motivation of Adult Learners

Cyril Houle (1961) identified three types of adult learners: learner-oriented, goal-oriented, and activity-oriented. Each type is associated with different motivational issues, which in turn affects how learning takes place (Draper, 1993).

Consider the example of two fire fighters who are attending a CPR training class. A fire fighter who is a goal-oriented learner might attend the class in effort to improve her job skills, whereas the fire fighter who is an activity-oriented learner might take the class to spend time with his friends and possibly learn CPR at the same time. Although the learning agendas and goals of these fire fighters are quite different, both are in the same class. If the goal-oriented learner has an auditory learning style (i.e., prefers listening to a lecture), she may have to alter her learning style during that class to a kinesthetic style (i.e., physically performing the skills) to meet the goal of learning CPR. By contrast, the activity-oriented learner does not have to adjust his style to achieve his goal and can readily learn CPR by actually performing the skill. Learner-oriented adults focus on the act of being involved in learning. While often adult learners are able to adjust their learning styles for a class, this type of transformation is not always successful or even possible for some learners (McClincy, 1995).

Another factor that greatly influences success in the classroom is the adult learner's grasp of how the material relates to the real world and how this learning can be beneficial to job performance. This factor is well documented (Wlodkowski, 1999).

Yet another key factor is the adult learner's psychological state. Mild anxiety can help the adult learner to focus on a particular task. Conversely, intense negative emotions (e.g., anxiety, panic, rage, insecurity) may detract from motivation, ultimately either interfering with learning or contributing to a low performance.

The current position of adult learners in the life cycle also influences the quality and quantity of learning. Research suggests that each phase of life is characterized by the need for learning certain behaviors and skills. Because they are adults, adult learners also want to be treated as autonomous, independent learners.

Geographical changes (moving from one location to another), physical changes (reaction times), visual and auditory acuity, and intellectual functioning influence the adult learner's performance as well. The adult learner's internal world of ideas, beliefs, goals, and expectations for success or failure can either enhance or interfere with the quality and quantity of thinking and information processing. In addition, the adult learner's attitudes about his or her ability to learn and judgment as to whether learning is an important life goal affects motivation (Merriam & Caffarela, 1999). Motivational and emotional factors also influence the quality of thinking and the processing of information.

No matter where the adult learner is and what he or she believes about his or her abilities, the adult learner continues to acquire knowledge both intentionally and unintentionally (Butler & McManus, 1998). The effectiveness of the learning process relates directly to the learning environment and the adult learner's interactions with that environment. In short, the adult learner's mental and emotional processes for learning can either help or hinder the fire fighter during training. Understanding each adult learner's preferred learning style helps you to tap into the strengths of each and every adult learner in your classroom.

## Motivation as a Factor in Class Design

Motivation is the primary learning component for adult learning. It generates the question, "Why?" Why is the adult learner in the class? Why does the adult learner want to take (or not want to take) this class? When you can identify the *why*, it will allow you to tap into the *how*—the precise motivation and techniques that will help the adult learner achieve the goal of learning.

**Teaching Tip**

An adult learner's motivation may change throughout a session. The number of factors that affect motivation is limitless. You need to be able to recognize the changing motivational needs of adult learners.

For example, knowing that attendance at a training class is mandatory allows you to design the lesson plan for the captive audience to maximize motivation. Unfortunately, most adult learners who are mandated to attend a class and forced to learn the information do poorly, simply because they are unmotivated. Some adult learners may not know why they must learn the information or what is expected from the training or educational program, whereas other adult learners may have high expectations of the class. The unmotivated adult learners, the questioning adult learners, and the adult learners with high expectations all share the same classroom.

Find out from your students why they are in your class and what they expect. This information will allow you to design the lesson plans to meet every students' needs and sow the seeds of self-motivation in every student. You will inevitably need a variety of motivational techniques to stimulate students. To learn, a student must be stimulated by the subject (McClincy, 2002).

If you are motivated, your enthusiasm may be contagious in the classroom. When you are dynamic, outgoing, charismatic, and generally interested in the topic, your lessons will be taught with energy. Your personality will shine through in the classroom and will serve as a motivational factor for your students.

The use of audiovisual aids, teaching strategies, teaching methods, and even the classroom itself may serve as motivational factors. How you incorporate any or all of these motivational factors into the classroom will be reflected in the success of the learning process.

**Safety Tip**

While a lack of sleep may be problematic for motivation, it is imperative for safety. Do not allow students to participate in training without proper sleep. Sleeping through part of a presentation could prove deadly.

## The Learning Process

Learning is an interactive process in which the adult learner encounters and reacts to a specific learning environment. That learning environment encompasses all aspects of the adult learner's surroundings—sights, sounds, temperature, other students, and so on. In addition, each adult learner has a distinct and preferred way of perceiving, organizing, and retaining these experiences—that is, a **learning style** (Merriam & Caffarela, 1999). Some adults learn by listening to and sharing ideas with others, other adults by thinking through ideas themselves, and still other adults by testing theories and/or synthesizing content and context.

Effective learning follows a natural process:

1. Adult learners begin with what they already know, feel, or need, based on the groundwork that was laid prior to what is currently taking place. In other words, learning does not take place in a vacuum.
2. Seeing that "newly" learned information has real-life connections prepares the adult learner for the next step—learning something new.

3. Adult learners use the new content, practicing how it can be applied to real life.
4. Adult learners take the material learned in the classroom and apply it to the real world, including in situations that were not always covered in the original lesson.

## What Is Adult Learning?

Fire service training and educational programs, for the most part, are designed with the adult learner in mind. An adult learner is defined as someone who has passed adolescence and is out of secondary school. Self-motivation is rarely a problem for these individuals, who typically see the courses as a way both to advance their careers and to better themselves as people FIGURE 3.1.

According to Bender (1999), adult learning is the integration of new information into existing values, beliefs, and behaviors. New information is learned and synthesized so that students will be able to consider alternative answers to never-before-asked questions, new applications of knowledge, and creative ways of thinking. Adult students need realistic, practical, and knowledgeable lessons that deliver information in a variety of ways to accommodate their differing learning styles. In particular, adult learners may have difficulty imagining a piece of equipment, responding to a written scenario, or performing a skill if they do not have a realistic application for the learning. The more realistic the lesson, the more likely the adult learner will generalize the information and be able to apply the learning in the field.

## Similarities and Differences among Adult Learners

The theory of andragogy (adult learning) is an attempt to identify the way in which adults learn. According to this theory's originator (Knowles, 1990), although adult learners differ, some commonalities can be found. For example, adults typically need to know *why* they need to know something and *how* they can use it. Adult learners also tend to focus on life-centered (practical) tasks and activities. Also, adult learners like to learn at their own pace and using their preferred learning style. Successful learning also takes advantage of adult learners' rich history of life experiences.

FIGURE 3.1 Fire service training and educational programs, for the most part, are designed with the adult student in mind.

Not all adult learners are alike. Two early theorists in behavioral science, Erikson and Levinson, focused on age to help understand how adult learners differ. According to these researchers, young adults seem to be primarily focused on issues surrounding the development of a sense of identity, establishing a career, pursuing a life dream, having an intimate relationship, getting married, and achieving parenthood. Middle-aged adults focus more on caregiving for children and perhaps older family members, career changes, physical and mental changes, role changes, and planning for retirement. Older adults, by contrast, often face a series of upheavals—loss of their jobs (via retirement), the transition from parent to grandparent, age-related physical and mental changes, and the deaths of friends, spouses, and eventually themselves. Stage of life directly affects the adult learner in the classroom.

As discussed previously, Houle (1961) identified three types of adult learners:

- For learner-oriented adults, the content is less important than the act of being involved in learning.
- Goal-oriented learners acquire knowledge with the goal of improving their job prospects or learning a new skill. Members of this group view the instructor as being responsible for disseminating knowledge that they need.
- Activity-oriented learners learn by doing. To succeed in learning, they require personal productive time, with control over content and learning style, with the goal of improving social contact rather than the acquisition of knowledge.

In all of these orientations, the adult learners set goals, identify objectives, select relevant resources, and use the instructor as a facilitator for learning.

## Influences on Adult Learners

Adult learning theories are based on the premise that adults have had different experiences than children and adolescents, and that these differences are relevant to learning. Although age is certainly a factor that can affect learning, other factors—such as motivation, prior knowledge, the learning context, and influences exerted by situational and social conditions—are also influential. For example, in the classroom an adult learner might not pay close attention to the topic because a previous experience left the impression that the subject is not important. This view is merely reinforced for the adult learner when others in the class have the same feelings about the subject.

Adult learners are more apt to learn and retain information when it is meaningful to their lives. For instance, information directly related to a new piece of equipment will

## Applying the JOB PERFORMANCE REQUIREMENTS (JPRs)

Effective training has many components. The audience, the learning environment, the instructor, and the expectations of the training process are all pieces of the puzzle that combine to make training effective. The methods of instruction chosen by the instructor are among the most vital parts of the puzzle, as they have to match the audience, the environment, and the expectations or outcomes of the training session. The wrong method may have disastrous effects on the outcome and lead to poor performance or even safety and survival issues on the fire ground. Students' motivation can be enhanced by varying approaches to instructional delivery and choosing the appropriate setting for training delivery.

### Instructor I

The Instructor I is responsible for delivering training from prepared lesson plans. Knowing how to identify the factors relating to the appropriate methods of instruction for the lesson plan will assist in the delivery of effective training.

### Instructor II

Conducting an audience analysis and evaluating the learning environment help the Instructor II develop lesson plan materials that are effective for teaching students. The best learning process is interactive and student centered. Varied approaches to your training methods make this work.

### JPRs at Work

Deliver the prepared lesson plan with the ability to adjust and adapt the lesson plan to the learning conditions presented by the students and the training location. Correct unproductive student behavior and control the delivery of training relating to the students performance.

### JPRs at Work

Be able to adjust the delivery of a lesson plan by varying the methods of instruction to keep students motivated and interested while ensuring that all objectives prepared for the lesson plan are covered and the student performance is improved.

### Bridging the Gap Between Instructor I and Instructor II

The lesson plan content and the instructor's ability both play large roles in the method of instruction that is used to deliver a training session. Certain methods are required for specific types of objectives and intended outcomes. The Instructor II will prepare the lesson plan with a specific setting in mind for the training delivery. The Instructor I, based on his or her experience, comfort, and background, may vary the training delivery according to these factors.

### Instant Applications

1. Identify the most effective and least effective training sessions you have attended. Determine which methods of instruction were appropriate or inappropriate and may have influenced your impressions of those training sessions.
2. Review a list of upcoming training sessions scheduled for your department and determine the best methods of instruction that should be used to achieve the best outcomes for them.
3. Rearrange your typical classroom seating arrangement to accommodate various methods of discussion such as a lecture, demonstration, or group discussion.

capture the interest of fire fighters working on that truck. Material that is outside the realm of the adult learner's experience and knowledge is less meaningful. The prior knowledge and experiences of the adult learner is more diverse than those of a child or adolescent, yet you must always be aware that some adult learners may lack the expected prerequisite knowledge needed to learn new information.

What influences adult learning? Here is one answer to that question:

> Adults often engage in learning as one way to cope with the life events they encounter, whether that learning is related to or just precipitated by a life event. Learning within these times of transition is most often linked to work and family. (Merriam & Caffarela, 1999, p. 107)

Fire fighters work in a very intense environment. What is learned during times of intense stress inevitably influences learning in the classroom. If a fire fighter was just on an intense call, he or she may have trouble concentrating or sitting still in the classroom. Coping mechanisms may affect behavior in the classroom.

Other influences on adult learners include social cultural perspectives, such as social roles, race, and gender. Some adult learners, for example, may feel uncomfortable with classmates who are more educated or who hold a higher rank. You must ensure that in the classroom *everyone* is equal.

A unique feature of adult learning is that adults often desire to be self-directed learners—that is, they want to take their own direction within the learning process. Self-directed learning is related to humanism, the concept that the adult learner can make "significant personal choices within the constraints imposed by heredity, personal history, and environment" (Elias & Merriam, 1980). While adults may have the desire to be self-directed in their learning, their prior experience with didactic classroom-based instruction may not have prepared them sufficiently for self-management in the learning environment. As a fire service instructor, it is up to you to provide the necessary guidance to keep them on track.

Critical reflection—that is, the ability to reflect back on prior learning to determine whether what has been learned is justified under present circumstances—is often a unique advantage attributed to adult learners, and a key asset that you should exploit. This kind of reflection involves "metathinking" about the strategies required to achieve a goal. Such reflection often occurs when adult learners compare their ideas to those of their mentors or those of classmates. Over a period of time, an adult acquires basic cultural beliefs and ways of interpreting the world. Often such learning and beliefs are assimilated uncritically. In the classroom, however, adult learners may reflect critically on the assumptions and attitudes that underlie their knowledge. Such reflection helps them become deeper thinkers and can enhance their processes of thinking.

The benefits of this evolution in thinking are readily evident in the fire service. How many times have fire fighters gone to training with an attitude of "I really don't want to be here"? These students would rather be doing something else, but they go because the education is mandated. This disdain for training may represent an assimilated attitude—perhaps others in the station feel the same way. That is, it may not represent the way that a particular fire fighter truly feels about training or education. Fire fighters naturally want to fit in socially, so they tend to adopt the attitudes that others around them have. As the fire service instructor, part of your job is to help all students become motivated to learn the information presented.

Factors such as participation in formal educational programs or training programs since high school will also affect the adult learner's ability to learn. For many adult learners, traditional learning skills (e.g., reading and understanding textbook material, taking notes during lectures, and studying for tests) may have been forgotten or, in some cases, were never developed properly. Many adult learners have developed a learning style that is limited to "doing" (McClincy, 1995). Adult learners with this learning style learn best when they can perform the skill. Difficulty with developing study skills and problems with unlearning and relearning present challenges to both the adult learner and to you, as the fire service instructor.

## Today's Adult Learners: Generations X, Y, and the Boomers

### Baby Boomers

The baby boomers were born between 1946 and 1964. At the end of World War II, soldiers returned home, married, and began families in the midst of an economic boom. The baby boom ended when the effects of birth control caused a sharp decline in the birth rate in 1964. Characteristics of baby boomers include the following:

- The baby-boomer generation was raised on television, which had a tremendous impact on them and caused them to see world events from a single source FIGURE 3.2.
- This generation was raised on rock and roll.
- This generation spoke out against social injustice.

FIGURE 3.2 The baby-boomer generation was raised on television.

How do you teach to baby boomers? This generation of individuals makes up the backbone of most organizations today. Look at the senior members of your fire department and observe how you interact with them. To teach baby boomers, you must first gain their trust and show them respect. As a fire service instructor, use them to assist you in teaching your course. Baby boomers will want to share their experiences and knowledge. This generation is known for both accepting change and resisting it because it pushes them out of their comfort zones.

## Generation X

As B. L. Brown points out in *New Learning Strategies for Generation X*, the "gap between Generation X and earlier generations represents much more than age and technological differences. It reflects the effect of a changing society on a generation" (Brown, 1997, p. 4). The term Generation X has been used to describe the generation consisting of those people whose teenaged years were touched by the 1980s, although this definition excludes the oldest and youngest Gen Xers. This was the first generation where both parents worked outside of the home. Their parents' absence required children to learn to take care of themselves and their siblings until Mom and Dad returned home. While their physical development was the same as earlier generations, the psychological and learning development of Generation X was not. This generation was the first to abandon the traditional family models, to consider divorce to be the norm, and to promote single-parent families. Many Gen Xers prefer to stay childless FIGURE 3.3.

Many members of this generation were raised in single-parent homes, growing up with "fast food, remote controls, entertainment and quick response devices such as automatic teller machines and microwave ovens, all of which provide instant gratification" (Brown, 1997, p. 3). These life experiences shaped the way they learned and resulted in certain learning characteristics that require some modifications to traditional classroom tactics and new teaching strategies. Generation X is typically being educated and trained by people from earlier generations. The cultural gap between the generations reflects the diverse life experiences of the individuals in each generation. One of the major challenges facing instructors is the need to bridge the generation gap.

FIGURE 3.3 Generation X was the first generation to abandon traditional family models.

Generation X learners typically have the following characteristics:

- Independent problem solvers and self-starters. These individuals want support and feedback, but they don't want to be controlled.
- Technologically literate. Gen Xers are familiar with computer technology and prefer the quick access of Internet, CD-ROMs, and the World Wide Web as their sources for locating information.
- Conditioned to expect immediate gratification. Gen Xers crave stimulation and expect immediate answers and feedback.
- Skeptical of society and its institutions. Members of this generation want their work to be meaningful to them. They want to know *why* they must learn something before they will take the time to learn the subject.
- Lifelong learners. Gen Xers do not expect to grow old working for the same company, so they view their job environments as places to grow. They seek continuing education and training opportunities; if they don't get them, they will look for new jobs where they can get what they want.
- Ambitious. Members of Generation X often seek employment at technology start ups, found their own small businesses, and even adopt worthy causes.
- Fearless. Adversity—far from these discouraging youths—has given them a harder, even ruthless edge. Most believe, "I have to take what I can get in this world because no one is going to give me anything" (Brown, 1997).

In the classroom as well as in the workplace, the Xers often clash with baby boomer or traditionalist instructors. As an instructor, you may be the baby boomer trying to teach the Gen Xer. The common value that both generations share is the high priority they place on learning and developing new skills. This poses a challenge to instructors, who must continually raise the bar in the classroom and adapt to cross-generational training. To accommodate the unique characteristics of Generation X learners, instructors need to step outside the box and promote learning that has applications in the school, work, and community settings that are part of the Generation Xers' experience.

## Generation Y

Generation Y (also known as the Millennial Generation) is the name given to those people who were born immediately after Generation X. There is no consensus as to the exact range of birth years that are included in Generation Y, or if this term is geographically specific. The only consensus

# Voices of Experience

We have all heard and maybe used the quote, "You can lead a horse to water, but you cannot make it drink." Training officers can provide great learning opportunities for their personnel, but there is no guarantee that students will show up ready to learn. Effective learning is often accompanied by a reward of some type. Recognition or rewarding great performance can provide a source of motivation for fire service personnel.

I remember a training officer who announced a new twist to annual company evaluations. Not only would crews perform supervised evolutions on the drill grounds, but a comprehensive written test would also be included. The written test was to be developed using all classroom materials presented during the entire calendar year. This test would be presented in a closed-book format to all 120 personnel attending the regional training center from five different fire departments.

*"There is no guarantee that students will show up ready to learn."*

As you can imagine, the response to this initiative was mixed. Some personnel were reluctant to participate and risk embarrassing themselves. Many fire fighters seemed to embrace the competition and the challenge. As the year progressed, the prize for the top individual and top crew was announced: a gift certificate and a seven-course evening meal to be prepared by all of the local fire chiefs. Following this announcement of the prize, the competitive nature of fire fighters really began to surface. Teamwork increased as study groups began to form and personnel began showing up at drill ready to take notes and learn.

In the end, test scores provided documentation to support above-average retention of the year's training material. In addition, the dinner was a big hit with the winning crew. Photos of fire fighters being served dinner by the fire chiefs solidified this moment in time. I can't remember how I spent that gift certificate, but I can remember sitting in that recliner watching those fire chiefs cooking us a tasty meal. Anytime you can engage the competitive nature of fire fighters in the learning environment, good things will occur as long as safety is never compromised.

*Scott Weninger*
Clackamas County Fire District #1
Clackamas, Oregon

appears to be that Generation Y includes those individuals born after Generation X.

The "Y" in Generation Y refers to the namelessness of a generation that has only recently come into an awareness of its existence as a separate group. Members of this generation typically feel dwarfed and overshadowed by the baby boomers and Generation X. Nevertheless, members of Generation Y are known for holding wide-ranging opinions on political and social issues, religion, marriage, gay rights, and behavior **FIGURE 3.4**. This lack of a common perspective may have occurred because these individuals have not encountered a personal situation where their actions/reactions have forced them to consciously choose sides on issues and maintain consistent positions. Most Generation Yers are, for the most part, more tolerant of alternative lifestyles and unconventional gender roles. At the same time, Generation Y tends to be more spiritual and religious than the generation that includes their parents, and disagreement on social issues between the more liberal and more conservative members of this generation is commonplace.

Generation Yers are young, smart, and brash. They wear flip-flops to the office and listen to their iPods at their desks. While they want to work, they don't want their work to be their life. The members of Generation Y have been pampered, nurtured, and programmed with a slew of activities since they were toddlers, meaning they are both high performers and high maintenance (Armour, 2005).

Learning in virtual classrooms and virtual labs is ideally suited to Generation Y, as is teaching by technical professionals, because these adult learners can interact with others and get hands-on experience. Virtual labs suit the "click on it and see what happens" experiential approach to learning and have the added benefit that the learner remains in a safe environment.

**FIGURE 3.4** Members of Generation Y are known for holding wide-ranging opinions on political and social issues, religion, marriage, gay rights, and behavior.

### Ethics Tip

While fire fighters unite as one of the greatest family structures known, chastising those with diverse social views is often considered acceptable practice. Part of team development entails exploring one another as individuals and then creating bonds to form the group. Historically, fire fighters were fairly homogenous in their beliefs, so the practice of ostracizing those who were different was relatively infrequent. That situation is changing with today's generations, as their values are often much different than those held by many older, more traditional fire fighters. How do you promote team development and cohesion without obstructing diverse thinking?

As a fire service instructor, you have a responsibility to ensure that the learning environment is appropriate for all of your students—even those who do not fit in the mainstream. You cannot expect, or desire, to change the social beliefs of students. You must provide an environment that is tolerant of all views and immediately address actions by students that are designed to ostracize other classmates.

The Generation Yers in the classroom present a challenge to the fire service instructor who is either a baby boomer or a Generation Xer. The fire service instructor needs to have a general understanding of members of this generation and support their learning styles and needs.

### Theory into Practice

Understanding generational differences will aid in creating an environment that is conducive to learning for all students. To see how you can capitalize on these differences, have students break out into small groups. Ask them first to identify the primary differences in the generations and then to identify themselves with one of the generations. If time and facilities allow, have the students use other resources, such as the Internet, to learn more about the generation assigned to them. Have them discuss what might work in the classroom to teach all of the generations about a subject such as basic fire behavior, and develop a lesson plan.

## Professionalism during the Learning Process

No matter which type of teaching methods you use, which combination of methods you employ, or which generation you teach, the following communications techniques will help you to improve the delivery of your information to your class.

*Dress professionally.* Students should see you as a professional and recognize you instantly as a subject matter expert. Even though we like to believe that we don't "judge a book

FIGURE 3.5 A good way to make a positive impression is to greet your students as they enter the classroom.

by its cover," you will be sized up by your students. A professional appearance lays the groundwork for making a good first impression.

*Welcome each student.* Your first impression begins the moment the students enter the classroom. Shake students' hands as they arrive, introduce yourself, and have a warm smile on your face—this will immediately convey that you are a likable, approachable person who cares about your students FIGURE 3.5.

*Call students by name.* Making every student feel important is essential in creating motivation and maintaining attention. Calling students by their name is one of the most effective methods of demonstrating to them that they are valued. While it may not be possible in large, lecture-style courses that meet only once, most fire service courses are small enough to allow you to learn the student's names. Most all of us struggle to learn names, but there are methods that can help you with this task. One of the most basic strategies is to use table tents with the students' names. Another technique is to keep a seating chart in front of you. Name association may also work well. Each method has its own advantages and disadvantages, but finding an effective method will go a long way toward helping you manage your class.

*Use eye contact.* In our culture, it is important to make eye contact with people to whom you are speaking. Making eye contact with each fire fighter for a few moments in the course of your presentation will help students to feel that you are actually communicating directly with them. Resist the temptation to look at the wall, at the clock, or over students' heads. Lack of eye contact makes people think you are not willing to be straight with them or that you are not confident in what you are saying. In fact, it is probably one of the most disturbing mannerisms you can display.

*Use your normal speaking voice.* Sound interested by being interested in the subject and in what students have to say. As a fire service instructor, you have a strong motivation to build a strong relationship with your class. If you are committed to the importance of what you are teaching, then that commitment will be revealed in your voice and by the enthusiasm that you display for the topic. Remember that enthusiasm is contagious—let it show!

## Safety Tip

Eye contact is an important tool for engaging students and demonstrating an interest in them. However, avoid the overuse of eye contact where it might be misconstrued as conveying your sexual interest in a student.

*Speak with respect.* You are teaching adults, and it is important that they perceive that they are respected as adults. When you are respectful to them, they will be respectful to you.

*Use pauses when you speak.* A short pause can be used to signal an important point or transition in your lecture. A pause can also indicate the conclusion of one topic and the beginning of another. It also allows you the opportunity to view your students and visually check for comments, interest levels, or questions.

*Use information appropriately.* It is important to gauge your students' level of understanding correctly. For example, Fire Fighter I students will have a different understanding of information than seasoned fire officers. Make sure that you speak at a level that students will understand and explain concepts or information on the same level. You may have to spend time ensuring that the level of training or education is appropriate for all participants in the course.

*Use gestures sparingly, for emphasis.* Avoid meaningless gestures—for example, jingling keys or change in your pocket, playing with a pencil, tapping mindlessly on a podium or desk. Try to avoid playing with things, chewing on the ends of pencils, or repeatedly tapping your foot. Hide those tempting items if playing with them has become an unconscious habit for you. Eventually, these habits will become intrusive—and distracting to your students—and fire fighters will be placing bets on whether you'll swallow the pencil or how many times you will tap your foot in a minute.

*Use posture and movement to signal a transition of topic or point of view.* Make your movements conscious and meaningful, and try to avoid pointless, nervous pacing. Think about where you want to move and when.

*Practice positive mannerisms.* Well-placed mannerisms can increase your effectiveness in presentation of your lectures, and appropriate gestures can add to the verbal messages. Vary your position and movement, because your ability to change the pace of the lesson will increase involvement. Your body language and facial expression can give a fire fighter support and encouragement even when you are not saying anything aloud.

*Choose your teaching aids appropriately.* The technology that you have at your fingertips in the classroom is simply a way to communicate with students and reinforce information. Eraser boards and paper pads have some unique advantages in communicating, as they can be used to convey a variety of information during a single lesson. If you use these tools, make sure that your writing is legible and readable. These teaching aids also provide a visual reinforcement of your verbal message. PowerPoint presentations, charts, posters, models, slides, films, and DVDs are other means of communicating information to students and can reinforce information being presented verbally. Technology changes rapidly, and your use of it can help to enrich your classroom and assist you in teaching. But don't rely on technology alone to teach: It can break, so always have a backup plan.

*Know how to use your chosen teaching aids in the classroom.* Your communication with the students is interrupted when you do not handle technology smoothly. Each time you are guilty of mishandling your teaching aids, you will weaken the impact of the lesson by sending an unintended message ("I don't know what I'm doing").

## Effective Communication

Effective communication is the key to ensuring that the students learn—that they receive and remember the information you intended to transmit to them. As a general guideline, consider these findings by educator Edgar Dale, in 1969:

- We remember 10 percent of what we read.
- We remember 20 percent of what we hear.
- We remember 30 percent of what we see.
- We remember 50 percent of what we see and hear.
- We remember 70 percent of what we say.
- We remember 90 percent of what we say while doing.

It is important to choose the right kind or kinds of communication when you are preparing for class. It is also important to use the selected communications methods effectively. Understanding how the communication process works will help you develop lesson plans or adapt lesson plans based on your students' learning styles. For example, if your students are visual learners, then you may need to include more visual presentations. Chapter 5, Communication Skills for the Instructor, describes the elements and role of the communication process.

## Disruptive and Unsafe Behaviors in the Classroom

Disruptive students come in many shapes and your strategy for handling the problematic situation will vary depending on the precise behavior. The most important thing to remember

### Theory into Practice

Have students identify the components of communications and demonstrate those communications skills in short individual presentations. Students should be allowed time to prepare prior to making their presentations. Skills sheet developed from the previous list could be used to have students provide feedback to presenters. The filled-out skills sheets might then be given to students to enhance their personal learning.

### Teaching Tip

When preparing your lesson plans, remember the following points:

- Adult learning should be problem centered.
- Learning must be experience centered with meaningful goals.
- Fire fighters must have feedback regarding their progress toward their goal.
- Adults learn at different speeds and in different ways.

is that *you are in charge of the classroom.* If you don't control your classroom, you may have some serious problems. You may be perceived by your students, peers, and administration as being a less competent instructor. Disruption in the classroom can significantly reduce or limit the amount of learning time for students **FIGURE 3.6**. It can also lead to unsafe and negative learning that could have serious implications when students apply what they have learned in the field. You are responsible for providing a safe learning environment and proper, safe instruction to all students.

Disruptive students come in many guises:

- The *Monopolizer* has a tendency to take control of classroom discussions and monopolize class time. To deal with this type of student, make certain to call on other students during classroom discussions.
- The *Historian* has some experience and wants to make sure that everyone is aware of it. Real Historians have the benefit of years of experience and share their experiences to help illustrate an instructor's lesson. Real Historians focus on helping the instructor and other students. By contrast, Artificial Historians focus on helping themselves and interject their opinions to show other students how much they know. Real Historians can be disruptive if their stories become distracting for the other students. Carefully guide the discussion back to the main topic. Another way to redirect Real Historians is by pairing them up

**FIGURE 3.6** You are responsible for providing an appropriate learning environment to all of your students.

with struggling students. Artificial Historians are almost always distracting. Assertively redirecting the discussion back to the lesson plan generally corrects this issue. If not, take the student aside and discuss the need for the class to stay focused on the lesson plan. Treat your adult students as adults—with respect.
- The *Day-dreamer* is preoccupied with anything other than the lesson. He or she may have been forced to attend the training class or may be participating as part of a job assignment. Many times this student is focusing on life outside of the classroom. Day-dreamers may not directly disrupt other students, but are more of a distraction to you, the instructor. Try to draw such students into the class by using direct questioning techniques and making maximum eye contact.
- The *Expert* is a student who may become confrontational or monopolize the class in an attempt to show off his or her brain power and perceived expertise, often as a measure of competitiveness. The Expert will provide up-to-date information, factoids, applications, and case information about the topic and can easily subvert your position as the classroom authority. You must corral the Expert and may even have to discuss his or her contributions to the class during a private break.

Some behaviors are clearly inappropriate in the classroom because they are unsafe or illegal. Chapter 2 describes many of these issues, including harassment (sexual or otherwise) and discrimination. Other worrisome behaviors may include actions or words that are threatening or offensive, or that create risks to students, instructors, or property. These behaviors need to be dealt with immediately and without hesitation.

There are many reasons for behavior problems in the classroom, ranging from very complex to very simple. For example, the student who falls asleep in class may have been on a working fire during all or part of the previous night and is just plain tired. This issue is relatively easy to identify and is not a persistent problem. By contrast, the student who falls asleep in class every day may have a medical disorder that causes him or her to be unusually drowsy. Learn more about human behavior, behavioral problems, and their causes to become a more effective fire service instructor.

Behavioral problems displayed by the disruptive student may be rooted in other problems as well. As the fire service instructor, you may be put in the position where you must deal with behaviors that are the result of situations outside the instructional environment. These problems may be as simple as hunger or thirst, illness, or medical needs. On other occasions, disruptive students may be experiencing personal problems. Your first priority is to create a positive learning environment for the whole class so all behavioral expressions can be addressed. Your second priority is to ensure the existence of a positive learning environment for the individual student. If you are able to determine the reason for the behavioral issue, and it is one that you can help to address, this will allow the individual student to learn as well.

Other behaviors that may prove disruptive to the class include calling out, asking irrelevant questions, or giving excessive examples. In these cases, students may be seeking attention from the class or from you. Some students may seek to gain power by being argumentative, lying, displaying a temper, or not following directions. These students need to be dealt with quickly and directly before their behavior gets out of hand.

Some of these behaviors can be easily addressed. The hungry student may need to take a break so he or she can eat. The bored student may need a more stimulating lesson to motivate him or her to learn. No matter what the problematic student behavior is, understanding the sources of some basic behaviors will help you control the classroom environment.

As the fire service instructor, you have the opportunity to make positive behavioral changes possible for your students. Here are some simple ways to do so:

- *Lead by example.* You set the tone for the class through your own behavior; in doing so, you set a standard for the class. Act as a professional.
- *Take steps to prevent undesired behaviors.* By preventing the behavior from occurring, you will have changed it before it happens and disrupts the class. For example, in your opening discussion with the class, you can outline the rules that govern the classroom and tell students what kind of behavior is (and is not) expected of them (Bear, 1990).
- *Know what to expect.* By putting the rules in writing, you will have anticipated what the behaviors may be. When formulating these policies, it is important to include the consequences for their violation and to let the students know what rights they have in the classroom situation.
- *Stick to the rules.* Don't contradict the facility's or administration's rules.
- *Communicate.* Talk to problem-causing students and explain that it is unfair for their behavior to continue to disrupt the class.
- *Reward good and appropriate behavior.* This can be done through special assignments or may be as simple as a pat on the back—in front of the class, of course (Walker & Shea, 1991).
- *Act like a professional,* which includes being prepared, approachable, and good natured.
- *Don't overreact or under react.* Some things may bother some that don't bother others. Conversely, problems left unaddressed seldom resolve themselves.
- *Discuss the problem in the class.* Be open and honest about what is happening in the classroom.
- If all else fails, *call a time out.*

**Teaching Tip**

If you cannot gain control of the class, it may be time for someone else to take over.

**Teaching Tip**

In small groups, students should be encouraged to discuss what influences them to learn, including identifying their personal agendas and goals for the class. It certainly would be appropriate to have students present their group's results to their class. This exercise is a good way to encourage discussion and for you to learn more about the students.

## Lesson Plans

All students have special needs in the classroom that may require you to adjust some part of the lesson plan at some juncture. When you develop your plan, it should be detailed and complete enough to achieve the identified goals and desired outcomes, yet flexible enough that it can be adjusted to meet students' needs on the fly. Be prepared to repeat information in different ways or to perform demonstrations more than once in different ways. Make sure that you follow a detailed, sequential plan when demonstrating a skill; explain and outline the information to be explained; and discuss the results, including key questions to guide the discussion. These simple yet flexible components will enable you to develop an effective, yet concise lesson plan. Chapter 6, Lesson Plans, details this process.

Ask yourself three basic questions before building your lesson plan:

1. *What are the goals of the lesson?* These goals are the purpose, aims, and rationale for the class period. You and your students will be working to achieve these goals.
2. *How will you achieve those goals?* Identify the objective of the lesson and focus on what the students will do to acquire the knowledge and skills of the lesson.
3. *How will you know when those goals are achieved?* Through some kind of assessment, you need to ensure that the students have learned, at a minimum, the desired information. Be sure to provide students with an opportunity to practice what you will be evaluating.

**Safety Tip**

Always be aware of what is going on around you. As the instructor, it is your responsibility to ensure the safety of your class. Modifications to exits, changes in the environment, and unusual activities or commotions in the area should prompt a careful investigation to keep your learning environment safe. Items out of place on the floor, people not sitting with all of the legs of the chair on the floor, or furniture that is unstable may seem benign, but are actually dangers that can cause injury to the students in the classroom setting. Use common sense and practice safety.

# Wrap-Up

## Chief Concepts

- Understanding how adult learning takes place, which factors affect motivation, and how different generations approach the educational process will help you become a better instructor.
- Meeting the needs of the students is the key to motivating them and helping them to learn.
- Effective learning requires instructors to step outside the realm of personal experiences and into the world of the learner.
- Students' leaning requirements differ because their experiences and development have differed.
- When you can communicate to students that you know the information, and that you understand their unique needs and motivations, you will find that it is much easier to achieve the goal of teaching the information and having the students learn.

## Hot Terms

**Adult learning** The integration of new information into the values, beliefs, and behaviors of adults.

**Andragogy** The identification of characteristics associated with adult learning.

**Baby boomer** The generation born after World War II (1946–1964).

**Generation X** People born after the baby boomers; they are today's adult learners.

**Generation Y** People born immediately after Generation X.

**Learning style** The way in which the individual prefers to learn.

**Motivation** The activator or energizer for an activity or behavior.

**Motivational factors** States of the person that are relatively temporary and reversible and that tend to energize or activate the behavior of the individual.

## References


Armour, S. (2005). http://www.usatoday.com/money/workplace/2005-11-06-gen-y_x.htm.

Bear, G. G. (1990). Models and techniques that focus on prevention. In A. Thomas & J. Grimes (Eds.), *Best practices in school psychology* (p. 652). Silver Spring, MD: National Association of School Psychologists.

Bender, D. (1999). Unique characteristics of adult learning. www2.bw.wdu/~dbender/csu/adultlearn.html.

Brown, G. L. (1997). New learning strategies for Generation X. ERIC Clearinghouse. http://www.ed.gov/databases/ERIC.

Butler, G., & McManus, F. (1998) *Psychology*. Oxford, UK: Oxford University Press.

Draper, J. (1993). *The craft of teaching adults*. Toronto: Culture Concepts.

Elias, J. L., & Merriam, S. (1980). *Philosophical foundations of adult education*. Malabar, FL: Krieger.

Houle, C. (1961). *The inquiring mind*. Madison, WS: University of Wisconsin Press.

Knowles, M. (1990). *The adult learner: A neglected species*, rev. ed. Houston, TX: Gulf.

McClincy, W. D. (1995). *Instructional methods in emergency services*. Englewood Cliffs, NJ: Brady Prentice Hall.

McClincy, W. D. (2002). *Instructional methods in emergency services* (2nd ed). Englewood Cliffs, NJ: Brady Prentice Hall.

Merriam, S. B., & Caffarela, R.S. (1999). *Learning in adulthood*. San Francisco, CA: Jossey-Bass.

Walker, J. E., & Shea, T. M. (1991). *Behavior management: A practical approach for educators* (5th ed.). New York: Macmillan.

Wlodkowski, R. J. (1999). *Enhancing adult motivation to learn: A comprehensive guide for teaching all adults*, rev. ed. San Francisco, CA: Jossey-Bass/Pfeiffer.

## Fire Service Instructor *in Action*

As a fire service instructor, you will face the challenge of teaching many different kinds of students and will need to cover many different topics. It is important that you remember to take the time to get to know the students in your classroom by using introductions, interviews, student surveys, or just general discussions. Identify their strengths and weaknesses based on their behaviors, their personal learning styles, and the generation to which they belong. These characteristics will help you figure out how your students learn and how you can effectively control the class.

If some of your students have disabilities, you may need to alter or adjust your lesson plan to help meet those individuals' learning needs. If students are disruptive, you may need to discuss their behavior with them, remove the problem students from the classroom, or possibly refer them to a superior for further action.

Getting to know the students is vital to your success, your students' success, and your ability to maintain control over the classroom.

**1.** When working with students, the instructor should recognize
- **A.** All fire fighters are pretty much alike, so they can be motivated in the same way.
- **B.** It is impossible to know how to motivate students.
- **C.** Adult learners are motivated differently than children.
- **D.** The instructor is not responsible for student motivation.

**2.** When working with adult learners, the instructor should recognize
- **A.** All adult learners are pretty much alike and, therefore, have much the same motivation.
- **B.** Members of different age groups may be motivated differently.
- **C.** The instructor cannot create motivation in adult learners.
- **D.** Adult learners will already be motivated.

**3.** According to Houle, which of the following adult learner type places the responsibility of disseminating knowledge upon the instructor?
- **A.** Goal-oriented
- **B.** Activity-oriented
- **C.** Action-oriented
- **D.** Learner-oriented

**4.** Motivation is defined as
- **A.** The integration of new information into the existing values, beliefs, and behaviors of adults.
- **B.** The way in which the individual prefers to learn.
- **C.** Behaving in a certain way.
- **D.** The activator or energizer that calls the behavior into action.

**5.** Which generation was the first where large numbers had both parents working?
- **A.** Baby boomer
- **B.** Generation X
- **C.** Generation Y
- **D.** Generation Z

**6.** Which of the following characteristics is *not* a trait of Gen Xers?
- **A.** Self-starters
- **B.** Problem-solvers
- **C.** Prefer structure
- **D.** Expect immediate gratification

**7.** How might an instructor improve communication with students?
- **A.** Learn their names.
- **B.** Use gestures routinely.
- **C.** Avoid pauses.
- **D.** Speak loudly.

**8.** How might an instructor correct inappropriate behavior by a student?
- **A.** Avoid eye contact.
- **B.** Move away from the student.
- **C.** Address the behavior.
- **D.** Chastise the student in front of the class.

# The Learning Process

# CHAPTER 4

## NFPA 1041 Standard

**Instructor I**

4.3.2* Review instructional materials, given the materials for a specific topic, target audience and learning environment, so that elements of the lesson plan, learning environment, and resources that need adaptation are identified.

(A) **Requisite Knowledge.** Recognition of student limitations, methods of instruction, types of resource materials, organization of learning environment, and policies and procedures. [p. 58–69]

(B) **Requisite Skills.** Analysis of resources, facilities, and materials.

4.4.3 Present prepared lessons, given a prepared lesson plan that specifies the presentation method(s), so that the method(s) indicated in the plan are used and the stated objectives or learning outcomes are achieved.

(A) **Requisite Knowledge.** The laws and principles of learning, teaching methods and techniques, lesson plan components and elements of the communication process, and lesson plan terminology and definitions. [p. 58–62]

(B) **Requisite Skills.** Oral communication techniques, teaching methods and techniques, and utilization of lesson plans in the instructional setting.

4.4.5 Adjust to differences in learning styles, abilities, and behaviors, given the instructional environment, so that lesson objectives are accomplished, disruptive behavior is addressed, and a safe learning environment is maintained. [p. 65–69]

(A)* **Requisite Knowledge.** Motivation techniques, learning styles, types of learning disabilities and methods for dealing with them, and methods of dealing with disruptive and unsafe behavior. [p. 65–69]

(B) **Requisite Skills.** Basic coaching and motivational techniques, and adaptation of lesson plans or materials to specific instructional situations.

**Instructor II**

NFPA 1041 contains no Job Performance Requirements for this chapter.

## Knowledge Objectives

After studying this chapter, you will be able to:

- Describe the laws and principles of learning.
- Identify the three types of learning domains.
- Define learning styles and discuss the effects of learning styles on the classroom.
- Describe learning disabilities and methods of dealing with learning disabilities and disruptive behavior in adult learners.

## Skills Objectives

After studying this chapter, you will be able to:

- Analyze student learning styles and preferences.
- Demonstrate methods of dealing with learning disabilities.

## You Are the Fire Service Instructor

You are preparing to teach an Introduction to Firefighting class. After working for several weeks by reviewing the curriculum and identifying learning objectives and outcomes, you have developed lesson plans, identified the resources needed to teach the class, and prepared the classroom for the first day. Your challenge is not just to master the curriculum, but also to teach all of the adult learners in the class, and to offer all students the best possible learning environment. Although you have been a fire service instructor for several years, you teach only occasionally. Wanting to do the best job you can, you consult some of the more experienced fire service instructors at the training center.

In your conversations with those fire service instructors, you are reminded about some of the basic laws and principles of learning and the ways in which that information can be used to assist you. You discuss learning domains, learning styles, and student behavior, and your colleagues offer some practical guidelines to help you meet the challenges of teaching the adult learner.

1. Is it valuable for you to know about your students' learning styles?
2. Why should you seek more information about the various learning domains?
3. What does knowing about learning styles and learning domains have to do with being a fire service instructor?

## Introduction

**Learning** is a change in a person's ability to behave in certain ways. This change can be traced to two key factors—past experience with the subject (e.g., in the field) and practice (e.g., training in the classroom). Learning can occur both formally (inside the classroom) and informally (around the dinner table) (Connick, 1997). Formal learning does not occur by accident—it is the direct result of a program designed by an instructor (Butler & McManus, 1998). An adult learner may intentionally set out to learn by taking classes or by reading about a subject. He or she may also gather information through the experience of living that changes the learner's behavior. Informal learning occurs spontaneously and continually changes the adult learner's behavior. Ideally, learning is created through the blending of individual curiosity, reflection, and adaptation (Stewart, 2003).

Learning takes time and patience. As the fire service instructor, you have the opportunity to positively influence fire fighters at all levels of the fire department FIGURE 4.1. You are a leader and have a responsibility to teach information, hone skills, and promote positive values and motivation in your students.

FIGURE 4.1 As the fire service instructor, you have the opportunity to positively influence fire fighters at all levels of the fire department.

## The Laws and Principles of Learning

Numerous theorists, including Edward L. Thorndike and his contemporaries Edwin R. Gutherie, Clark L. Hull, and Neal E. Miller, have studied learning and developed ideas that have become recognized as the basic laws of learning. According to Thorndike, there are six laws of learning:

1. *The law of readiness*: A person can learn why physically and mentally he or she is ready to respond to instruction.
2. *The law of exercise*: Learning is an active process that exercises both the mind and the body. Through this process, the learner develops an adequate response to instruction and is able to master the learning through repetition. For

FIGURE 4.2 Adult learners use the five senses (hearing, seeing, touching, smelling, and tasting) to obtain information.

example, a fire fighter can master ropes by practicing tying knots every night for a month.

3. *The law of effect*: Learning is most effective when it is accompanied by or results in a feeling of satisfaction, pleasantness, or reward (internal or external) for the student—for example, when the student receives an A on a pop quiz.
4. *The law of association*: In the learning process, the learner compares the new knowledge with his or her existing knowledge base. For example, suppose a new type of nozzle arrives at the fire department. During training, fire fighters will compare the new nozzle with the nozzles with which they are already very familiar, having used them countless times at incidents.
5. *The law of recency*: Practice makes perfect, and the more recent the practice, the more effective the performance of the new skill or behavior. Running drills on new skills will reinforce and perfect training in fire fighters.
6. *The law of intensity*: Real-life experiences are more likely to produce permanent behavioral changes, making this type of learning very effective. A student can hear your lecture on the importance of always wearing full personal protective equipment at an incident a dozen times, but the lesson might not really sink in until the helmet prevents a concussion during an incident (Thorndike, 1932).

Adult learners use the five senses (hearing, seeing, touching, smelling, and tasting) to obtain information FIGURE 4.2. The type and number of senses used to learn determines how learning occurs and what is remembered. Seeing (e.g., a visual demonstration of a skill) is generally the most effective means of learning, followed by hearing (e.g., listening to a lecture), smelling (e.g., smelling smoke), touching (e.g., handling fire equipment), and tasting (rarely used in training fire fighters).

## The Behaviorist and Cognitive Perspectives

Two schools of thought have developed to explain how behavior evolves: behaviorist and cognitive. According to the **behaviorist perspective**, learning is a relatively permanent change in behavior that arises from experience. Not all changes in behavior reflect classroom learning. Many changes in behavior take place owing to a person's maturation and physical changes as he or she ages (Rathus, 1999).

According to the **cognitive perspective** (mental), learning is an intellectual process where experience contributes to relatively permanent changes in the way individuals mentally represent their environment. Cognitive theories stress the use of mental capabilities to change behavior. Cognition is defined as the acquisition of knowledge through the use of perception, encounters with ideas, and obtaining of experiences (Lefrancois, 1996). Thus cognitive learning is demonstrated by recall of knowledge and use of intellectual skills. It entails the comprehension of information, organization of ideas, analysis and synthesis of data, application of knowledge, choice of alternatives in problem solving, and evaluation of ideas or actions. According to this perspective, learning is an internal process that is demonstrated externally by behavioral changes.

## Competency-Based Learning Principles

Learning that is intended to create or improve professional behavior is based on performance, rather than on content. Such **competency-based learning** is usually tied to skills or hands-on training. Not surprisingly, much of what is accomplished on the fire ground occurs through hands-on activities that require proficiency with skills—for example, the ability to force entry on a door. Competency-based learning is based on what the adult learner will do.

Competency-based learning originates with the identification and verification of the needed competencies to perform particular skills. The competencies are actually job requirements, and ideally they will have been identified through a job analysis. Once the job requirements have been identified, they can be used as course goals and put into written form as lesson objectives.

To ensure the success of a competency-based course, it is essential to follow a standardized process. This process ensures that each successive step has the proper foundation. Without the proper foundation, what may look like an excellent course might not address the real skills issues of the fire department. The design of a competency-based course requires the following components:

- Competencies exist.
- Job analysis is performed.
- Standards are identified and set.
- Course goals are identified and written.
- Lesson objectives are identified and written.
- Competency-based instruction is delivered.

The most critical facet of competency-based learning is the notion that the skills must be mastered by the adult learners for each of the competencies. To ensure that this requirement is met, students must be given enough time and appropriate instruction to both learn and master the skill. You and the students must understand that with mastery comes the responsibility of maintaining a particular performance level. Thus learning objectives, instruction, and the evaluation of

FIGURE 4.3 Immediate feedback needs to be given on the training ground to ensure that each student masters the skill.

### Teaching Tip

Instructors always need to be thinking about what might help their students learn. The creation of an environment that addresses the physical, emotional, and cognitive needs of all students provides for a successful learning experience. To achieve this kind of supportive environment, make sure that your classroom accommodates differences in students' learning styles.

### Teaching Tip

The use of teams can be an especially helpful strategy in the classroom setting. It allows students to work with people who differ from them in terms of how they learn, think, or respond.

the skills must be tailored to performance in the field, not just to the acquisition of abstract knowledge.

A competency-based learning course must be flexible enough to be adapted to the needs of all students. Each student inevitably learns at his or her own pace and in his or her own individual fashion. Some students may take longer than others to learn the skill, so you must take each individual's preferred learning style into account when teaching skills. Also, appropriate and immediate feedback needs to be provided to students to allow them to achieve mastery level of the skill FIGURE 4.3.

## Forced Learning

There is an important fact of which you need to be aware: Adults cannot be forced to learn. Unfortunately, there are too many adult learners who attend training programs or engage in educational situations only because they are required to do so. Owing to the dynamics of the fire service, there always seem to be new continuing education credits to be accumulated for recertification or some new required certification to be obtained. Mandating learning often results in a lack of motivation in classroom activities and a poor attitude toward the learning process in general, which present quite a challenge to the fire service instructor.

The presence of students who are mandated to attend a class or training session should alert you that the lesson plans might need to be altered. For example, if you are assigned to teach a CPR recertification course to paramedic students, then you might look for ways to involve students in both the teaching process and the learning process. The creation of such dual roles will help capture their attention.

### Areas of Interest

All adults have certain areas of interest. These areas may help reinforce the desire to learn. The source is less important than its mere presence; because all students need to have encouragement if learning is to take place. This positive reinforcement can serve to increase the motivation of students who are forced into your classroom. You may need to acknowledge the

### Ethics Tip

One method that is routinely used to encourage learning from forced participants is tapping into the participants' knowledge and letting them teach from their experiences. Is it ethical to have someone other than the assigned fire service instructor present the instructional material? Would it be ethical for a college business professor to have a student with equal business experience teach other students in an effort to maintain that student's interest in the class?

Like all ethical dilemmas, both situations have positive benefits. Having an otherwise disengaged student present instructional material will increase the learning for *that* particular student. Also, other students often learn much from the student presenter, with who they can identify more easily. However, if the other students do not receive the necessary instruction, or if there is the perception that they may not receive proper instruction, this practice can be detrimental. Having the assigned fire service instructor teach the material ensures quality and quantity.

When an experienced student is forced to attend your class and you are thinking about taking advantage of that student's value as a peer-teacher, consider the material, the experience of the student, the reputation of the experienced student among his or her classmates, and the reaction of the other students to a "guest" presenter from their own class. Most fire service instructors would agree that the student should not do the majority of the teaching because students have paid to benefit from the instruction of the professional fire service instructor; however, allowing the experienced student to assist in teaching small components might be acceptable to both the class and the experienced student.

## Applying the JOB PERFORMANCE REQUIREMENTS (JPRs)

Every student in each class you instruct learns at a different rate with different levels of comprehension and understanding. Likewise, each student responds differently to the methods of instruction that are used in a class. An experienced fire service instructor knows how to adapt a lesson plan while the presentation is under way. The need for such flexibility may be one of the more difficult tasks for new fire service instructors to understand. The goal of any presentation is to improve the student's understanding of the objectives. To do your job correctly, you must have a working knowledge of the learning process. Your understanding of these factors is critical to the success of your entire class.

### Instructor I

The Instructor I will deliver a prepared lesson plan using a variety of instructional methods. The appropriate method to be used might be identified after review of the content of the lesson plan, and perhaps even more by the analysis of the audience who will participate in the training. Student-centered delivery with the students involved in the learning will normally make the longest-lasting impressions on the students' learning.

### JPRs at Work

Review your assigned lesson plan and determine the best methods of presenting the material based on the objectives and evaluations developed for the class. Adjust the lesson plan delivery based on the students' understanding and progress through the material.

### Instructor II

When developing objectives and lesson plan content, the Instructor II must be aware of the various domains of learning and the levels of comprehension intended for the students. The objective must be written to the appropriate level and applications in the lesson plan must be developed to help students apply what they have learned. Such exercises check the students' understanding of course content.

### JPRs at Work

NFPA does not identify the JPRs that relate to this chapter. Nevertheless, it is important to understand how students learn and the most effective ways of instructing so that you can be the most effective instructor possible.

### Bridging the Gap Between Instructor I and Instructor II

Often a coffee-table discussion between instructors at different levels can assist in determining the best methods of delivering a prepared lesson plan. The Instructor II will typically have more experience in the development and delivery of training and should mentor the Instructor I in ways to engage the students in the learning process so as to achieve better outcomes. Select appropriate methods of instruction by understanding the learning process.

### Instant Applications

1. Review the laws and principles of learning, and identify an instructional method that incorporates each of these laws and principles.
2. Make a list of the students you will have in upcoming classes. Based on their knowledge, experience, and other instructional factors, identify the ways each of these students might learn and determine how you can make adaptations appropriate for the entire class.
3. Review how you would try to increase your students' understanding of basic material by adjusting the lesson plan to meet individual needs.

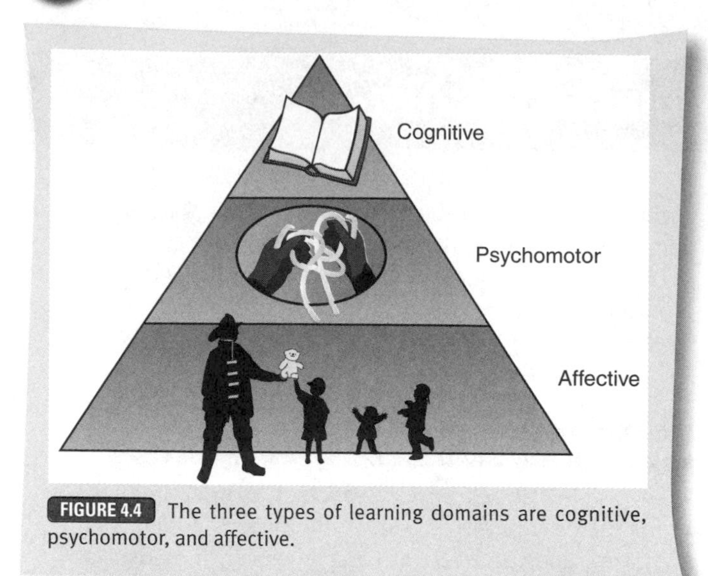


FIGURE 4.4 The three types of learning domains are cognitive, psychomotor, and affective.

## Teaching Tip

To be a great fire service instructor, you must truly know yourself. Some of us are excellent speakers, while others are effective writers. What are your strengths? What do you need to work on? Taking a personal inventory assessment such as the Myers–Briggs Type Indicator can clue you in to behaviors you may need to improve.

## Safety Tip

Sometimes you will find that the fire fighters in your department are ready and willing to learn and experience training—as a group. When it comes down to individual fire fighters, however, blocks to learning may spring up. For example, a fire fighter may be tired and unable to concentrate because he worked an incident all night. When basic needs are not met, learning becomes a second priority, which can in turn create a safety issue. A student may perform an unsafe act due to his or her physical or mental state, or a student may miss information that he or she must be able to perform on actual incidents because of inattention.

areas of special interest of reluctant students to guide them through the learning process and keep them engaged in your class. Knowing your students and their areas of interest allows you to tap into their knowledge base and become a more effective fire service instructor.

## Learning Domains

In 1956, Benjamin Bloom and his research team identified three types of **learning domains** (categories in which learning takes place) FIGURE 4.4:

- **Cognitive domain**: knowledge
- **Psychomotor domain**: physical use of knowledge
- **Affective domain**: attitudes, emotions, or values

Today, Bloom's classification system is widely known as Bloom's Taxonomy of the Domains of Learning. **Bloom's Taxonomy** has had a widespread influence in the field of learning research. Researchers have since developed new strategies of classifications and identified additional learning domains to build upon Bloom's original taxonomy.

While Bloom identified three general learning domains, the learning process seldom takes place solely in one domain. Additionally, within each of the domains, the levels of learning build one upon the other. These levels, which have been expanded and refined by subsequent researchers, serve as the building blocks of the learning process. Within each level are found three basic sublevels: knowledge, application, and problem solving. Being aware of these levels and Bloom's three original learning domains will assist you in developing lesson objectives and course material that will address a specific level of learning.

## Cognitive Learning

The most commonly understood of the learning domains is cognitive learning, which results from instruction. Bloom's Taxonomy distinguishes six levels of cognitive learning:

- Knowledge: remembering knowledge acquired in the past
- Comprehension: understanding the meaning of the information
- Application: using the information
- Analysis: breaking the information into parts to help understand all of the information
- Synthesis: integrating the information as a whole
- Evaluation: using standards and criteria to judge the value of the information

Each of these levels builds on the previous one. The most basic is the knowledge level; the highest is the evaluation level. By having an understanding of the cognitive domain of learning, you will be able to build more effective lesson plans and utilize students' thinking processes to encourage them to learn the information and apply that new knowledge. Adult learners should be able to fit the abstract thinking, facts, and information you present into the context of their jobs.

Consider these examples of cognitive learning and its application for fire fighters:

- Learning terms, facts, procedures, or principles: The fire fighter uses cognitive learning to learn the standard operating procedures of the fire department.
- The ability to translate and explain facts and principles: The fire fighter is taught the basic principles of fire and then asked to identify fuel types.
- The ability to apply facts and principles to a new situation: The fire fighter is able to size up the scene, formulate a plan based on that analysis, and evaluate all possible outcomes.

## Psychomotor Learning

For fire fighters, psychomotor learning is the most common learning domain. The term "psychomotor" refers to the use of the brain and senses (*psycho*) to tell the body what to do and the use of the muscles (*motor*) to tell the body how to move. The psychomotor learning domain involves the ability to physically manipulate an object or move the body to accomplish a task or perform a skill. This kind of learning is also referred to as **kinesthetic learning**.

As in the cognitive domain, the learning phases in the psychomotor domain build progressively one upon the other. According to Bloom's Taxonomy, there are six levels:

- Observation: watching the skill or activity being performed
- Imitation: copying the skill or activity in a step-by-step manner
- Manipulation: performing the skill based on instruction
- Precision: performing the skill or activity until it becomes habit
- Articulation: combining multiple skills together
- Naturalization: performing multiple skills correctly all the time

Each level of learning in the Bloom's Taxonomy psychomotor domain builds upon the previous level: observation → imitation → manipulation → precision → articulation → naturalization. By understanding that the physical use of knowledge is the basis for students learning to perform the skill, you will be able to develop a lesson plan that involves demonstrating the skill to students, who then practice until they are able to flawlessly perform that skill. As students progress from one level to the next within the psychomotor domain, they may experiment with new ways to perform the skill. This trial-and-error process is completely natural as students learn how to adapt the skill and become more familiar with the activity.

Awareness of the following aspects will assist you in teaching the psychomotor skill and developing an effective lesson plan:

- Gross body movement: the ability to move the arms, legs, and shoulders in a coordinated controlled manner
- Fine motor control: the ability to use hands, fingers, hand–eye coordination, and hearing
- Verbal behaviors: using sound to communicate
- Nonverbal behaviors: using facial expressions and body gestures to communicate

In your development of lesson plans, ideally you will identify how a student should perform a skill or activity by carrying out a task analysis and identifying the steps involved in performing the activity. Here are some examples of psychomotor learning and its application:

- Comprehensive skill approach: The skill is demonstrated by you and the fire fighter watches.
- Step-by-step approach: The fire fighter performs the skill in a precise step 1, step 2 manner until the entire skill activity is accomplished.
- Repetition of the skill or activity until it becomes habitual and is correct: The fire fighter performs the skill until it becomes second nature and he or she executes the skill flawlessly. The instructor then presents "what if" questions to help the student determine when to perform the skill.
- Combining multiple skills together: The fire fighter is able to adapt to changing situations and defend his or her choice of skills for the activity.
- Performing multiple skills correctly all the time: The fire fighter can perform the skill and it is truly a natural activity despite the environment or circumstance in which the skill is performed.

FIGURE 4.5 Practice makes perfect.

For students to properly learn skills or activities, they need to practice, practice, practice FIGURE 4.5. As the brain processes the attempts at performing the skill, it also learns to avoid what does not work or what was in error.

## Affective Learning

Affective learning is the feeling or attitude domain. It focuses on those characteristics that make each person unique—that is, an individual's preferences, perceptions, and values. Many of these characteristics will have evolved in individual fire fighters over long periods of time, so attitudes may not change immediately after the introduction of a new concept. Instead, learning within this domain progresses from simple awareness, to acceptance, to internalization, to finally acting out the attitude. Bloom's Taxonomy identifies five learning levels within the affective domain:

- Receiving: becoming aware of the skill or concept
- Responding: acknowledging the implications of the skill or concept and altering behavior accordingly
- Valuing: internalizing the skill or concept and having it become part of everyday life
- Organizing: comparing and contrasting skills or concepts
- Characterizing: adopting and personalizing the skill or concept

Each level of learning in the Bloom's Taxonomy affective domain builds on the previous level: receiving → responding → valuing → organizing → characterizing. Because much of

# Voices of Experience

As a new fire service instructor, I held the belief that I could apply the same methods of instruction that I had experienced in grammar school, high school, and college to the fire service instructional process. I quickly realized that in a volunteer organization, a majority of which was staffed with senior members who had 20 years of volunteer service, that lectures, written tests and overhead transparencies wouldn't translate well. These students wanted to be involved and share experiences.

*"These students wanted to be involved and share experiences."*

Firefighters with various educational backgrounds and learning abilities need basic lesson plans and training drills to be taught at a uniform level but with applications of the class content adapted to the skill level of each person. I found that using a blend of several instructional methods achieved this. For example, newer members responded to visually based instruction while more experienced members responded to participatory learning and skill demonstration.

Identification of skills and weaknesses and knowing your audience before you begin any learning process helps the instructor be better prepared to present the material. When preparing for a class, I try to remember how I learned the topic I am about to teach, what presentation techniques worked and what did not along with how today's learning environment and learning needs have changed. I consider empathy, or what it's like to be the student as an important consideration in the learning process. I also consider who the audience will be and then try to identify the best methods of instruction and any barriers to the instructional process. Other things I consider are the experience levels of the students, the level of the objectives or course content, the presentation methods necessary to deliver the material and any physical characteristics of the training area. If the training is to be practical or hands-on in nature, always consider the time needed to give all the participants an opportunity to perform the skill and an opportunity to review their performance with the instructor. Each participant will perform the skills at different levels and each participant will require instructor attention and feedback.

*Forest Reeder*
Pleasantview Fire Protection District
LaGrange Highlands, Illinois

this learning takes place inside the mind, it is important to look for subtle changes in students to confirm that affective learning is occurring. Here are some examples of affective learning:

- Acquisition of new values: The fire fighter is willing to learn the information and attends class regularly.
- Acknowledgment of the concept: The fire fighter studies for tests and participates in class activities.
- Internalization of values: The fire fighter volunteers to participate in an extra-credit activity.
- Internalization of the organization of information: The fire fighter decides to pursue continuing education because the additional training can keep him or her abreast of new developments and technologies.
- Full adoption of the new values: The fire fighter decides to pursue a degree in fire service management.

The affective domain is difficult to measure, but it should not be overlooked. The instructor must be observant to identify a student's reaction to the learning situation and changes in his or her value system.

**Teaching Tip**

Exercises that force students to compare two events involving similar problems demonstrate how different values may be applied to similar situations as well as how values may differ from one situation to another. For example, ask students to consider two different room and content fires: one that involves entrapment of children in their home, and one that involves entrapment of adults in a crack house. How do students perceive each of these scenarios?

## Learning Styles

### What Is a Learning Style?

When preparing for a class, you must ask, "What is learning?" as well as "How does a particular individual learn?" To answer these questions, you need to recognize that each student has a different **learning style**—that is, a way in which he or she prefers to learn. During the internal process of learning, a person encounters new information, analyzes it, and rejects or adopts it. Each and every adult learner has a distinct and consistently preferred way of perceiving, organizing, and retaining information. Learning styles are habitual modes of processing information; they show an individual's predisposition to adopt a particular learning strategy, regardless of the specific demands of the class. To use the terminology introduced by Bloom's Taxonomy, learning styles comprise characteristic cognitive, psychomotor, and affective behaviors.

The process of learning directly relates to the learning environment and interactions that are influenced by an individual's preferred learning style. These interactions are the central elements of the learning process and show a wide variation in terms of their pattern, learning style, and quality. Although learning is an invisible mental and emotional process, the results of the learning process are frequently visible in that they may be traced back to certain experiences (e.g., a training session) and are evident in behaviors (e.g., performance of a skill).

### Learning Styles' Effects on Training and Educational Programs

If you teach in a manner that favors a student's less preferred learning style, you need to be aware that the student's discomfort level may interfere with learning. Therefore, it is beneficial to recognize and understand all student learning styles. This knowledge comes from knowing your students. If you are new to the organization or are serving as guest speaker, then discovering students' preferred learning styles could present quite a challenge. Nevertheless, members of a profession or group often show a decided tendency toward certain learning styles. Part of your presentation preparation may be to contact the agency ahead of time and gather specific information about the audience's demographics, their learning backgrounds, and experience levels. Of course, it is inappropriate to make assumptions about a group simply based on one single group identifier. For example, if you were assigned to teach a group of senior citizens in an assisted living center and assumed that all of them would be in wheelchairs, you would be very surprised to learn that most of them cook or play golf and lead very active lives.

The ability to learn is not solely dependent on your use of the student's preferred learning style. Most adult learners will adapt their learning styles and use nonpreferred learning styles if necessary. For example, a study by Mel Silberman in 1998 revealed that in every group of 30 learners, 22 are able to learn effectively as long as the instructor provides a blend of visual, auditory, and kinesthetic activity. The remaining eight learners have a much stronger preference for a certain learning style and may struggle to understand the information unless special care is taken to present the material in their preferred learning style.

To see how this works in practice, consider a group of fire fighters in a class on fireground support operations. Although the majority of the class wants to immediately start placing ladders and practicing ventilation, a minority prefer to watch a video and discuss the skills before actually practicing the operations **FIGURE 4.6**. Both groups want to learn, but the process is different for each. Always take everyone's learning style preferences into account.

A learning style is a characteristic indicator of how a student learns, likes to learn, and learns most effectively. Given the importance of understanding the fire service and those who work in it, you can imagine how your appropriate use of this knowledge might affect fire fighters' training and educational opportunities. An understanding of the preferred learning styles of fire service personnel is a valuable tool for fire service instructors and curriculum designers. Identifying any patterns of learning that are common among firefighting personnel may serve to improve instructional methods

**FIGURE 4.6** Both fire fighters want to learn how to raise the ladder; they simply have different learning styles.

and curriculum content for educational courses and training geared toward fire fighters. In short, a thorough understanding of the audience's preferred learning style makes for better curriculum developers, trainers, and fire service instructors. As in any field, the development of a well-balanced program encourages learning by optimizing the learning styles of the learners and enhancing their learning experiences.

By understanding learning styles, you will gain some basic understanding of the strengths and weaknesses of the average fire service student. This information is essential given the trend toward more continuing education for fire personnel. A myriad of new education and training programs for the fire service are being developed to meet the demands of the industry, increase skills, and elevate the level of professionalism. These courses are designed to include not only the cognitive domain for recall, but also application of new skills, problem solving, and the psychomotor skills.

To accommodate this expanded scope of learning, course designers and fire service instructors need to have an even deeper understanding of their audience. By understanding learning style preferences and determining how those learning styles form patterns based on the audience, you should be able to increase the amount of learning that actually occurs in the classroom.

Training and education for the fire service must change to meet the needs and wants of these fire fighters. The increasing diversity of fire personnel—both in terms of backgrounds and learning styles—presents an especially thorny challenge for today's fire service instructors, who must strive to deliver increasingly complex information to a broader range of students. Traditionally, the fire service focused on the physical activities associated with extinguishing fires and providing rescue services. In the past, training was the key to developing skills. In the last 15 years, the scope of the fire service has expanded dramatically to include responding to terrorism, dealing with chemical and biological warfare threats, and managing new medical technology. These more sophisticated demands require similarly more sophisticated training that goes beyond hands-on activities to encompass education in explosives, chemistry, and computers—and, likewise, new methods of instruction and learning techniques.

**Safety Tip**

Using the appropriate teaching style for each adult learner will increase safety.

Although there has been speculation about the preferred learning styles of fire service personnel, little scientific research has been done to support or investigate that speculation. Three decades of experimenting with learning styles of non-fire service personnel has convinced hundreds of administrators and educators of the effectiveness of teaching by first identifying, and then complementing, how each person begins to concentrate on, process, internalize, and retain new and difficult information and skills. Once the learning styles of students have been identified, you can select the best teaching approach.

## Measurement of Learning Styles

Generally speaking, the researchers that are focused on learning styles are divided into two distinct camps: those supporting a narrower view that emphasizes the cognitive learning domain, and those supporting a broader view that encompasses other learning-related factors, such as motivation and personal preferences (Biggs, 1993). In the former group, David Kolb suggested that individuals are likely to feel most comfortable in one of four learning modes: using either abstract or concrete perception and either active or reflective processing. His Learning Style Inventory (LSI) instrument, which is designed to determine a person's placement along these dimensions, is a model of cognitive processing—that is, how we process learning in the brain (Kolb, 1995).

In general, distinctions in three or four different learning styles are well accepted as more or less prototypes of learning styles. A variety of instruments have been developed to measure an individual's affinity for particular styles (Semeijn & van der Velden, 1999). With any of these instruments, caution needs to be taken to ensure that what is being inventoried is learning style and not personality type.

Some researchers have focused on the **visual, auditory, and kinesthetic (VAK) characteristics** of learning styles based on the idea that we all seem to have a learning style preference based on sensory intake of information. Adult learners use all three of these sensory functions to receive information, although one or more of these receiving styles is normally dominant and hence filters the information received. The most common form of information exchange is speech, which arrives in the adult learner's ear and is considered to be an auditory means of learning. In a visual characteristic, information arrives in the form of graphs, charts, pictures, color and layouts, maps, or patterns. An adult learner using the kinesthetic characteristic would include the senses of touch, hearing, smell, taste, and sight. These adult learners want concrete, multisensory experiences as they learn. The VAK learning style focuses on the *how* of learning; it does not concern itself with the *why* of learning styles.

Even though the eyes take in all visual information, information is perceived differently, which in turn leads to subtle differences in students' learning styles based on the visual component of the VAK model (Fleming & Mills, 1992). For example, sometimes the information is largely composed of printed words, from which some students appear to get a greater or lesser degree of understanding; at other times the information consists of mostly graphic elements. Fleming (1995) suggested a modification to the basic VAK scheme in which the visual characteristic is divided into iconic (symbolic) and textual characteristics. Iconic characteristic visualizers learn more effectively when they are exposed to diagrams, symbols, and other graphic matter. Textual characteristic visualizers prefer the printed word, which includes both reading and writing.

The **VARK Preferences** instrument is a learning styles inventory designed to help students identify how they prefer to learn in terms of the previously described visual, auditory, read/write, and kinesthetic characteristics (Fleming, 2001). Developed in 1997 by Fleming at Lincoln University in New Zealand, VARK contains 13 questions whose answers provide students with an indication of their personal learning preferences.

The learning preferences identified via the VARK Preferences questionnaire indicate the ways the adult learner wants to take in or give out information in a learning context. The four categories of learning preferences seem to reflect the experiences of the students when they are taking in or giving out information fairly accurately. The categories overlap to some extent, but Fleming has defined the following dimensions:

- **Visual (V)**. This perceptual mode emphasizes a preference for the depiction of information in charts, graphs, flow charts, and all of the symbolic arrows, circles, hierarchies, and other devices that instructors use to represent what could have been presented in words. This definition does not include the use of television, videos, films, or computers. Instead, most of these media are considered primarily aural and kinesthetic because of their presentation of sound and reality.
- **Aural (A)**. This perceptual mode describes a preference for information that is "heard." Students with this preference report that they learn best from lectures, tutorials, tapes, and talking to other students.
- **Read/write (R)**. This preference is for information displayed as words. Many academics have a strong preference for this modality.
- **Kinesthetic (K)**. This modality includes a preference for experience and practice with the information, such as hands-on training or use of a simulator. Although such an approach may invoke other modalities, the key is that the student is connected to reality, either through experience, example, practice, or simulation.

A fifth category (multimodals) was added to the VARK Preferences questionnaire when it was found that the majority of adult learners actually have multiple preferences for learning styles. Some multimodal students may need to process information in more than one mode to learn effectively (Fleming, 1992).

VARK is structured specifically to have practical implications—namely, to improve learning and teaching. The VARK Preferences inventory, which is available for free on the Web, has been widely accepted as both practical and thought provoking (Fleming, 1992). Its results can be used in both course design and classroom activities, as students often find their scores highly provocative. In particular, you may be able to better design lesson plans and present information if you have an idea of the learning styles of your students. It is also beneficial to take the VARK questionnaire yourself, because your teaching style often matches your preferred learning style—which may not match the preferred learning styles of your students. It is important to focus on students' learning styles, employ a particular method of instruction, and not neglect the needs of adult learners who are different or who require an alternative method of instruction.

## Learning Disabilities

As a fire service instructor, you may encounter students with a wide variety of learning disabilities. The Americans with Disabilities Act (ADA—federal legislation that was originally passed in 1990) classifies learning disabilities into these major categories:

- Reading disabilities range from the inability to understand the meaning of words to the inability to read or comprehend to dyslexia. This disability could become apparent in class if a student is asked to read a section of course material aloud or to summarize a paragraph from a textbook.

### Teaching Tip

The VARK questionnaire can be downloaded from the Internet (www.vark-learn.com), filled out by students, and graded by both the students and you. The results could be used as a point of discussion to learn about preferred learning styles. It is self-defining and directed activity.

- Some students lack the ability to write or to spell or to place words together to complete a sentence. This type of disability, which is referred to as **dysphasia**, might present itself when a student submits work for a course and the work is poorly written.
- Other students might suffer from **dyscalculia**, or difficulty with math and related subjects. Students with this disability might struggle on a hydraulics course or test.
- **Dyspraxia** is the inability to display physical coordination of motor skills. This condition could be observed on the training ground with a student's inability to complete a task such as climbing a ladder.
- A disorder that affects children but can be carried into adulthood is known as **attention-deficit/hyperactivity disorder (ADHD)**. A person with ADHD has a chronic level of inattention and an impulsive hyperactivity that affects his or her ability to function on a daily basis. This disorder is a documented condition identified by the Centers for Disease Control and Prevention (CDC). Although it can be treated, there is no known cure.
- Other types of disabilities could also affect the learning process, such as color blindness or poor vision. Some adult learners may have poor hearing and cannot properly hear material presented in the classroom.

Knowing and understanding each of these learning disabilities will help you teach. Unfortunately, in most cases you will not know that students have these problems until a class has already begun. If you are a guest lecturer, you will not get the opportunity to know your students until after the first few hours of instruction. Once you have identified students with some type of learning disability, then you need to adjust your teaching as necessary. There are many ways to help students in your class who are struggling to learn. For example, a few positive words in private or suggestions on a written assignment might help a struggling student to understand a concept that is just beyond his or her reach. Students who have learning disabilities are not unintelligent or unable to learn; they simply learn in different ways. They usually have average to high intelligence, but they may perform poorly on tests because of their disabilities if accommodations are not made for their special needs.

## Disruptive Students

Other types of students may pose challenges as well. For example, some students, for one reason or another, feel the need to establish their presence in the class by acting out in other ways. This was also discussed in Chapter 3.

- The class clown feels the need to make comments or jokes about the course material or perhaps others in the classroom.
- The class know-it-all has been there and done that already, regardless of the topic.
- The gifted learner is very familiar with the course material or has the ability to read quickly. This individual becomes bored because he or she is so far ahead of the rest of the class.

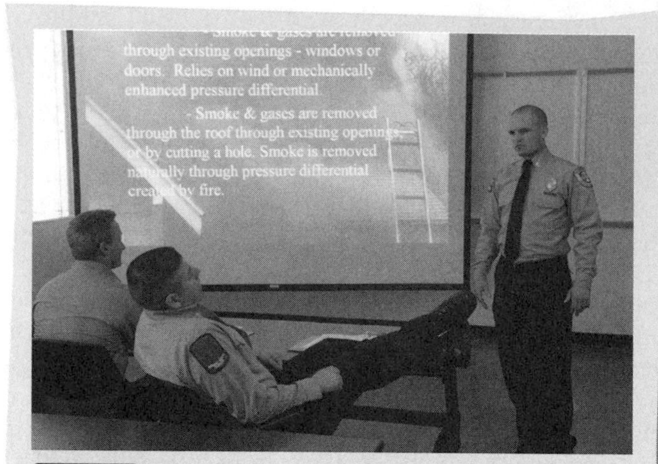


**FIGURE 4.7** Remember how your second-grade teacher silenced the class with one look? That same look works with adults, too.

Most often, these types of students have other underlying issues that cause them to seek out attention in a negative way. A variety of methods may be used to deal with these types of students, but in general the most effective method is to minimize their impact on the class by not giving into the same type of behavior. Most often a stern look is enough to get a class clown to stop cracking jokes **FIGURE 4.7**. If this doesn't work, then engage the student with a question that is meant to bring him or her into the class discussion. If this technique doesn't work, then at the next break a one-on-one discussion might be appropriate. The last step in dealing with a student who wants to continue to disrupt the learning environment would be to ask him or her to leave the class. In using this form of progressive discipline, the goal is to bring the student back into the classroom in a positive way that does not embarrass the student or you as the instructor.

A student who is a hesitator may be suffering from a learning disability. Such a student is often shy, reluctant, at a

### Ethics Tip

Is it ethical to explain an unfamiliar term to a student during an exam? For example, suppose a student is asked to "select a type of ritual" and one of the answers is "tenure." If the student does not know the term "tenure," is it ethical to explain it?

Again, a dilemma occurs when both instructor and student are trying to achieve a positive result. Not allowing any help during an exam achieves the goal of a level playing field for all students in the class. At the same time, the goal of learning is to see what the student knows. If the point of the exam is to see if the student knows what a ritual is, and not what tenure is, explaining the term "tenure": would allow for the exam to determine whether the student understands the term "ritual."

Ethical dilemmas have no easy answer. Using sound judgment and reasoning is required to determine the appropriate action.

## Theory into Practice

To identify which kinds of learners—visual, auditory, or kinesthetic—are present in your classroom, try the following exercise:

Tell students to sit quietly for 5 minutes—no reading, no talking, and no moving around. Observe what the students are doing. You may want to jot down some notes about each student so you can share your observations with them at the end of the 5 minutes.

- The visual learner will look around and absorb or read everything he or she sees.
- The auditory learner will mumble or make some kind of sound.
- The kinesthetic learner will fidget, may get out something to doodle on, or slump comfortably in his or her seat.

This may not be the most accurate or fun activity, but your observations will provide some valuable information about your students. You can provide feedback to the students about their behavior, and over time you should be able to see how accurate your observations are. Use this exercise as a springboard for your discussion on learning styles and behavior.

## Teaching Tip

When teaching adults, you are really more of a facilitator than a traditional teacher. Here are some core qualities that will help you become a good facilitator:

- Be genuine in your relationships with adult learners, rather than consistently adhering to the traditional role as "teacher." Show you really care about the students and their success in the classroom.
- Accept and trust in the adult learner as a person of worth. Provide positive reinforcement, and engage the learner with respect and value. Remember—people learn more from individuals whom they trust than from individuals whom they mistrust. Establishing a mutual climate of trust is important in fostering learning.
- Provide nonjudgmental understanding of adult learners' perspectives. Respond with empathy to both their intellectual and emotional perspectives. Remember that people are more open to learning when they are respected and feel supported rather than when they feel judged or threatened.
- Establish a climate for learning. Openness and authenticity are essential. People will be more likely to examine and embrace new ideas in an open and authentic environment.

loss for words, and quiet. Although he or she may know the material and have much to offer, the individual's shyness, fear, or lack of confidence may keep him or her from participating in class activities or answering questions. You can engage this student by asking nonthreatening questions and offering encouragement. You need to let a hesitator know that his or her contributions are worthwhile and important. Every student's participation in the classroom is important, so provide reassurance to this type of student as necessary.

The student who is bored or quiet may be displaying that behavior because of a particular circumstance. Sources of this problem may include a lack of interest in the subject, an objection to being forced to be in the class, unfamiliarity with the terminology in the class, boredom, and long and technical lectures. This student may just drift off mentally and refuse to participate in the classroom or become disruptive in other ways (see Chapter 3). As the fire service instructor, you need to keep all students interested and focused on the material, so keep your lectures interactive and concise.

Other students may not be interested in the topic. They may be in the class simply because they have to be. These students lack energy and attention, and you should try to determine whether this is a regular problem. If so, these students may have a disability or other problem that you may need to discover and address through counseling, tutoring, or other means.

A student who is a slow learner may be the most difficult for you to deal with. Such an individual has trouble keeping up and may not understand some or all of the material. One technique for handling this situation is to encourage input from other students—hearing the information differently may help. During a break, have a respectful one-on-one talk with the student to ensure understanding and comprehension.

Many students have some kind of impediment to learning, and most of those students have found unique ways to compensate for these obstacles. We all have times when our attention drifts or our energy is low, but if it is a reoccurring event it needs to be addressed because it will interfere with long-term learning. Some students may need help in the form of tutoring or individualized instruction, for example. These options should be discussed with students who have learning disabilities, and together you and the student should design a learning plan.

# Wrap-Up

## Chief Concepts

- In the education and training environment, you need ways of identifying what is helping the students to learn and what is not. What works for one student may not necessarily work for another.
- It is your responsibility to understand what learning is, which principles are associated with learning, how students learn, and how learning affects student behaviors.
- Teaching or instruction is the single greatest determinant of learning. The more you know about learning, learning styles, your students, and teaching methods and strategies, the more effectively you can influence student learning.
- Your effectiveness depends on being organized; communicating the goals, objectives, and expectations of the program clearly; and knowing the audience or group of students.
- Having an understanding of the learning process can help you to improve and continue to hone your instructional skills. It is helpful for students to understand the basics of learning theory, so that they will know what could potentially interfere with their own learning.
- Having an understanding of scientific research related to adult learners will help you become a better instructor and assist you in helping your students to learn.

## Hot Terms

**Affective domain** The domain of learning that affects attitudes, emotions, or values. It may be associated with a student's perspective or belief being changed as a result of training in this domain.

**Attention-deficit/hyperactivity disorder (ADHD)** A disorder in which a person has a chronic level of inattention and an impulsive hyperactivity that affects daily functions.

**Behaviorist perspective** The theory that learning is a relatively permanent change in behavior that arises from experience.

**Bloom's taxonomy** A classification of the different objectives and skills that educators set for students (learning objectives).

**Cognitive domain** The domain of learning that effects a change in knowledge. It is most often associated with learning new information.

**Cognitive perspective** An intellectual process by which experience contributes to relatively permanent changes (learning). It may be associated by learning by experience.

**Competency-based learning** Learning that is intended to create or improve professional competencies.

**Dyscalculia** A learning disability in which students have difficulty with math and related subjects.

**Dysphasia** A learning disability in which students lack the ability to write, spell, or place words together to complete a sentence.

**Dyspraxia** Lack of physical coordination with motor skills.

**Kinesthetic learning** Learning that is based on doing or experiencing the information that is being taught.

**Learning** A relatively permanent change in behavior potential that is traceable to experience and practice.

**Learning domains** Categories that describe how learning takes place—specifically, the cognitive, psychomotor, and affective domains.

**Learning style** The way in which the individual prefers to learn.

**Psychomotor domain** The domain of learning that requires the physical use of knowledge. It represents the ability to physically manipulate an object or move the body to accomplish a task or use a skill. This domain is most often associated with hands-on training or drills.

**VARK Preferences** A tool that measures a person's learning preferences along visual, aural, read/write, and kinesthetic sensory modalities.

**Visual, auditory, and kinesthetic (VAK) characteristics** Learning styles based on the idea that we all have a learning style preference based on sensory intake of information (visual, auditory, and kinesthetic).

## References


Biggs, J. (1993). What do inventories of student's learning processes really measure? A theoretical review and clarification. *British Journal of Educational Psychology*, *63*, 3–19.

Bloom, Benjamin S. (1984). *Taxonomy of Educational Objectives*. Allyn and Bacon, Boston, MA: Pearson Education.

Butler, G., & McManus, F. (1998). *Psychology*. Oxford, UK: Oxford University Press.

Connick, G. (1997). Beyond a place called school. *NLII Viewpoint*, Fall Winter.

Fleming, N. (1995). I'm different: Not dumb. Modes of presentation (V.A.R.K.) in the tertiary classroom. Research and Development in Higher Education, *Proceedings of the 1995 Annual Conference of the Higher Education and Research Development Society of Australasia (HERDSA)*, *18*, 308–313.

Fleming, N. (2001). A guide to learning styles: V.A.R.K. Retrieved 8/16/2002, from www.vark-learn.com.

Fleming, N. (2005). A Guide to Learning Styles: V.A.R.K. Retrieved 2/15/1995 from www.vark-learn.com/english/page.asp?p=whatsnew.

Fleming, N.D. & Mills, C. (1992). *Not Another Inventory, Rather a Catalyst for Reflection*. To Improve the Academy, 11, 137–155.

Felder, R. (1996). Matters of Style. ASEE Prism, 6(4), 18–23. Retrieved 6/15/1999 from www2ncsu.edu/unity/lockers/users/f/felder/LS-Prism.htm.

Goodman, P. (1962). The Community of Scholars. NY: Random House.

Hoetmer, G. (2000). *Fire services today, managing a changing role and mission*. Washington, D.C: International City/County Management Associations.

International Fire Service Training Association. (1994) *Fire service instructor*. 5th ed. Stillwater, OK: Fire Protection Publications.

Kolb, D. (1995). *Experiential learning: Experience as the source of learning and development*. Englewood Cliffs, NJ: Prentice-Hall.

LeFever, M. (1995). *Learning Styles—reaching everyone God gave you to teach*. Colorado Springs, CO: David C. Cook Publishing Co.

Lefrancois, G. (1996). *The lifespan* (5th ed.). Belmont, CA: Wadsworth.

Rathus, S. (1999). *Psychology in the new millennium* (7th ed.). For Worth, TX: Harcourt Brace College.

Semeijn, J., & van der Velden, R. (1999). Aspects of learning style and labor market entry: An explorative study. Maastricht, Maastricht University.

Silberman, M. (1998). *Active training* (2nd ed.). San Francisco, CA: Jossey-Bass/Pfeiffer.

Stewart, D. (2003). Computer and technology skills. Retrieved 10/23/2003, from http://www.ncwiseowl.org/Kscope/techknowpark/Kiosk/index.html.

Thorndike, E. (1932). *The Fundamentals of Learning*. New York: Teachers College Press.

Fire.jbpub.com

# Fire Service Instructor *in Action*

As the fire service instructor, you have been assigned to the training center. For the next several weeks, you will be teaching a skills recertification class. As you begin to plan your instruction materials, you think about the diverse group of people who will be in the classroom and wonder how the information you will be presenting can best be shared with them.

**1.** Which of the following questions would help you prepare for this class?
   **A.** On which day of the week will the class be given?
   **B.** What shifts are the students from?
   **C.** Which type of learning or relearning will be going on?
   **D.** What time will lunch be?

**2.** How should the information be presented to the students?
   **A.** You should present the information in the way that you prefer to learn.
   **B.** You should pick one way to present the information and stick with it throughout the course.
   **C.** It doesn't matter how the information is presented.
   **D.** You should provide critical thinking opportunities that appeal to a variety of student learning styles and preferences.

**3.** What is the purpose of evaluating students' skills performance?
   **A.** Provide students with feedback that will help them master the skill.
   **B.** Give students the opportunity to get a passing grade.
   **C.** Provide an exercise to fill up the allotted class time.
   **D.** Give students the opportunity to show how good they are.

**4.** Which of the following outcomes demonstrates that learning occurred?
   **A.** The student passes a written exam.
   **B.** The student passes a practical exam.
   **C.** The student correctly answers questions that he or she could not answer prior to the class.
   **D.** The student answers all of the answers on the exam correctly.

**5.** If you give the VARK Preferences questionnaire at the beginning of the class, you will be able to determine:
   **A.** whether learning has occurred.
   **B.** what the preferred learning styles of the students are.
   **C.** how effective your teaching style will be.
   **D.** how effective your teaching style was for the students.

**6.** According to Bloom, changes in attitudes would occur in the _______ learning domain.
   **A.** affective
   **B.** psychomotor
   **C.** cognitive
   **D.** behavioral

**7.** According to the behaviorist, perspective changes in behavior occurs due to:
   **A.** teaching.
   **B.** intellectual process.
   **C.** experience.
   **D.** conscious thought.

**8.** If you want to reach the greatest number of styles of learning, you should include material that:
   **A.** is delivered in the form of a lecture.
   **B.** is delivered in the form of a lecture and the students take notes.
   **C.** a student can see, hear, and touch.
   **D.** a student can hear, see, touch, smell, and taste.

# Practical Applications

PART III

# Communication Skills for the Instructor

# CHAPTER 5

## NFPA 1041 Standard

**Instructor I**

4.4.3 Present prepared lessons, given a prepared lesson plan that specifies the presentation method(s), so that the method(s) indicated in the plan are used and the stated objectives or learning outcomes are achieved.

**(A) Requisite Knowledge.** The laws and principles of learning, teaching methods and techniques, lesson plan components and elements of the communication process, and lesson plan terminology and definitions. [p. 76–77]

**(B) Requisite Skills.** Oral communication techniques, teaching methods and techniques, utilization of lesson plans in the instructional setting. [p. 80-81, 84-85]

**Instructor II**

NFPA 1041 contains no Job Performance Requirements for this chapter.

## Knowledge Objectives

After studying this chapter, you will be able to:

- Identify and describe the elements of the communication process.
- Describe the role of communication in the learning process.
- Compare and describe the different types and styles of communication.

## Skills Objectives

After studying this chapter, you will be able to:

- Demonstrate effective oral communication techniques.
- Demonstrate effective written communication techniques.
- Demonstrate the ability to use various communication styles in the classroom.

## You Are the Fire Service Instructor

You have a new class of ten volunteers to orient to the fire service and to your fire department. As you make your way around the room, you recognize that two of them are career fire fighters in the big city who are volunteering in their local community. You recognize another two recruits as mechanics from the public works department. In addition, there is a college professor and two recent high school graduates from your cadet program. The last three recruits have no firefighting experience outside of watching the fire hose truck go by during the Fourth of July parade.

The ability to speak and be understood by each of these recruits is a communication skill. This class will test your ability to read classroom cues and react in a positive, supportive manner. It is a test of your ability as a fire service instructor to communicate both verbally and in writing to such a diverse group of students.

1. Which communication barriers must you overcome?
2. How will you communicate effectively with such a diverse group?
3. How will the communication process enhance the students' learning?

## Introduction

Communication is the most important element of the learning process. Communication in and of itself is a process that you must master to become an effective fire service instructor. When you have mastered communication skills, the environment in which you are teaching becomes less of a factor in the learning process. Even if the environment is less than perfect, strong communication skills will enable you to be an effective and dynamic fire service instructor. Your goal is to make sure that the students leave the classroom with a greater knowledge base than they had when they came in.

This chapter addresses both the spoken and written communication processes as well as the environment's effect on communication. As mentioned in Chapter 4, students have different learning styles and use different cues when learning. Knowing how to access these diverse cues and use them to enhance the learning experience will make you a more effective fire service instructor.

The fire service classroom is so much more than four walls, desks, and a chalkboard. It also includes the back of an engine after the call, a burn building, and even a call itself. No matter what the environment, the end objective is the same: that the student learns. This requires you to have excellent communication skills.

### Teaching Tip

The fire service is steeped in tradition, but the means of communication used by fire fighters have certainly changed over the years. The first fire officers used speaking trumpets to communicate to the members of their department at a fire scene. Today we have more modern solutions for the transmission of the message, but the message itself remains the critical piece of the communication puzzle.

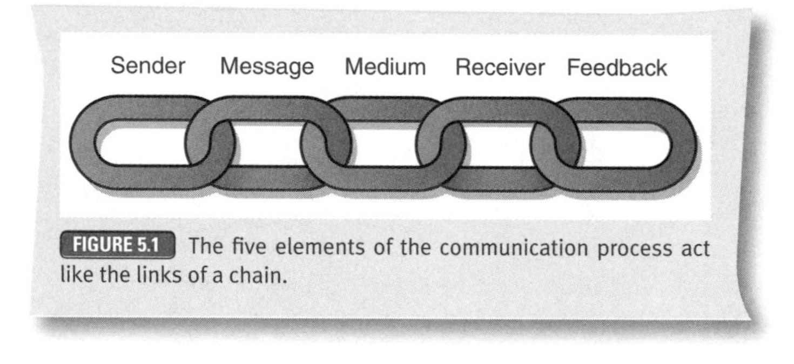


FIGURE 5.1 The five elements of the communication process act like the links of a chain.

## The Basic Communication Process

The basic **communication process** consists of five elements: the sender, the message, the medium, the receiver, and feedback FIGURE 5.1. Surrounding these elements is the environment. The five elements of the communication process act like the links of a chain. For the chain to function properly, all of the links must be attached to one another. If one of the links breaks, then the chain falls apart.

### Sender

In the communications chain of the fire service classroom, the **sender** is you, the fire service instructor. As the sender, you must know which style of communication to use. The choice of the style of communication is based on many factors, including your comfort level. If you are more comfortable using oral or verbal communication, then you are likely to use that style most often. Sometimes, however, this means of communication is not the most appropriate style. For instance, when you are giving feedback on a research paper, writing notes in the margins of the paper allows the student to carefully review and process your comments, as opposed to just saying, "Nice job," and handing the student the paper.

## Message

The next link in the communication chain is the **message**. What, as the sender, are you trying to communicate to your students? This may sound simple, but it is actually the most complex part of the communication process. For students to understand your message, they must hear it, put it into their own terms, and then ensure that their terms match your terms. Consider the command, "Put out that fire now." A student might hear this message as "Get a hose and spray some water right now," when the message really was "Get a fire extinguisher and suppress those sparks."

## Medium

How the message is conveyed is as important as the message itself. How a message is conveyed constitutes the **medium**. For example, your tone of voice, volume, speed of voice, and other qualities all convey information. If you give a lecture about a new piece of personal protective equipment (PPE) while sitting behind a desk, fiddling with a paper clip, and speaking in a very relaxed manner, you will convey to the students the message "This new piece of PPE really isn't that important," even though the message you are stating verbally might be "This piece of equipment will save your life."

## Receiver

The next link in the communications chain is the **receiver**. In the learning environment, this is the student. The receiver plays an important, active role in the communication chain; that is, the student needs to actively listen and maintain focus on your message. On the first day of class, remind students of their role in the communications chain. Fulfilling this role means leaving all distractions at the door and focusing solely on the information you present. Ensure that all students understand their active role in the communications process.

## Feedback

**Feedback** is the link that completes the communication process. It allows the sender to determine whether the receiver understood the message. In the fire service, the feedback process is used when orders are given over the radio. Consider this example:

> **Incident Commander:** Engine 2, take a line to Division C and provide exposure protection.
> **Engine 2:** Engine 2 copies—a line to Division C for exposures.

Feedback can also be nonverbal. For example, blank stares are feedback indicating that the students do not understand the message. Collectively asking students if they understand is not the best method of determining whether the message was properly received. The natural response is "yes," because no student wants to admit that he or she does not understand. Asking a student to describe your message in his or her own words is a much better way to make sure that the student understands your message, and it will also reinforce the learning process for the student.

### Teaching Tip

A student's previous experience and knowledge base plays a major role in determining whether the student is able to receive your message. Adapt your style and presentation for your students. If you "talk down" to students or if the presentation goes over their heads, the message is not in a form that they can receive.

## The Environment

Surrounding all of the links of the communication chain is the environment. The environment—which includes physical, social, and environmental factors—can dramatically affect the communication process.

The physical environment is the room or area in which communication is taking place. It could be a classroom, the training ground, or even cyberspace. The physical environment can either inhibit or enhance the learning process. Teaching in a classroom with radios blaring and phones ringing directly outside, for example, will inhibit the communication process. When students cannot hear or see you, your message has a limited chance of being understood. If you are in a classroom or a busy training ground, make sure that all of your students can see and hear you.

The social environment is the context in which the communication takes place. If the class is not enthusiastic about the message, the message will have a limited chance of being received. Mandatory training usually has this effect. You must make sure that your students want or need to receive your message—this is the art of instruction. Show your students *why* they need your message.

The final component is the environmental factor. The receiver must first feel that his or her needs are being met. An environment that is physically difficult for its inhabitants distracts students, for example. If the room is too cold or too hot, then your message will not be received properly. See Chapter 7, The Learning Environment for more information.

### Teaching Tip

Select an appropriate location for instruction. Teaching in a truck bay with heaters running may distract students and cause them to quickly lose interest in your message.

# Advanced Communication Process

## Nonverbal Communication

According to Charles R. Swindoll, American writer and clergyman, "Life is 10 percent what happens to you and 90 percent how you react to it." The same can be said about nonverbal communication: Communication is 10 percent what you say and 90 percent how you say it.

Nonverbal communication is a difficult communication skill to control because you generally don't watch yourself in the mirror all day. This type of communication includes the tone of your voice, your eye movement, your posture, your hand gestures, and your facial expressions. For example, eye contact can show sincerity. If you can look a student in the eye, you are demonstrating trustworthiness and caring. If you cannot maintain eye contact, you lessen the feeling of trust. You can also use your eyes to elicit a reaction: If you continue to look at a student, then that student is compelled to focus on you.

Standing straight and tall is a sign of confidence. You should work on your posture for this reason alone **FIGURE 5.2**. In addition to showing confidence, such a stance helps your back health during long days of instruction.

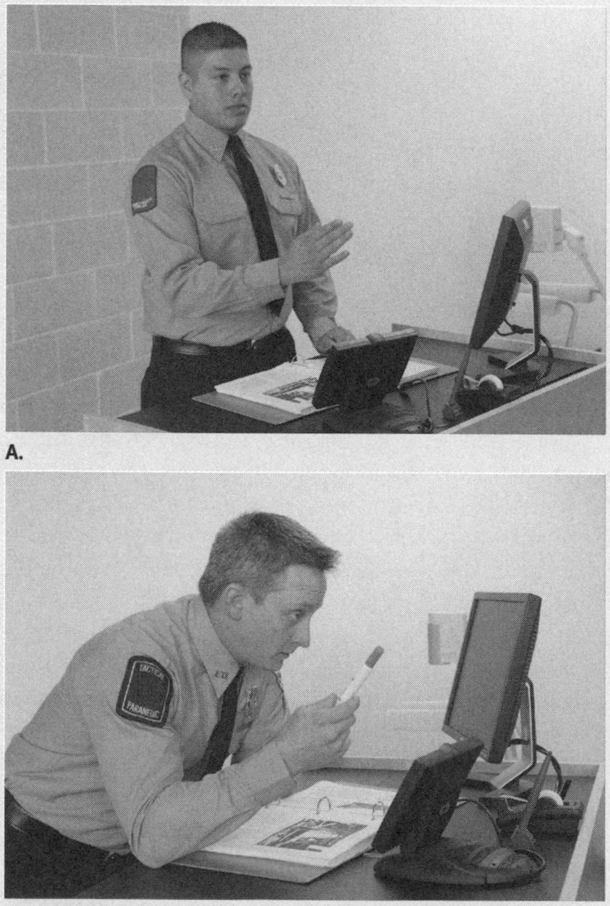

A.

B.

**FIGURE 5.2** **A.** Good posture is critical for good communication. **B.** Poor posture is unprofessional.

Hand gestures can express either aggression or enthusiasm. Rapid hand movements or repeated movements of the hands to the face can be signs of deception or uncertainty.

## Active and Passive Listening

As an instructor, you must be well versed in both active and passive listening. **Active listening** is the process of hearing and understanding the communication sent and demostrating that you are listening and have understood the message. It requires you to keep your mouth closed and your ears wide open. As part of active listening, it is also very important that you hear the entire message. All too often, instructors formulate an answer before the student finishes asking the question. If you find yourself jumping in with an answer, try restating the student's question before answering. "Why do we use foam on a chemical fire? We use foam on a chemical fire because . . ."

**Passive listening** is listening with your eyes and other senses without reacting to the message. Observe the student's body language and facial expressions **FIGURE 5.3**. Nervous movements can be indicative of misunderstanding or confusion, while a relaxed brow and alert eyes can indicate comprehension and interest.

### Safety Tip

Carefully monitor what you are communicating nonverbally. Unknowingly, you may communicate that students should take unnecessary risks on the training ground with your body language or how you are dressed.

### Teaching Tip

One of the best ways to develop your communication skills is straightforward: Practice them. One method is to have another fire service instructor sit through your presentation and evaluate you. This evaluation should include how you communicated your message, which nonverbal cues you gave, and how your tone of voice affected the delivery of your message. Through such a peer evaluation, you can see the student's perspective of you.

Another technique is to video your presentation. Watch the recording and look for any nonverbal cues that do not match your message. Many of the nonverbal cues you project become more obvious when you play the recording back at a faster than normal speed. Finally, close your eyes and listen to your voice at normal speed. Is your voice really sending your intended message?

## Applying the JOB PERFORMANCE REQUIREMENTS (JPRs)

Being able to communicate effectively is key to being a great fire service instructor. The most engaging fire service instructors are truly great communicators. A great communicator uses many senses beyond just hearing (i.e., his or her speaking). Good listeners complement the learning process by participating in the course delivery. When an instructor builds a relationship with his or her students, the bond forged improves the chances that class objectives will be achieved. Your communication skills are the key to the whole learning process because they tie together instruction methods and learning methods. The more you teach, the more you will appreciate this linkage—and the more you will understand why you must continually improve your communication skills throughout your instructional career.

### Instructor I

The Instructor I is assigned the duties of presenting (using communication skills) information from a prepared lesson plan. Understanding the elements in the communication process and knowing how to interpret formal and informal communication signals from your students will allow you to present better training sessions. Your communication skills are always being observed by your students, so continual improvement of those skills must be a professional development goal.

### JPRs at Work

You must understand the elements of the communication process if your presentation is to be effective. Know how to engage your students using a variety of communication skills throughout the training process.

### Instructor II

Always keep in mind the methods of instruction and the learning process when developing lesson plans for the Instructor I to deliver. Awareness of various instructors' communication skills may influence who is assigned to teach certain topics or the particular type of lesson plan to be presented. New instructors should be monitored and advised about ways to improve their communication skills.

### JPRs at Work

NFPA does not identify any JPRs for this chapter. However, the Instructor II must have knowledge of how instructors present material and the best methods for delivery. Some lesson plans will require instructors with specific communication skill levels.

### Bridging the Gap Between Instructor I and Instructor II

Work together to identify better communication skills. Experienced instructors should always audit new instructors' training sessions to see if they can give any pointers about ways to eliminate distractions, poor communications, or better ways of getting the objectives across. An Instructor I should attempt to utilize a variety of communication methods and be able to interpret the communication process throughout the delivery of a lesson plan. The Instructor II can help in this area by suggesting methods of instruction and providing input into the delivery methods being used.

### Instant Applications

1. Make a list of common distracters that you have witnessed when being taught by your instructors. How did these distracters impair the communication process? How could they have been avoided by using a different communication method?
2. Which great communicator traits can you relate to?
3. Watch a political candidate speak to an audience. Identify which communication skills the candidate uses to sway public opinion in an effort to gain votes. How can you translate that information to your own training delivery?
4. Watch news broadcasts and identify methods used to communicate the day's news. Which stories held your interest? Why? Why did you select the channel? Are there lessons you can learn from your selection?

FIGURE 5.3 Pay attention to what your students are telling you with their body language.

## Communications Background for the Instructor

According to Leah Davies,

> Being able to communicate is vital to being an effective educator. Communication not only conveys information, but it encourages effort, modifies attitudes, and stimulates thinking. Without it, stereotypes develop, messages become distorted and learning is stifled.

As a fire service instructor, it is important for you to be an effective communicator. One of the keys to the learning process is the conveyance of information. With effective communication skills, you can take the learning process to even greater heights by encouraging effort, modifying attitudes, and stimulating thinking.

## Verbal Communication

Verbal or oral communication is more than just talking. Many people talk, yet never actually communicate. Great communicators, by contrast, use many tools to convey their message. Watch and listen to the speeches of John F. Kennedy and Martin Luther King, Jr.—both were highly effective communicators, inspiring their followers to take action. At his inauguration in 1961, Kennedy said, "And so, my fellow Americans, ask not what your country can do for you; ask what you can do for your country." The inflection in his voice added emphasis to his words and reinforced his message. In August 1963, King said in his famous speech at the Lincoln Memorial, "I have a dream that my four little children will one day live in a nation where they will not be judged by the color of their skin but by the content of their character." In his speech, King used personalization so that the entire audience would put themselves in his shoes for a moment.

Many factors affect verbal communication, including language and tone. The volume of your speech is important. By lowering or raising your voice, you can emphasize a point. You can also use volume to gain the attention of your students. A loud, forceful tone will let the class know that you are very serious. In cases of imminent danger to a student, your voice must be loud and forceful.

Likewise, the importance of speaking softly cannot be understated. In some cases, a simple whisper in a student's ear will let the student know that he or she needs to pay closer attention because you are watching. When speaking to a student privately, maintain both a level of decorum and a softer tone to your voice. In addition, a softer voice is less threatening and does not put the student on the defensive. If the student is defensive, he or she is less likely to receive your message.

Bringing your voice to a loud crescendo and then making it softer can serve to emphasize your message. This technique is especially effective when you are using compare and contrast strategies. For example, during a discussion of tone of voice with fire officers, you might want to compare a dominant tone versus a supportive tone. You would speak loud and forcibly to show dominance, and then speak calmly and softly to show support. The same technique can also be used when describing an escalating event. For example, when describing a flashover, you would speak at normal volume to describe the fire building. As the fire builds, your volume would increase until the flashover occurs, with your voice booming.

The learning environment is also a factor when it comes to voice volume. If you are instructing outside or in a noisy environment, the volume of your voice may have to increase to project it over any **ambient noise**. Sometimes ambient noise is chattering students. In those cases, you should use volume to gain your students' attention and silence.

Projecting your voice is important. If you are stationary or do not have the ability to walk around, be aware of making your voice carry to the back of the room to each student. This feat can be accomplished by using a microphone FIGURE 5.4. If a microphone is not available, use your voice as a tool to convey your message. If you are more mobile, you may actually be behind or among the students, so be sure to project your voice so that everyone in the class can hear you.

### Theory into Practice

Great instructors can hold the attention of a class even after students have pulled an all-nighter. Unfortunately, not-so-great instructors can put even a triple-shot espresso enthusiast into a coma. The difference between the two types of instructors can sometimes be as simple as the level of enthusiasm projected through the voice.

Enthusiasm breeds enthusiasm. It creates the interest in the topic and in you. Focus on teaching the courses that fascinate you—this will add passion to your presentation. For those topics that may not be as exciting, research why the topic is important. Understanding the underlying value of the topic will help you make it more interesting for the students.

Dynamic use of voice is one of the most effective methods of holding students' attention. Knowing when and how to change your voice is a skill learned through practice. Practice making a *conscious* effort of really using your voice to convey your message until it becomes automatic.

FIGURE 5.4 Microphones can help project your voice in large lecture halls.

### Language

The fire service has its own unique language. We use terms that mean something completely different to the general public. For example, when we say "company," we mean a group of personnel assigned to perform a task; by contrast, the general public defines "company" as a business. Within the fire service, terminology can change from department to department and from region to region. A rescue in one department is called an ambulance in another. It is imperative that you use the terminology of the department when discussing equipment, apparatus, or procedures.

The use of appropriate language is critical. Avoid language that does not put you or your department in the best light. Ask yourself if you would use the same language around the family dinner table. Once spoken, words cannot be taken back. Exercise restraint and caution when speaking. Always be on guard with your speech: As a fire service instructor, you are held to a higher level. Many students have expectations of what a fire service instructor should say and do, so never let your guard down.

The tone of your voice can express more than your words. The purpose of tone is to express emotion. Your tone should always be positive, while expressing the passion of conviction. If your words say that you believe in something and your tone expresses apathy, the students will hear apathy, no matter what your words say. Watch an actor on stage. He may be having a bad day, but when he steps on stage, he portrays good humor and joy. As a fire service instructor, you must put your personal feelings aside and use your tone of voice to properly express the message.

#### Ethics Tip

In firehouses across the country, fire fighters routinely use language that they would never use in front of their grandmothers. Is it ethical to use that same language when instructing?

While your first instinct may be to communicate directly in the students' everyday language, it is inappropriate to do so in the classroom. As a fire service instructor, you are judged by your actions. Always take the high road.

#### Teaching Tip

Review departmental policies prior to teaching on a subject to ensure that your instruction reinforces policy.

#### Teaching Tip

The fire service classroom can be as simple as a room with chairs, tables, and any number of audiovisual devices to enhance the learning experience, or it can be as complex as a state-of-the-art "techno-heaven" with multimedia consoles that respond to your requests with only a touch of a button. Sometimes the classroom may be a bunkroom that serves as a multipurpose room with an overhead projector and slide projector, but it can also be the back of a fire engine or the base of the training tower. Many facilities have "dirty" classrooms—that is, designated areas allowing fire fighters in full turnout gear to discuss the training session before or after it occurs.

The purpose of the classroom is to facilitate the learning process. The physical location is less important than the communication process. The back of an engine can be a great learning environment if the proper communication skills are used. Many company officers take time after a call to discuss how the call went. This mini-critique can be of great value to the company, even without the traditional classroom or the audiovisual devices; only good communication is necessary.

## Written Communications

### Reading Levels

Most newspapers in the United States are written at approximately a fourth-grade reading level to allow for understanding by the general public. The newspaper does not intend to talk down to its readers who have a higher level of education; instead, it is simply ensuring that the majority of readers can easily understand the information presented.

Do you know the reading levels of all of your students? Most fire fighters are required to have a high school education or a GED, so the expectation is that all students can read and write at the high school level. Most educational textbooks are written at a reading level that the majority of fire fighters should understand. Technical editors and select fire service personnel review textbooks to establish its readability before it is printed and bound.

Most word processing programs provide an evaluation of the document's reading level based on the Flesch–Kincaid Readability index. With this index, a higher number indicates that the document is easier to read. Therefore, the higher Flesch–Kincaid number, the lower the reading level. The index is based on the average number of words per sentence

One of my first classes was a fire fighter academy class that was run through the local community college. It was the first time the class had been taught through the college, so I was on fresh ground. My class consisted of two fire fighters from the neighboring volunteer fire company, three paid-on-call members of a local department, one part-time fire fighter who traveled about 30 miles each day to attend class, and several paid-on-call fire fighters.

I had never taught a class that was 240 hours long to a group of fire fighters I knew little about, in a place that was not my normal fire station. I had some students with years of experience, but many who had none. I had students who came from firefighting families, but many who had not. I had students with higher education, but many with none. By experimenting with various communication styles, I was able to develop a common language and communicate effectively with each student.

Four of my students now work with me at our Fire District. Three others are career fire fighters and fire officers. Several students have returned and become fire service instructors for subsequent classes. By using good communication skills, I was able to get my message across to the students; this is the greatest job in the world.

*Terry Vavra*
Lisle–Woodridge Fire District
Lisle, Illinois

*"By experimenting with various communication styles, I was able to develop a common language and communicate effectively with each student."*

and the number of syllables per 100 words. As a fire service instructor, you must be aware of the length of your sentences and the complexity of the words you use. Always keep your students' reading levels in mind when writing.

### Writing Formats

It is also important to know which format you should use when writing. In general, when writing to individuals within your organization, you can use a memo (short for "memorandum"). Such a document is designed for in-house communications only. For example, a memo from the chief of your training division to all fire service instructors might inform them of an upcoming training drill. Written communications for outside the organization should be in letter format, using departmental letterhead, and should be signed by the sender. A letter would be the appropriate form of communication to Mr. Jones confirming a preincident planning appointment at his coffee shop, for example.

### The Five W's of Writing

The basic rule of thumb for writing is to include the "five W's":

- Who
- What
- Where
- When
- Why

Add an "H"—*How*—to expand the discussion further.

By including these points in your writing, you will ensure that you address the information necessary for readers to properly understand your message. For example, suppose you need to write a memo about a live burn that will take place at a donated structure on the outskirts of town **FIGURE 5.5**. You want your students and additional fire service instructors to know where and when to meet on the live burn day. You also want the entire fire department to know when the live burn is taking place, who is scheduled to be on site, and whether permission must be granted if additional fire fighters want to be on the live burn site.

This memo contains all five W's (plus H) of writing:

- Who: Lavalle Fire Department
- What: live burn training exercise
- Where: 1015 Willow Way
- When: July 10 at 10:00 am
- Why: permission must be granted to be on the live burn site
- How: fire fighters must see Officer Archie Reed for a permission request form

### The Rules of Writing

Good, clear, concise writing is critical to getting your message across. Fortunately, it's easy to improve your writing: If you follow the rules of good writing step-by-step and practice for 15 minutes each day, you can become an effective written communicator. As a fire fighter, you train on a regular basis to maintain your firefighting skills. Similarly, as fire service instructor, you need to practice your writing skills on a regular basis. This step involves actually writing and allowing fellow fire service instructors to read your writing to ensure that your message is coming through loud and clear.

To: Lavalle Fire Department
Cc: Chief Michael Deforge
From: Officer Archie Reed
Date: May 23
Subject: Live Burn on July 10

On July 10, the Lavalle Fire Department will conduct a live burn at 1015 Willow Way. Fourteen students from the Fire Suppression Course will be present. Students will participate in supervised training using medium-diameter hose lines during defensive operations.

Students and Fire Service Instructors Thomas, Kelly, Hinkler, and Andrews will assemble at the South Street Fire Station at 10:00 a.m. to be transported to the live burn structure. Safety procedures will be reviewed at the site prior to the live burn. In addition to the students and the fire service instructors, Engine Company 3 will be on site.

If you are not a student in the Fire Suppression Course, an assigned fire service instructor, or a member of Engine Company 3, you may not be on the live burn site without written permission. Please see Officer Archie Reed for a permission request form.

**FIGURE 5.5** A sample memo using the five W's of writing.

The rules of good writing begin with the paragraph. Each paragraph begins with an introduction, a sentence or two that states what you plan to discuss in the paragraph. Then comes the main body of the paragraph—the frame. Like the frame of a structure, the frame of the paragraph is the support that holds everything together. The conclusion reviews what was discussed in the paragraph and reinforces the frame.

For example:

> The fire officer must first consider the intended audience for the report before writing. For example, a fire officer may be assigned to prepare a study that proposes closing three companies and using quints that can respond as either engine or ladder companies. If the report is intended for internal use, the technical information would use normal fire department terminology. Conversely, if the report is intended for the fire chief to deliver to the city council, with copies going to the news media, many of the terms and concepts would need to be explained in simple terms for the general public. Always consider who your audience is before writing a report.

To practice constructing a solid paragraph, consider writing a two- to three-paragraph summary after every class. What was your message for the class? Did your students receive it?

Once you have mastered the building of a paragraph, you are on your way to constructing a complete written structure—anything from a two-paragraph memo, to a full-page letter, to a 10-page report for the city council. The paragraph is your basic building block of writing. All good writing has an introduction, a frame, and a conclusion. In an eight-paragraph report that is intended for the city council and explains why the fire department's training budget should be increased by 10 percent, for example, the report would be broken down in the following way:

- Paragraphs 1 and 2: Introduction. These paragraphs give an overview of why the fire department needs to increase the training budget.
- Paragraphs 3 through 7: Frame. These paragraphs discuss in depth the specific reasons why the training budget should be increased. For example, the city might have grown by 20 percent in three years and needs more fire fighters to provide good service.
- Paragraph 8: Conclusion. This paragraph summarizes the arguments for increasing the fire department's training budget.

### Style

Most of the writing you do as a fire fighter and a fire service instructor will be technical in nature. In such a case, your sole objective is to convey information—not to express emotion or to inspire passionate feelings. Writing in the fire service should not be boring or uninspiring, but it should remain true to the premise that you write to impart facts.

Once words are printed, they are difficult to retract. Read an editorial column in a newspaper to see the effect writing can have on people's thoughts and emotions. Words can be construed in different ways by different people, so it is extremely important to choose your words carefully. The meaning of words can change based on geographic location. A person's individual experiences also can change the meaning of a word. Know your audience before you begin to write. Your goal is to make certain that your message is clearly and properly conveyed to every reader.

> **Safety Tip**
>
> Choose your words carefully when writing. Make every effort to ensure that alternative interpretations cannot occur. Failure to do so can mislead students and ultimately lead to undesired performances.

## Communication Styles in the Classroom

Many types of speeches exist, including persuasive, informative, and special speeches. A keynote speech is generally a persuasive speech. The purpose of such a speech is to motivate and inspire the audience to move forward with the message given. Most presentations for a fire service instructor involve informative speeches, or lectures. As a good fire service instructor, you need to determine how your message needs to be communicated before determining the correct format for your presentation. You also need to evaluate your audience to determine how they will best receive your message.

At the most basic level of firefighting, the lecture is usually the most appropriate presentation method. Because students in these types of classes have very little experience to draw upon, it is best to use an informative lecture to communicate concepts and objectives. You need to pass the information to the student in a very direct manner. It will be repetitive in nature—some would say almost rote memorization.

In speaking to more experienced fire fighters, the lecture is more persuasive in nature. With these students, you may be taking a skill learned in basic training and discussing it in finer detail at a higher level. This type of class requires students to be convinced about the importance of the message; otherwise, they might tune out the lecture because they mistakenly believe that they know everything there is to know about the skill.

The third type of audience encountered by fire instructors is made up of experienced fire fighters and fire officers. These students tend to require discussions rather than informative lectures. Discussions engage these experienced students and encourage them to incorporate their knowledge and experiences into the learning process. This concept, which is sometimes referred to as a "tailboard chat," is used with more experienced fire fighters because it takes their existing knowledge and experience and melds it to a new concept **FIGURE 5.6**. By examining and reexamining the way we do things, and by discussing experiences and options, new ideas and new ways of doing things are developed. This sort of exchange can be very exciting and rewarding because you can witness an idea develop in front of you. In fact, many of the tools used in the fire service today evolved out of such discussions, when fire fighters saw a problem or a situation that needed to be improved and made adaptations to rectify the need. This style of instruction is sometimes referred to as facilitation.

Alternative communication methods should also be considered. For example, role playing can often reinforce learning objectives. In role playing, you can take the position of someone with whom the student interacts. This method requires you to have an idea of how a student will respond in this situation and to be prepared for that response. You should be

**FIGURE 5.6** The tailboard chat can provide an opportunity for experienced fire fighters to develop and adapt procedures.

prepared for several potential responses from the student and be able to lead the student toward the learning objective. At the end of the role playing, you should explain the learning objective of the exercise.

Another alternative communication method focuses on the use of group exercises. With this method, you use the collaborative efforts of the group to benefit all members of the group. Such a technique exploits the natural tendency of fire fighters to work in groups. With group work, you can evaluate whether students can work collaboratively to solve a problem. In most cases, this is the way a fire company solves problems in the field. It is imperative that you listen to the conversations in the groups. Walk around to each group and stand near or actually sit down with the group and listen. You can see if the group is moving in the right direction or needs to be redirected. If necessary, help the group along by saying, "Have you considered. . . ." Steering the discussion is not giving students the answer, but rather is assisting them in remaining focused on the message.

Your communication style will be situational, such that you may have to use several different styles during the same presentation. When teaching a class on fire extinguishers, for example, you might begin with a lecture style and then move into a psychomotor session. When giving the lecture, you are simply presenting facts: There is no pressure and no sense of danger. When you move into the psychomotor session, however, it may be necessary to speak authoritatively to ensure that the students understand the issues of safety.

Issues of safety will inevitably affect your communication style. When you explain to new recruits that they must stay down in a fire situation, they generally nod their heads as you explain the science of thermal layering. When you get on the fire ground, be more forceful in your tone and action to keep your students focused on safety.

When an important objective needs to be reinforced, you can highlight it by changing your communication style. For example, by adding a group exercise in the middle of a lecture, you highlight the learning objective.

Time is another factor when considering communication styles. If students spend two hours straight sitting and listening to lecture, retention of the material can diminish. Nothing is worse for a student than to have a monotone instructor lulling him or her to sleep with an endless lecture. Retention of material is increased by adding activities. For example, if a student has to speak and do something, the information will be retained longer and more completely. This consideration is especially important after a meal. The more activities you can provide to students after a lunch break, the more likely they are to retain the information.

### Safety Tip

**When having students work in teams, it is essential to carefully monitor each student to ensure that he or she masters the material. Without close monitoring during group activities, one student may not be able to perform the task when others are not there to provide assistance.**

### Teaching Tip

Use words that are familiar to your audience and explain new terms well. Mentioning the problems of fighting a fire in a "group home," for instance, may not be effective if the students do not understand what a group home is.

# Wrap-Up

## Chief Concepts

- Communication is the most important element of the learning process.
- The basic communication process consists of five elements: the sender, the message, the medium, the receiver, and feedback.
- Be aware of the message that you are sending to students nonverbally through your body language.
- Your communication style must be adapted to match the situation. You may have to use several different styles during a presentation.

## Hot Terms

**Active listening** The process of hearing and understanding the communication sent; demonstrating that you are listening and have understood the message.

**Ambient noise** The general level of background sound.

**Communication process** The process of conveying an intended message from the sender to the receiver and getting feedback to ensure accuracy.

**Feedback** The fifth and final link of the communication chain. Feedback allows the sender (the instructor) to determine whether the receiver (the student) understood the message.

**Medium** The third link of the communication chain. The medium describes how you convey the message.

**Message** The second link of the communication chain; the most complex link. The message describes what you are trying to convey to your students.

**Passive listening** Listening with your eyes and other senses without reacting to the message.

**Receiver** The fourth link of the communication chain. In the fire service classroom, the receiver is the student.

**Sender** The first link of the communication chain. In the fire service classroom, the sender is the instructor.

## References


Davies, L (2001). Effective Communication. Kelly Bear website [www.kellybear.com]. Retrieved 04/16/08 from http://www.kellybear.com/TeacherArticles/TeacherTip15.html.

## Fire Service Instructor *in Action*

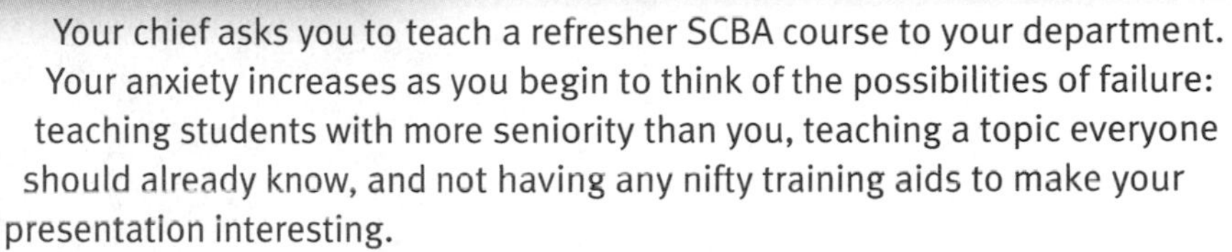

Your chief asks you to teach a refresher SCBA course to your department. Your anxiety increases as you begin to think of the possibilities of failure: teaching students with more seniority than you, teaching a topic everyone should already know, and not having any nifty training aids to make your presentation interesting.

As you prepare, you remember some of the keys to effective instruction: Be enthusiastic and use good communication skills. You search for the latest information in the trade publications, you review departmental policies and procedures, and you go over the equipment. As you develop your teaching outline, you focus your attention on creative ways to cover the basics as you teach information that your students might not otherwise know.

**1.** Which of the following is not a form of communication?
  **A.** Verbal
  **B.** Nonverbal
  **C.** Written
  **D.** Audiovisual

**2.** Which of the following is *not* part of the communication process?
  **A.** Sender
  **B.** Receiver
  **C.** Message
  **D.** Verbal or written communication

**3.** Imparting or interchange of thoughts, opinions by speech, writing or signs is called
  **A.** verbal communication.
  **B.** written communication.
  **C.** nonverbal communication.
  **D.** communication.

**4.** Enthusiasm is what type of communication?
  **A.** Verbal communication
  **B.** Written communication
  **C.** Nonverbal communication
  **D.** Communication

**5.** Which type of communication are departmental policy and procedure manuals?
  **A.** Verbal communication
  **B.** Written communication
  **C.** Nonverbal communication
  **D.** Communication

**6.** What is the primary reason for adding an activity into your presentation?
  **A.** To increase retention of information
  **B.** For evaluation purposes
  **C.** To fill the time requirements
  **D.** To rest your voice

**7.** During your presentation, a student asks a question to clarify what you said. Which part of the communications process is this follow-up?
  **A.** Sender
  **B.** Receiver
  **C.** Message
  **D.** Feedback

# Lesson Plans

CHAPTER

# 6

## NFPA 1041 Standard

**Instructor I**

4.2.2 Assemble course materials, given a specific topic, so that the lesson plan and all materials, resources, and equipment needed to deliver the lesson are obtained. [p. 92–111]

(A) **Requisite Knowledge.** Components of a lesson plan, policies and procedures for the procurement of materials and equipment, and resource availability. [p. 92–98]

(B) **Requisite Skills.** None required.

4.3 **Instructional Development.**

4.3.1* **Definition of Duty.** The review and adaptation of prepared instructional materials. [p. 99–104]

4.3.2* Review instructional materials, given the materials for a specific topic, target audience and learning environment, so that elements of the lesson plan, learning environment, and resources that need adaptation are identified. [p. 92–111]

(A) **Requisite Knowledge.** Recognition of student limitations, methods of instruction, types of resource materials, organization of the learning environment, and policies and procedures.

(B) **Requisite Skills.** Analysis of resources, facilities, and materials.

4.3.3* Adapt a prepared lesson plan, given course materials and an assignment, so that the needs of the student and the objectives of the lesson plan are achieved. [p. 99–104]

(A)* **Requisite Knowledge.** Elements of a lesson plan, selection of instructional aids and methods, origination of learning environment. [p. 92–98]

(B) **Requisite Skills.** Instructor preparation and organizational skills. [p. 98–102]

4.4.3 Present prepared lessons, given a prepared lesson plan that specifies the presentation method(s), so that the method(s) indicated in the plan are used and the stated objectives or learning outcomes are achieved. [p. 94–98]

(A) **Requisite Knowledge.** The laws and principles of learning, teaching methods and techniques, lesson plan components and elements of the communication process, and lesson plan terminology and definitions. [p. 92–111]

(B) **Requisite Skills.** Oral communication techniques, teaching methods and techniques, and utilization of lesson plans in the instructional setting.

4.4.4* Adjust presentation, given a lesson plan and changing circumstances in the class environment, so that class continuity and the objectives or learning outcomes are achieved. [p. 99–104]

(A) **Requisite Knowledge.** Methods of dealing with changing circumstances. [p. 99–104]

(B) **Requisite Skills.** None required.

**4.4.5** Adjust to differences in learning styles, abilities, and behaviors, given the instructional environment, so that lesson objectives are accomplished, disruptive behavior is addressed, and a safe learning environment is maintained.

(A)* **Requisite Knowledge.** Motivation techniques, learning styles, types of learning disabilities and methods for dealing with them, and methods of dealing with disruptive and unsafe behavior.

(B) **Requisite Skills.** Basic coaching and motivational techniques, and adaptation of lesson plans or materials to specific instructional situations. [p. 99–104]

**Instructor II**

**5.3** **Instructional Development.**

**5.3.1** **Definition of Duty.** The development of instructional materials for specific topics. [p. 104–111]

**5.3.2** Create a lesson plan, given a topic, audience characteristics, and a standard lesson plan format, so that the JPR's for the topic are achieved, and the plan includes learning objectives, a lesson outline, course materials, instructional aids, and an evaluation plan. [p. 104–111]

(A) **Requisite Knowledge.** Elements of a lesson plan, components of learning objectives, instructional methods and techniques, characteristics of adult learners, types and application of instructional media, evaluation techniques, and sources of references and materials. [p. 92–98]

(B) **Requisite Skills.** Basic research, using JPRs to develop behavioral objectives, student needs assessment, development of instructional media, outlining techniques, evaluation techniques, and resource needs analysis. [p. 101–111]

**5.3.3** Modify an existing lesson plan, given a topic, audience characteristics, and a lesson plan, so that the JPR's for the topic are achieved and the plan includes learning objectives, a lesson outline, course materials, instructional aids, and an evaluation plan. [p. 111]

(A) **Requisite Knowledge.** Elements of a lesson plan, components of learning objectives, instructional methods and techniques, characteristics of adult learners, types and application of instructional media, evaluation techniques, and sources of references and materials. [p. 92–98]

(B) **Requisite Skills.** Basic research, using JPR's to develop behavioral objectives, student needs assessment, development of instructional media, outlining techniques, evaluation techniques, and resource needs analysis. [p. 101–111]

## Knowledge Objectives

After studying this chapter, you will be able to:

- Identify and describe the components of learning objectives.
- Identify and describe the parts of a lesson plan.
- Describe the four-step method of instruction.
- Describe the instructional preparation process.
- Describe the lesson plan adaptation process for the Fire Service Instructor I.
- Describe how a Fire Service Instructor II creates a lesson plan.
- Describe how a Fire Service Instructor II modifies a lesson plan.

## Skills Objectives

After studying this chapter, you will be able to:

- Utilize the four-step method of instruction.
- Review a lesson plan and identify the adaptations needed.
- Create a lesson plan that includes learning objectives, a lesson outline, instructional materials, instructional aids, and an evaluation plan.
- Adapt a lesson plan so that it both meets the needs of the students and ensures that learning objectives are met.
- Modify a lesson plan so that it both meets the needs of the students and ensures that all learning objectives are met.

# You Are the Fire Service Instructor

Your officer asks you to conduct a forcible-entry class for a group of new fire fighters. As part of this class, you have to tell the new fire fighters everything you know about getting in through a locked door. You have two hours to deliver the class. As your officer walks away, you think about everything that is involved in teaching a class.

1. What does your officer expect the outcome of this class to be?
2. Which resources and equipment will you need?
3. How will you know whether the new fire fighters have learned what is expected?

## Introduction

When most people think about the job of a fire service instructor, they picture the actual delivery of a presentation in the front of the classroom. Although lectures are an important aspect of instruction, they are not the only part of the job. Most fire service instructors spend many hours planning and preparing for a class before students ever arrive in the classroom. There are many details to address when planning a class:

- How much time will the class take?
- How many students will attend the class?
- What must the students know in order to understand what is being taught in the class?
- Which equipment will be needed?
- In what order will the instructional material be presented?

All of these questions and more are answered during the planning and preparation for the class. This information is compiled into a document called a lesson plan. A **lesson plan** is a detailed guide used by the fire service instructor for preparing and delivering instruction to students. A fire service instructor who uses a well-prepared and thorough lesson plan to organize and prepare for class greatly increases the odds of ensuring quality student learning. A Fire Service Instructor I uses a lesson plan that is already developed. A Fire Service Instructor II may develop his or her own lesson plan.

FIGURE 6.1 Attempting to deliver instruction without a lesson plan is like driving in a foreign country without a map. Don't waste valuable class time searching for directions.

### Teaching Tip

At first, lesson plans may seem awkward and disorganized, but there is a logical thought process behind the design. You simply need to learn its methodology.

## Why Use a Lesson Plan?

Most people without experience in the field of education do not understand the importance of a lesson plan. Attempting to deliver instruction without a lesson plan is like driving in a foreign country without a map FIGURE 6.1. The goal in both situations is to reach your intended destination. In a lesson plan, the learning objectives are the intended destination. Without a map (the lesson plan), you most likely will not reach the destination. Also, without a lesson plan that contains learning objectives, you may not even know what the destination for the class is. In other words, if you do not have clearly written learning objectives for your class and a plan for how to achieve them, odds are that you will not be successful.

Written lesson plans also ensure consistency of training across the fire department. When a class is taught multiple times, especially by different fire service instructors, a common lesson plan ensures that all students receive the same information. Lesson plans are also used to document what was taught in a class. When the class needs to be taught again in the future, the new fire service instructor will be able to refer to the existing lesson plans and achieve the same learning objectives.

## Teaching Tip

If you are using a common lesson plan, carefully review it, and write your comments and thoughts in the margins as you prepare for the class. It is particularly useful to note illustrative examples that you can use in class during the preparation stage so you do not have to think up examples under pressure. Good examples and local applications include your own firsthand experiences, and they make the material come alive for your students.

## Learning Objectives

All instructional planning begins by identifying the desired outcomes. What do you want the students to know or be able to do by the end of class? These desired outcomes are called objectives. A **learning objective** is defined as a goal that is achieved through the attainment of a skill, knowledge, or both, and that can be observed or measured. Sometimes these learning objectives are referred to as performance outcomes or behavioral outcomes, for a simple reason: If students are able to achieve the learning objectives of a lesson, they will achieve the desired outcome of the class.

## Components of Learning Objectives

Many different methods may be used for writing learning objectives. One method commonly employed in the fire service is the **ABCD method**, where ABCD stands for Audience (Who?), Behavior (What?), Condition (How?), and Degree (How much?). (Learning objectives do not need to always be written in that order, however.) The ABCD method was introduced in the book *Instructional Media and the New Technologies of Education* written by Robert Heinich, Michael Molenda, and James D. Russell (Macmillan, 1996).

The *audience* of the learning objective describes who the students are. Are your students experienced fire fighters or new recruits? Fire service learning objectives often use terms such as "fire fighter trainee," "cadet," "fire officer," or "students" to describe the audience.

Once the students have been identified, then the *behavior* is listed. The behavior must be an observable and measurable action. A common error in writing learning objectives is using words such as "know" or "understand" for the behavior. Is there really a method for determining whether someone understands something? It is better to use words such as "state," "describe," or "identify" as part of learning objectives—these are actions that you can see and measure. It is much easier to evaluate the ability of a student to identify the parts of a portable fire extinguisher than to evaluate how well the student understands the parts of a portable fire extinguisher.

The *condition* describes the situation in which the student will perform the behavior. Items that are often listed as conditions include specific equipment or resources given to the student, personal protective clothing or safety items that must be used when performing the behavior, and the physical location or circumstances for performing the behavior. For example:

- ". . . in full protective equipment including self-contained breathing apparatus."
- ". . . using the water from a static source, such as a pond or pool."

The last part of the learning objective indicates the *degree* to which the student is expected to perform the behavior in the listed conditions. With what percentage of completion is the student expected to perform the behavior? Total mastery of a skill would require 100 percent completion—this means perfection, without missing any steps. Many times knowledge-based learning objectives are expected to be learned to the degree stated in the passing rate for written exams, such as 70 percent or 80 percent. Another degree that is frequently used is a time limit, which can be included in learning objectives dealing with both knowledge and skills.

ABCD learning objectives do not need to contain all of the parts in the ABCD order. Consider the following example:

> In full protective equipment including SCBA, two fire fighter trainees will carry a 24-foot extension ladder 100 feet and then perform a flat raise to a second-floor window in less than one minute and thirty seconds.

Here the audience is "the fire fighter trainees." The behaviors are "carry a 24-foot extension ladder" and "perform a flat raise." Both carrying and raising are observable and measurable actions. The conditions are "full protective equipment including SCBA," "100 feet," and "to a second-floor window"; they describe the circumstances for carrying and raising the ladder. The degree is "less than one minute and thirty seconds." The fire fighter trainees must demonstrate the ability to perform these behaviors to the proper degree to successfully meet this learning objective.

Strictly speaking, well-written learning objectives should contain all four elements of the ABCD method. Nevertheless, learning objectives are often shortened because one or more of the elements are assumed to be known. If a lesson plan is identified as being used for teaching potential fire service instructors, for example, every single objective may not need to start with "the fire service instructor trainee." The audience component of the ABCD method may be listed once, at the top of all the objectives, or not listed at all.

The same principle applies to the condition component. If it is understood that a class requires all skills to be performed in full personal protective gear, it may not be necessary to list this condition in each individual learning objective. It is also common to omit the degree component, as many learning objectives are written with the assumption that the degree will be determined by the testing method. If the required passing grade for class written exams is 80 percent, it is assumed that knowledge learning objectives will be performed to that degree. Similarly, if the skill learning objectives for a class are required to be performed perfectly, a 100 percent degree for those learning objectives can be assumed.

Learning objectives should be shortened in this way only when the assumptions for the missing components are clearly stated elsewhere in the lesson plan. Of course, a learning objective is unlikely to omit the behavior component, because this component is the backbone of the learning objective.

## Parts of a Lesson Plan

Many different styles and formats for lesson plans exist. No matter which lesson plan format is used, however, certain components should always be included. Each of these components is necessary for you to understand and follow a lesson plan FIGURE 6.2.

### Lesson Title or Topic

The **lesson title or topic** describes what the lesson plan is about. For example, a lesson title may be "Portable Fire Extinguishers" or "Fire Personnel Management." Just by the lesson title, you should be able to determine whether a particular lesson plan contains information about the topic you are planning to teach.

### Level of Instruction

It is important for a lesson plan to identify the **level of instruction** because your students must be able to understand the instructional material. Just as an elementary school teacher would not use a lesson plan developed for high school students, you must ensure that the lesson plan is written at an appropriate level for your students. Often the level of instruction in the fire service corresponds with NFPA standards for professional qualifications. If you are teaching new recruits or cadets, you would use lesson plans that are designated as having a Fire Fighter I or Fire Fighter II level of instruction. If you are teaching fire service professional development classes, you may use lesson plans that are specified as having a Fire Officer I or a Fire Service Instructor III level of instruction. Another method of indicating the level of instruction is by labeling the lesson plan with terms such as "beginner," "intermediate," or "advanced." No matter which method is used to indicate the level of instruction, you should ensure that the material contained in a lesson plan is at the appropriate level for your students.

Another component of the level of instruction is the identification of any prerequisites. A **prerequisite** is a condition that must be met before the student is permitted to receive further instruction. Often, a prerequisite is another class. For example, a Fire Service Administration class would be a prerequisite for taking an Advanced Fire Service Administration class. A certification or rank may also be a prerequisite. Before being allowed to receive training on driving an aerial apparatus, for example, the department may require a student to hold the rank of a Driver and possess Driver/Operator—Pumper certification.

### Behavioral Objectives, Performance Objectives, and Learning Outcomes

As mentioned earlier, learning objectives are the backbone of the lesson plan. All lesson plans must have learning objectives. Many methods for determining and listing learning objectives are available. The specific method used to write the learning objectives is not as important as ensuring that you understand the learning objectives for the lesson plan that you must present to your students.

> **Safety Tip**
>
> You should ensure the proper prerequisites are met by each of your students. Failure to do so may mean that a student performs tasks that he or she is not qualified to perform.

### Instructional Materials Needed

Most lesson plans require some type of instructional materials to be used in the delivery of the lesson plan. Instructional materials are tools designed to help you present the lesson plan to your students. For instance, audiovisual aids are the type of instructional material most frequently listed in a lesson plan—that is, a lesson plan may require the use of a video, DVD, or computer. Other commonly listed instructional materials include handouts, pictures, diagrams, and models. Also, instructional materials may be used to indicate whether additional supplies are necessary to deliver the lesson plan. For example, a preincident planning lesson plan may list paper, pencils, and rulers as the instructional materials needed.

### Lesson Outline

The **lesson outline** is the main body of the lesson plan. This is discussed in detail on page 105.

### References/Resources

Lesson plans often simply contain an outline of the information that must be understood to deliver the learning objectives. Fire service instructors who are not experts in a subject may need to refer to additional references or resources to obtain further information on these topics. The references/resources section may contain names of books, Web sites, or even names of experts who may be contacted for further information. By citing references in the lesson plan, the validity of the lesson plan can also verified.

### Lesson Summary

The **lesson summary** simply summarizes the lesson plan. It reviews and reinforces the main points of the lesson plan.

### Assignment

Lesson plans often contain an **assignment**, such as a homework-type exercise that will allow the student to further explore or apply the material presented in the lesson plan. Be prepared to explain the assignment, its due date, the method for submitting the assignment, and the grading criteria to be used.

## Applying the JOB PERFORMANCE REQUIREMENTS (JPRs)

The lesson plan is the tool used by a fire service instructor to conduct a training session. It is as essential as personal protective equipment (PPE) is to a fire fighter. The lesson plan details the information necessary to present the training session, which includes everything from the title of the class to the assignment for the next training session. In between are the resources needed, behavioral objectives, the content outline, and various teaching applications used to complete the training. As a fire service instructor, you must review and practice the delivery of the lesson plan, check the materials needed for the class, and be ready to present the materials. Using the lesson plan, you must present a structured training session by taking advantage of appropriate methods of instruction to engage the students and use a variety of communication skills to complete the learning objectives.

### Instructor I

The Instructor I will teach from a prepared lesson plan using appropriate methods of delivery and communication skills to ensure that the learning process is effective. The instructor must understand each component of the lesson plan. It may be necessary to adapt the lesson plan to the needs and abilities of the audience and the teaching environment.

### Instructor II

The Instructor II will prepare the lesson plan components and determine the expected outcomes of the training session. The four-step method of instruction should be defined within the lesson plan and all instructional requirements outlined for the presenter's use.

### JPRs at Work

Present prepared lesson plans by using various methods of instruction that allow for achievement of the instructional objectives. Adapt the lesson plan to student needs and conditions.

### JPRs at Work

Create and modify existing lesson plans to better satisfy the student needs, job performance requirements, and objectives developed for the training session.

### Bridging the Gap Between Instructor I and Instructor II

A partnership must exist between the developer of the lesson plan and the instructors who will deliver that lesson plan. In many cases, they may be the same person. If another person must deliver your lesson plan, however, you must be sure that all components of the lesson plan are clear and concise and that the material and instructional methods match the needs of the students. Your communications skills and knowledge of the learning process will be used at both instructor levels in the development and delivery of this content.

### Instant Applications

1. Using a sample lesson plan included in supplemental course material, identify the components of the lesson plan.
2. Using the same sample lesson plan, adjust the lesson plan based on the needs of different audiences.
3. Analyze an existing lesson plan from your department. Are the components complete and accurate?

## Instructor Guide
## Lesson Plan

**Lesson Title:** Use of Fire Extinguishers ← **Lesson Title**
**Level of Instruction:** Firefighter I ← **Level of Instruction**

**Method of Instruction:** Demonstration
**Learning Objective:** The student shall demonstrate the ability to extinguish a Class A fire with a stored-pressure water-type fire extinguisher. (NFPA 1001, 5.3.16) ← **Learning Objective**
**References:** Fundamentals of Firefighter Skills, 2nd Edition, Chapter 7 ← **References**

**Time:** 50 Minutes

**Materials Needed:** Portable water extinguishers, Class A combustible burn materials, Skills checklist, suitable area for hands-on demonstration, assigned PPE for skill ← **Instructional Materials Needed**
**Slides:** 73–78*

**Step #1 Lesson Preparation:**
- Fire extinguishers are first line of defense on incipient fires
- Civilians use for containment until FD arrives
- Must match extinguisher class with fire class
- FD personnel can use in certain situations, may limit water damage
- Review of fire behavior and fuel classifications
- Discuss types of extinguishers on apparatus
  Demonstrate methods for operation

**Step #2 Presentation**

A. Fire extinguishers should be simple to operate.
  1. An individual with only basic training should be able to use most fire extinguishers safely and effectively.
  2. Every portable extinguisher should be labeled with printed operating instructions.
  3. There are six basic steps in extinguishing a fire with a portable fire extinguisher. They are:
     a. Locate the fire extinguisher.
     b. Select the proper classification of extinguisher. ← **Lesson Outline**
     c. Transport the extinguisher to the location of the fire.
     d. Activate the extinguisher to release the extinguishing agent.
     e. Apply the extinguishing agent to the fire for maximum effect.
     f. Ensure your personal safety by having an exit route.
  4. Although these steps are not complicated, practice and training are essential for effective fire suppression.
  5. Tests have shown that the effective use of Class B portable fire extinguishers depends heavily on user training and expertise.
     a. A trained expert can extinguish a fire up to twice as large as a non-expert can, using the same extinguisher.
  6. As a fire fighter, you should be able to operate any fire extinguisher that you might be required to use, whether it is carried on your fire apparatus, hanging on the wall of your firehouse, or placed in some other location.

B. Knowing the exact locations of extinguishers can save valuable time in an emergency.
  1. Fire fighters should know what types of fire extinguishers are carried on department apparatus and where each type of extinguisher is located.
  2. You should also know where fire extinguishers are located in and around the fire station and other work places.
  3. You should have at least one fire extinguisher in your home and another in your personal vehicle and you should know exactly where they are located.

**Step # 3 Application**
Slides 7–10

*Ask students to locate closest extinguisher to training area*

FIGURE 6.2 The components of a lesson plan.

**Step # 3 Application cont.**
*Review rating systems handout—Have students complete work book activity page #389*

C. It is important to be able to select the proper extinguisher.
   1. This requires an understanding of the classification and rating system for fire extinguishers.
   2. Knowing the different types of agents, how they work, the ratings of the fire extinguishers carried on your fire apparatus, and which extinguisher is appropriate for a particular fire situation is also important.
   3. Fire fighters should be able to assess a fire quickly, determine if the fire can be controlled by an extinguisher, and identify the appropriate extinguisher.
      a. Using an extinguisher with an insufficient rating may not completely extinguish the fire, which can place the operator in danger of being burned or otherwise injured.
      b. If the fire is too large for the extinguisher, you will have to consider other options such as obtaining additional extinguishers or making sure that a charged hose line is ready to provide back-up.

*What happens if wrong type or size extinguisher is used?*

   4. Fire fighters should also be able to determine the most appropriate type of fire extinguisher to place in a given area, based on the types of fires that could occur and the hazards that are present.
      a. In some cases, one type of extinguisher might be preferred over another.

D. The best method of transporting a hand-held portable fire extinguisher depends on the size, weight, and design of the extinguisher.
   1. Hand-held portable fire extinguishers can weigh as little as 1 lb to as much as 50 lb.
   2. Extinguishers with a fixed nozzle should be carried in the favored or stronger hand.
      a. This enables the operator to depress the trigger and direct the discharge easily.

*Display available types of extinguishers*

   3. Extinguishers that have a hose between the trigger and the nozzle should be carried in the weaker or less-favored hand so that the favored hand can grip and aim the nozzle.
   4. Heavier extinguishers may have to be carried as close as possible to the fire and placed upright on the ground.
      a. The operator can depress the trigger with one hand, while holding the nozzle and directing the stream with the other hand.
   5. Transporting a fire extinguisher will be practiced in Skill Drill 7-1.

E. Activating a fire extinguisher to apply the extinguishing agent is a single operation in four steps.
   1. The P-A-S-S acronym is a helpful way to remember these steps:
      a. Pull the safety pin.
      b. Aim the nozzle at the base of the flames.
      c. Squeeze the trigger to discharge the agent.
      d. Sweep the nozzle across the base of the flames.

*Have students demonstrate steps using empty extinguisher*

   2. Most fire extinguishers have very simple operation systems.
   3. Practice discharging different types of extinguishers in training situations to build confidence in your ability to use them properly and effectively.
   4. When using a fire extinguisher, always approach the fire with an exit behind you.
      a. If the fire suddenly expands or the extinguisher fails to control it, you must have a planned escape route.
      b. Never let the fire get between you and a safe exit. After suppressing a fire, do not turn your back on it.

*Complete skills sheet #7–9 for each student*

   5. Always watch and be prepared for a rekindle until the fire has been fully overhauled.
   6. As a fire fighter, you should wear your personal protective clothing and use appropriate personal protective equipment (PPE).
   7. If you must enter an enclosed area where an extinguisher has been discharged, wear full PPE and use SCBA.
      a. The atmosphere within the enclosed area will probably contain a mixture of combustion products and extinguishing agents.

*Review PPE required for extinguisher use*

**FIGURE 6.2** The components of a lesson plan *(continued)*.

F. The oxygen content within the space may be dangerously depleted.

**Step # 3 Application cont.**

*Discuss hazards of extinguishing agents*

**Step #4 Evaluation:**

1. Each student will properly extinguish a Class A combustible fire using a stored-pressure type water extinguisher. (Skill Sheet x-1)
2. Each student will return extinguisher to service. (Skill Sheet x-2)

**Lesson Summary:** ← **Lesson Summary**

- Classifications of fire extinguishers
- Ratings of fire extinguishers
- Types of extinguishers and agents
- Operation of each type of fire extinguishers
- Demonstration of Class A fire extinguishment using a stored pressure water extinguisher

**Assignments:** ← **Assignment(s)**

1. Read Chapter 8 prior to next class.
2. Complete "You are the Firefighter" activity for Chapter 7 and be prepared to discuss your answers.

FIGURE 6.2 The components of a lesson plan *(continued)*.

## The Four-Step Method of Instruction

While reviewing and preparing for class with your lesson plan, keep in mind the four-step method of instruction, which is shown in TABLE 6.1.

**Table 6.1 The Four-Step Method of Instruction**

| Step in the Instructional Process | Instructor Action |
| --- | --- |
| 1. Preparation | The instructor prepares the students to learn by identifying the importance of the topic, stating the intended outcomes, and noting the relevance of the topic to the student. |
| 2. Presentation | The presentation content is usually organized in an outline form that supports an understanding of the learning objective. |
| 3. Application | The instructor applies the presentation material as it relates to students' understanding. Often the fire service instructor will ask questions of the students or ask students to practice the skill being taught. |
| 4. Evaluation | The evaluation of students' understanding through written exams or in a practical skill session. |

## Preparation Step

The **preparation step** is the first phase in the **four-step method of instruction**, which is the method of instruction most commonly used in the fire service. The preparation step—also called the motivation step—prepares or motivates students to learn. When beginning instruction, you should provide information to students that explains why they will benefit from the class. Adult learners need to understand how they will directly benefit by attending the class, because very few adults have time to waste in sitting through a presentation that will not directly benefit them.

The benefit of a class can be explained in many ways:

- The class may count toward required hours of training.
- The class may provide a desired certification.
- The class may increase students' knowledge of a subject.

Whatever the benefit may be, you should explain it thoroughly during the preparation step. In a lesson plan, the preparation section usually contains a paragraph or a bulleted list describing the rationale for the class. During the preparation step, the Fire Service Instructor I needs to gain students' attention and prepare them to learn when the instructor begins presenting the prepared lesson plan. The Fire Service Instructor II, while developing the lesson plan, will include suggested preparation points, including safety- and survival-related information, local examples, and explanations of how the material will help improve students' ability to do their job

## Presentation Step

The **presentation step** is the second step in the four-step method of instruction; it comprises the actual presentation of the lesson plan. During this step, you lecture, lead discussions, use audiovisual aids, answer student questions, and perform other techniques to present the lesson plan. Chapter 3, Methods of Instruction, discusses the various methods of instruction used during this step. In a lesson plan, the presentation section normally contains an outline of the information to be presented. It may also contain notes indicating when to use teaching aids, when to take breaks, or where to obtain more information.

## Application Step

The **application step**, which is the third step in the four-step method of instruction, is the most important step because it is during this phase that students apply the knowledge presented in class. Learning occurs during this step as students practice skills, perhaps make mistakes, and retry skills as necessary. You should provide direction and support as each student performs this step. You must also ensure that all safety rules are followed as students engage in new behaviors.

In a lesson plan, the application section usually lists the activities or assignments that the student will perform. In the fire service, the application section often requires the use of skill sheets for evaluation purposes. The experienced fire service instructor uses the application step to make sure that each student is progressing along with the lesson plan. This step also allows students to actively participate and remain engaged in the learning process.

## Evaluation Step

The **evaluation step** is the final step of the four-step method of instruction. It ensures that students correctly acquired the knowledge and skills presented in the lesson plan. The evaluation may, for example, take the form of a written test or a skill performance test. No matter which method of evaluation is used, the student must demonstrate competency without assistance. In a lesson plan, the evaluation section indicates the type of evaluation method and the procedures for performing the evaluation.

### Teaching Tip

If you are creating a lesson plan, the evaluation should directly link back to the initial learning objectives outlined at the beginning of class.

## Instructional Preparation

Once you have a lesson plan, the instructional preparation begins. Which materials are needed for the class? Which audiovisual equipment will be utilized? Where will the class be conducted? How much time will be needed? These and many other questions must be answered during instructional preparation. The information contained in the lesson plan should be used as a guide for instructional preparation.

### Teaching Tip

Check all instructional materials prior to class and replace any missing or nonfunctional materials. This review includes ensuring that a video clip that plays on your home computer also plays on the computer that you will be using in the classroom.

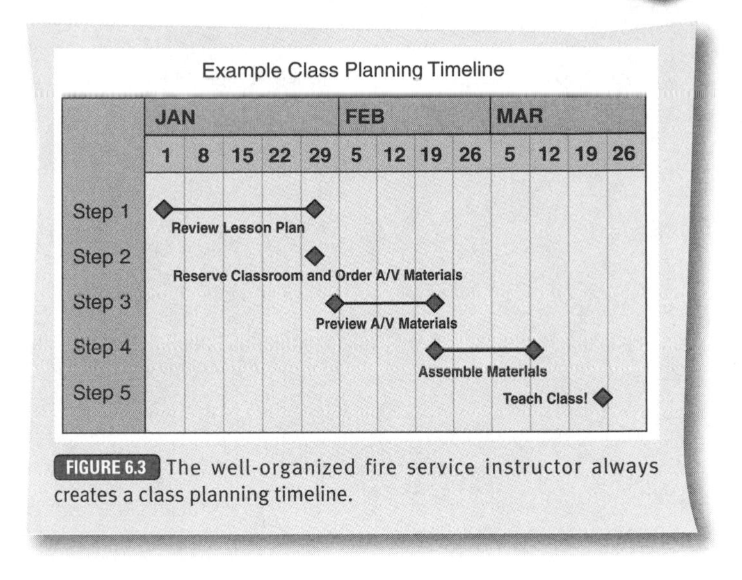


FIGURE 6.3 The well-organized fire service instructor always creates a class planning timeline.

## Organizational Skills

Taking the information from a lesson plan and transforming it into a well-planned class takes good organizational skills. First, you should organize the class planning timeline FIGURE 6.3. Identify the time available for you to plan and prepare for the class. The time available for preparation is usually the amount of time from when the lesson plan is identified until the day when the class is scheduled to be taught. Identify the milestones that must be accomplished as part of this timeline. Depending on the lesson plan, milestones may include obtaining audiovisual equipment, purchasing materials, reserving a classroom, or previewing audiovisual aids.

## Procuring Instructional Materials and Equipment

Most classes take advantage of instructional materials or equipment. The method of obtaining these instructional materials and equipment differs from fire department to fire department. A common method of procuring materials is for the fire service instructor to contact the person in the fire department who is responsible for purchasing training materials, such as a training officer or someone assigned to the training division. You may be required to provide a list of needed materials to the training officer. Often, this list of materials must be submitted long before the class is scheduled to begin. The training officer then compiles the materials either by purchasing new materials or by securing materials already available at the training division. The training officer contacts the fire service instructor when all class-related materials are available.

A common method for procuring class equipment is the equipment checkout process, which is typically managed by the fire department's training division. For example, if you need a multimedia projector for a class, you would submit a request for the projector in which you indicate the date and time the projector is needed. The training division would then reserve the projector for you. On the day of the class, the projector would be available for you. Depending on the

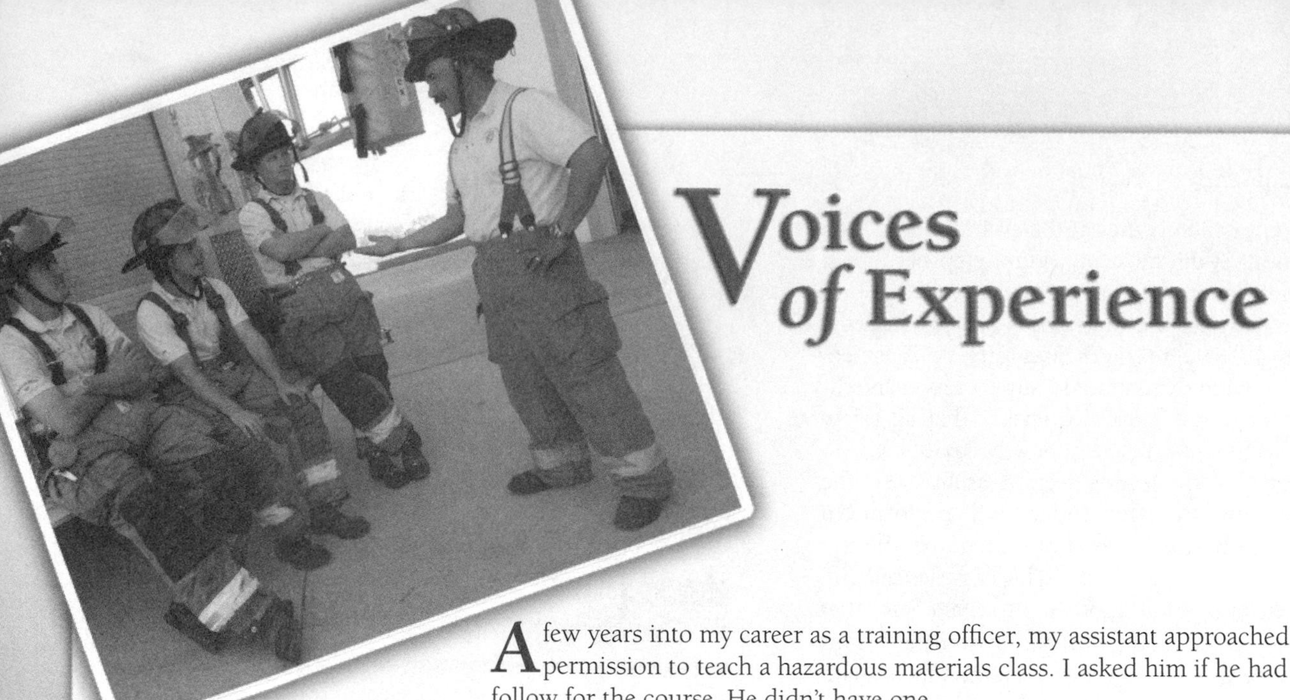

# Voices of Experience

A few years into my career as a training officer, my assistant approached me asking for permission to teach a hazardous materials class. I asked him if he had a lesson plan to follow for the course. He didn't have one.

I spent the next hour or so with my assistant explaining what exactly a lesson plan is and the components of a successful lesson plan. I also emphasized that his lesson plan and supporting documents should be usable by anyone with little modification.

*"I remember thinking to myself that I could even teach the course using his materials."*

A few weeks later we sat down and reviewed the lesson plan he had prepared. I didn't have any trouble following his lesson plan. Additionally, I saw that he had his course materials, instructional aids, and his evaluation plan ready to go. I remember thinking to myself that I could even teach the course using his materials. I gave my approval and the class was scheduled.

As fate would have it, about a week before the class, my assistant broke his leg. There was no way he would be standing for six hours and conducting practical exercises. It was also too late to reschedule, as several individuals had adjusted their time to be there. Knowing that I had a good lesson plan to work from, I stepped up to the challenge.

With the lesson plan in place, I simply modified it to accommodate my teaching style without changing the objectives. Having the references and resources in the lesson plan made my preparation simple.

Eventually the day came to present. Despite my experience as an educator, I still found myself extremely nervous. To make matters worse, there were several people in the audience who had a lot more overall experience in this area than I did. I stuck to my guns, however, and followed the lesson plan. Not only did the class go well, but with the reference and resources list, I could answer all questions asked, either during the class or shortly afterward. I also took care to make certain everyone knew who had prepared the lesson plan and was responsible for the class success.

*Philip J. Oakes*
Laramie County Fire District #6
Burns, Wyoming

organizational procedures, you might be required to pick up the projector and set it up, or the training division might set up the projector at the class location for you.

## Preparing for Instruction Delivery

The most important part of instructional preparation is preparing for actual delivery of the lesson plan in the classroom. If you obtain the necessary materials, equipment, and classroom, but you do not prepare to deliver the lesson plan, the class will not be successful. You should be thoroughly familiar with the information contained in the lesson plan, which may require you to consult the references listed in the lesson plan and research the topic further. If the lesson plan includes a computer presentation, then practice using this technology to deliver the presentation.

No matter which method of instructional delivery is used, you should always rehearse your presentation before delivering it to a classroom full of students. A class is destined for failure if you are seeing the presentation material for the first time in front of the class. Successful fire service instructors have a sound understanding of the information that they are delivering and can adapt to the particular needs of their class because they always know what is coming next.

## Adapting a Lesson Plan

One of the most important—yet confusing—distinctions between a Fire Service Instructor I and a Fire Service Instructor II is the Fire Service Instructor II's ability to modify a lesson plan. A lesson plan is a guide or a roadmap for delivering instruction, but it is rarely implemented *exactly* as written. To understand what can and cannot be modified by each level of fire service instructor, let's review what the NFPA job performance requirements (JPRs) say about modifying and adapting a lesson plan.

### Fire Service Instructor I

4.3.2 Review instructional materials, given the materials for a specific topic, target audience and learning environment, so that elements of the lesson plan, learning environment, and resources that need adaptation are identified.

4.3.3 Adapt a prepared lesson plan, given course materials and an assignment, so that the needs of the student and the objectives of the lesson plan are achieved.

The Fire Service Instructor I should not alter the content or the lesson objectives. Prior to the beginning of the class, the Fire Service Instructor I should be able to evaluate local conditions, evaluate facilities for appropriateness, meet local standard operating procedures (SOPs), and evaluate students' limitations. He or she should be able to modify the method of instruction and course materials to meet the needs of the students and accommodate their individual learning styles, including making adaptations as necessary due to the learning environment, audience, capability of facilities, and types of equipment available.

### Fire Service Instructor II

5.3.3 Modify an existing lesson plan, given a topic, audience characteristics, and a lesson plan, so that the JPRs for the topic are achieved, and the plan includes learning objectives, a lesson outline, course materials, instructional aids, and an evaluation plan.

To clearly understand the difference between adapting and modifying, you must understand the proper definitions of these terms:

- **Modify**: to make basic or fundamental changes
- **Adapt**: to make fit (as for a specific use or situation)

Put simply, a Fire Service Instructor II can make basic or fundamental changes to the lesson plan but a Fire Service Instructor I cannot. Fundamental changes include changing the performance outcomes, rewriting the learning objectives, modifying the content of the lesson, and so on.

So what can a Fire Service Instructor I do? He or she can make the lesson plan fit the situation and conditions. Conditions include the facility, the local SOPs, the environment, limitations of the student, and other local factors.

The NFPA standard specifically states that a Fire Service Instructor I may modify the method of instruction and course materials to meet the needs of the student and accommodate the individual fire service instructor's style. Here are a few real-life examples:

- A Fire Service Instructor I may modify a lesson plan's method of instruction from lecture to discussion if he or she determines that the latter method would be a better presentation format because of the students' level of knowledge.
- A Fire Service Instructor I may adapt the classroom setting if the facility cannot meet the seating arrangement listed in the lesson plan.
- A Fire Service Instructor I may adapt the number of fire fighters performing an evolution in a lesson plan from three to four to meet local staffing SOP requirements.
- A Fire Service Instructor I *cannot* modify a lesson plan learning objective that states a fire fighter must raise a 24-foot extension ladder because he or she feels the task is too difficult for one fire fighter.

**Teaching Tip**

All instructors routinely adapt and modify courses. While a Fire Service Instructor I may believe that a curriculum needs to be modified, those changes should be made only by a Fire Service Instructor II. This process will ensure that development of the curriculum is done correctly and that coverage of the lesson objectives is not reduced.

**Theory into Practice**

Lesson plans must remain dynamic in both the short term and the long term. In the short term, you should understand when it is appropriate to adjust a lesson plan during its delivery based on students' learning styles, changing conditions, timing considerations, and students' progress. In the long term, you should provide input to your supervisor regarding the success of the delivery. If problems occurred or improvements are needed, report this feedback as well.

One critical component of lesson plan adaptability is the break times. The break times should be adjusted to fit the environment. The length of breaks can also be adjusted.

If you make adjustments to the delivery of a lesson plan, it is critical that you ensure that all learning objectives are still covered. For example, many times activities must be scheduled around the activities of other courses. Scheduling resources in the field with other instructors can help reduce conflicts. The program coordinator should be advised if an instructor intends to move portions of the program around so that the coordinator can ensure the change doesn't affect other programs and shared resources.

- A Fire Service Instructor I *cannot* change the JPR of developing a budget in a Fire Officer lesson plan because he or she does not feel comfortable teaching that subject.

As with all other positions within the fire service, it is important that fire service instructors perform only those actions within their level of training. As a Fire Service Instructor I, you must recognize what you can and cannot do. Acting outside your scope of training may lead to legal liability. If you are ever unsure if you have the ability to do something, check with a superior.

## Reviewing Instructional Materials for Adaptation

There are many ways for a Fire Service Instructor I to obtain a lesson plan: fire service Web sites, commercially published curriculum packages, the National Fire Academy, your fire department's training library, or other fire departments. No matter which method is used, the lesson plan must be reviewed and any areas that need adaptation must be identified. This is true even for lesson plans that were originally developed within your fire department. Over time, standards and procedures change, so that a lesson plan that was completely correct for your department when it was created may be out-of-date in just a few months.

After obtaining a lesson plan, you must review the entire lesson plan and determine whether adaptations are needed to make the lesson plan usable for your class. As part of the class planning and preparation process, lesson plan adaptations must be scheduled and completed before you deliver the presentation to the class. A lesson plan might need adaptations for many reasons, such as differences related to the learning environment, the audience, the capability of facilities, and the types of equipment available.

**Teaching Tip**

The National Fire Academy maintains a Web site called TRADE's Virtual TRADEing Post that enables fire service instructors to share non-copyrighted information, PowerPoint presentations, lesson plans, and training programs or other downloadable materials. The training information is provided free of charge with the understanding that you must give credit to the department or agency that developed it.

This Web site is part of the Training Resources and Data Exchange (TRADE) program, which is a regionally based network designed to foster the exchange of fire-related training information and resources among federal, state, and local levels of government.

## Evaluating Local Conditions

The main focus when adapting a lesson plan is to make minor adjustments so it fits your local conditions and your students' needs. To accomplish this, you must be familiar with your audience. Which organizational policies and procedures apply to the lesson plan? What is the current level of knowledge and ability of your students? Which types of tools and equipment will your students use in performing the skills within the lesson plan? These are all questions a Fire Service Instructor I should be contemplating when reviewing a lesson plan for any adaptations needed to accommodate the intended audience.

The second area pertaining to the local conditions that must be considered is you, the Fire Service Instructor I. What is your experience level and ability? How familiar are you with the topic that will be taught? What is your teaching style? The answers to these types of questions will allow you to adapt the lesson plan so that you deliver the lesson in the most effective way given your own abilities.

## Evaluating Facilities for Appropriateness

You should review and adapt the lesson plan based on the facilities that will be used when delivering the class. Several factors—for example, the equipment available, student seating, classroom size, lighting, and environmental noise—must be considered as part of this evaluation. For example, a lesson plan may call for students to sit at tables that have been moved into a U-shaped arrangement. However, if the local classroom has desks that are fixed to the floor, you will not be able to arrange the seating as indicated in the lesson plan. The lesson plan would then need to be adapted to meet the conditions of the facility and the seating arrangement changed accordingly. You should make this adaptation, keeping in mind the reason for the indicated seating arrangement.

## Meeting Local Standard Operating Procedures (SOPs)

A lesson plan must be reviewed to ensure that it meets and follows local SOPs. This is one of the most important considerations when adapting a lesson plan. You should never teach information that contradicts a SOP. Not only would this lesson be confusing for the students, but it would also create a liability for you. If a student were to be injured or killed while performing a skill in violation of a SOP, you would be held responsible. At a minimum, you would be disciplined within the organization. It is also possible that you might be held legally responsible in either criminal or civil court.

When reviewing a lesson plan, make note of the SOPs that may cover this topic. After completely reviewing the lesson plan, research the SOPs and ensure that no conflicts exist. If your research turns up conflicting information, you should adapt the lesson plan to meet the local SOPs. If you are not familiar with your local SOPs, contact someone within the department who can assist you with ensuring that the lesson plan is consistent with local SOPs.

## Evaluating Limitations of Students

The lesson plan should also be reviewed based on student limitations and adapted to accommodate those limitations if possible. The lesson plan should be at the appropriate educational level for the students, and the prerequisite knowledge and skills should be verified. For example, if you were reviewing a lesson plan to teach an advanced hazardous materials monitoring class, students should have already undergone basic hazardous materials training. If you were training new fire fighters and reviewing this lesson plan, you may not be able to adapt it. Instead, you would most likely have to require additional training before the lesson plan would be appropriate to present to those students.

Sometimes it is possible to adapt a lesson plan to include information to accommodate the students' limitations. Many times, however, this mismatch indicates that the lesson plan should not be used.

## Adapting a Prepared Lesson Plan

Reviewing and adapting a lesson plan should be a formal process. For instance, you should document in writing which adaptations have been made. Many times it is appropriate for the Fire Service Instructor I to obtain approval for the adaptations. After completing the review and adaptation process, you should ensure that the adaptations are not really modifications. In other words, the minor adjustments you made while adapting the lesson plan should not significantly change the class or alter the learning objectives.

### Safety Tip

When adapting a lesson plan, closely evaluate the revised plan's safety implications. It is all too easy to omit important safety information that was previously included or to include information that may create a safety issue when combined with other material.

## Modifying the Method of Instruction

Method of instruction is the one area that a Fire Service Instructor I may readily modify. Such a modification may be needed to allow you to effectively deliver the lesson plan, but it should not change the learning objectives. For example, you may not be comfortable using the discussion method to deliver a class on fire service sexual harassment as indicated by the lesson plan. Instead, you might modify the lesson plan and change the method of instruction to lecture. This would allow the same information to be taught, just in a different format, and the same learning objectives would still be achieved.

## Accommodating Instructor Style

In addition to ensuring that the method of instruction best suits your abilities, lesson plans may be adapted to accommodate your style. A lesson plan often reflects the style of the fire service instructor who wrote it. When reviewing and adapting a lesson plan, consider whether the lesson plan—and especially the presentation section—fits your style. For example, a lesson plan may call for a humorous activity designed to establish a relationship between the instructor and the students. If you are teaching a military-style academy class, this may not be the best style, so you may need to adapt the presentation accordingly.

### Ethics Tip

Fire service instructors regularly adapt material to meet their departmental needs and to improve the curriculum. Is it ethical to modify material to reflect your personal opinions when those opinions run counter to the traditional way of thinking? The solution to this dilemma is not as simple as it may seem. Imagine where the fire service would be today if only a few years ago bold fire service instructors did not step up to the plate and refuse to present material that was not based on safe practices. "Doing the right thing" is what ethical decisions are all about—but there is a fuzzy line between "the right thing" and "my way is the only right way."

For example, modifying a course by eliminating the use of fog nozzles because you believe that these nozzles lead to hand burns is as dangerous as another fire service instructor eliminating the use of smooth-bore nozzles from a lesson plan. The reality is that students must understand the appropriate use, benefits, and dangers of each type of nozzle. What might seem like a simple modification could have serious consequences for a student who is not trained thoroughly and properly to department standards.

## Meeting the Needs of the Students

All adaptations should be done with one purpose in mind—namely, meeting the needs of the students. As with all lesson plans, the main goal is to provide instruction that allows students to obtain knowledge or skills. This goal should be verified after you review and adapt a lesson plan.

## Creating a Lesson Plan

The Fire Service Instructor II is responsible for creating lesson plans. Depending on the subject, this task can take anywhere from several hours to several weeks. No matter how big or small the lesson plan may be, the ultimate goal is to create a document that any fire service instructor can use to teach the subject and ensure that students achieve the learning objectives. A Fire Service Instructor II should ensure that the lesson plan is complete and clearly understandable so that any other fire service instructor can use it. Many fire departments have lesson plan templates for the Fire Service Instructor II to use as a starting point. Such a standard format makes it easier for all fire service instructors in the department to understand the lesson plan and ensures consistency in training.

**Safety Tip**

Once a lesson plan is modified, go back and confirm that all learning objectives are met.

## Achieving Job Performance Requirements

The first step of lesson plan development is to determine the learning objectives. What are students expected to achieve as a result of taking the class? Many times this desired outcome is obvious, because you are teaching a class to prepare students to perform a certain job or skill. For example, if you were to develop a lesson plan for a class to train fire fighters to drive a fire engine, you would start by listing the job performance requirements for a fire engine driver. On many other occasions, however, the learning objectives are not that clear.

It is very difficult to develop a lesson plan when the learning objectives are not clearly stated. Many fire service instructors have been in the unhappy position of being told to teach a certain class, such as one dealing with workplace diversity or safety, without clear direction on the intended learning objectives. Although the person requesting the class may have a general idea of what the class is intended to accomplish, he or she may not know the specific learning objectives that the Fire Service Instructor II needs to develop a lesson plan.

For example, the Fire Chief may want to improve fire fighter safety through training. Unless given specific learning objectives, the Fire Service Instructor II cannot develop a lesson plan to "improve fire fighter safety." When placed in this position, the Fire Service Instructor II should attempt to clarify the Fire Chief's vision of improving fire fighter safety. Would he like all fire fighters to understand the chain of events that leads to an accident and to know how to break that chain so that an accident is avoided? A learning objective can be written to achieve that goal. Does the Fire Chief expect all fire fighters to don their structural firefighting protective equipment properly within a time limit? A learning objective can be written to achieve that goal, too. Whenever you are asked to develop a lesson plan for a class, start by clarifying the intended outcome of the class with the person requesting the class.

## Learning Objectives

Once the Fire Service Instructor II has a clear outcome for a class, he or she should develop the learning objectives for the class. As described earlier in this chapter, learning objectives can be written utilizing the ABCD method.

### Audience

The audience should describe the students who will take the class. If the lesson plan is being developed specifically for a certain audience, the learning objectives should be written to indicate that fact. For example, if a Fire Service Instructor II is writing a lesson plan for a driver training class, the audience would be described as "the driver trainee" or "the driver candidate." Both of these terms indicate that the audience consists of individuals who are learning to be drivers. If the audience is not specifically known or is a mixed audience, the audience part of the learning objective could be written more generically, such as "the fire fighter" or even "the student."

### Behavior

As described earlier, the behavior part of the learning objective should be specified using a clearly measurable action word, which allows the evaluation of the student's achievement of the learning objective. Another important consideration is the level to which a student will achieve the learning objective. This level is most often determined using Bloom's Taxonomy (*Taxonomy of Educational Objectives*, 1965), a method to identify levels of learning within the cognitive domain. For the fire service, the three lowest levels are commonly used when developing learning objectives. In order of simple to most difficult, these levels are knowledge, comprehension, and application:

- *Knowledge* is simply remembering facts, definitions, numbers, and other items.
- *Comprehension* is displayed when students clarify or summarize important points.
- *Application* is the ability to solve problems or apply the information learned in situations.

To use this method, a Fire Service Instructor II must determine which level within the cognitive domain is the appropriate level for the student to achieve for the lesson plan. For example, if a Fire Service Instructor II is developing objectives for a class on portable extinguishers, the following objectives could be written for each level:

- Knowledge: "The fire fighter trainee will identify the four steps of the PASS method of portable extinguisher application." For this objective, the student simply needs to memorize and repeat back the four steps of pull, aim, squeeze, and sweep. Achievement of this very simple knowledge-based objective is easily evaluated with a multiple-choice or fill-in-the-blank question.

- Comprehension: "The fire fighter trainee will explain the advantages and disadvantages of using a dry-chemical extinguisher for a Class A fire." This objective requires the student to first identify the advantages and disadvantages of a dry-chemical extinguisher and then select and summarize those that apply to use of such an extinguisher on a Class A fire. This higher-level objective may be evaluated by a multiple-choice question but is better evaluated with a short-answer-type questions.
- Application: "The fire fighter trainee, given a portable fire extinguisher scenario, shall identify the correct type of extinguisher and demonstrate the method for using it to extinguish the fire." This is the highest level of objective because it requires the student to recall several pieces of information and apply them correctly based on the situation. This type of objective is often evaluated with scenario-based questions that may be answered with multiple-choice or short answers.

There is no one correct format for determining which level or how many learning objectives should be written for a lesson plan. Typically, a lesson plan will contain knowledge-based learning objectives to ensure that students learn all of the facts and definitions within the class. Comprehension objectives are then used to ensure that students can summarize or clarify the material. Finally, application objectives are used to ensure that the student can actually use the information learned in the lesson.

## Converting Job Performance Requirements into Learning Objectives

Often, a Fire Service Instructor II needs to develop learning objectives to meet **job performance requirements (JPRs)** listed in an NFPA professional qualification standard; a JPR describes a specific job task, lists the items necessary to complete the task, and defines measurable or observable outcomes and evaluation areas for the specific task. Matching of learning objectives to JPRs occurs when a lesson plan is being developed to meet the professional qualifications for a position such as Fire Officer, Fire Instructor, or Fire Fighter. The JPRs listed in the NFPA standards of professional qualifications are not learning objectives per se, but learning objectives can be created based on the JPRs. Each NFPA professional qualification standard has an annex section that explains the process of converting a JPR into an instructional objective, including examples of how to do so FIGURE 6.4. By following this format, a Fire Service Instructor II will be able to develop learning objectives for a lesson plan to meet the professional qualifications for NFPA standards.

**Safety Tip**

When writing objectives in an application format, be sure to address any safety issues required to meet the objective.

## Lesson Outline

After determining the performance outcomes and writing the learning objectives for the lesson plan, the next step for the Fire Service Instructor II is to develop the lesson outline FIGURE 6.5. The lesson outline is the main body of the lesson plan and is the major component of the presentation step in the four-step method of instruction.

One method for creating a lesson outline involves brainstorming the topics to be covered and then arranging them in a logical order. Begin listing all of the information that needs to be taught to achieve the learning objectives. Which terms do students need to learn? Which concepts must be presented? Which skills need to be practiced? Which stories or real-life examples would demonstrate the need to learn this material?

Once you have listed all of the topics that should be covered in the lesson outline, organize them into presentation and application sections. Arrange the listed topics you will lecture on in a logical and orderly fashion in the presentation section. Topics should be presented in order starting from the basic and then moving on to the more complex. Ensure that the topics flow together and that the presentation does not contain any gaps that might confuse a student. If you identify a gap, you may need to create a new topic to bridge it.

In the application section, list the topics that require students to apply the information learned in the presentation section. Most often the topics in the application section will be activities or skills practice. If the lesson does not include actual hands-on activities, the application should at least consist of discussion points for you to talk about with the students to ensure the information in the lecture was learned and can be applied.

Many lesson outlines utilize a two-column format. The first column contains the actual outline of the material to be taught. If this lesson outline is to be used by experienced fire service instructors, a simple outline of the material may suffice. For less experienced fire service instructors or to ensure consistency between multiple instructors, the outline may be more detailed. The second column of the lesson outline contains comments or suggestions intended to help a fire service instructor understand the lesson outline. It is also a good practice to indicate in the second column which learning objectives are being achieved during the presentation or application sections. This information is especially helpful when you are developing a lesson plan to teach an established curriculum that uses a numbering system to identify learning objectives.

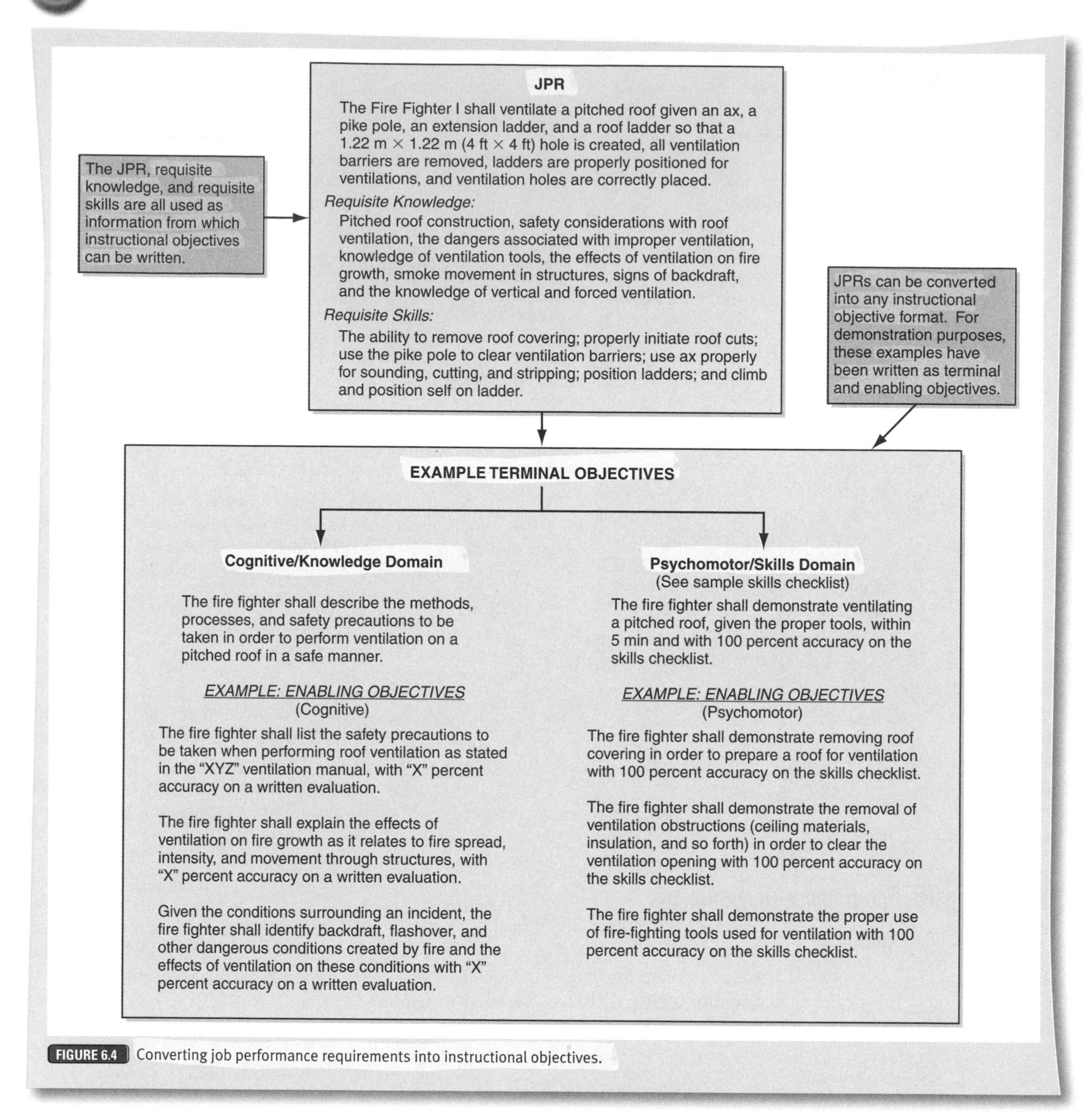


**FIGURE 6.4** Converting job performance requirements into instructional objectives.

## Instructional Materials

Once the lesson outline is developed, all instructional materials needed to deliver instruction should be identified and listed in the lesson plan. This list should be specific so that the exact instructional aid can be identified. For example, if the lesson plan is a fire safety lesson for children that incorporates a DVD as an instructional aid, just listing "Fire Safety Video" in the lesson plan does not provide enough information for the fire service instructor. Instead, give the exact title of the video, such as "*Sparky Says: Join My Fire Safety Club*," by the NFPA. This information will allow any fire service instructor who uses the lesson plan to obtain the correct instructional aid.

## Pre-Lecture (Preparation Step)
## I. You Are the Fire Fighter

**Time:** 5 Minutes
Small Group Activity/Discussion

Use this activity to motivate students to learn the knowledge and skills needed to understand the history of the fire service and how it functions today.

**Purpose**
To allow students an opportunity to explore the significance and concerns associated with the history and present operation of the fire service.

**Instructor Directions**

1. Direct students to read the "You Are the Fire Fighter" scenario found in the beginning of Chapter 1.
2. You may assign students to a partner or a group. Direct them to review the discussion questions at the end of the scenario and prepare a response to each question. Facilitate a class dialogue centered on the discussion questions.
3. You may also assign this as an individual activity and ask students to turn in their comments on a separate piece of paper.

## Lecture (Presentation Step)
## I. Introduction

**Time:** 5 Minutes
**Slides:** 1–6
**Level:** Fire Fighter I
Lecture/Discussion

**A.** Training to become a fire fighter is not easy.
1. The work is physically and mentally challenging.
2. Firefighting is more complex than most people imagine.

**B.** Fire fighter training will expand your understanding of fire suppression.
1. The new fire fighter must understand the roots of the fire service, how it has developed, and the fire service "culture" in order to excel.
2. This course equips fire fighters to continue a centuries-old tradition of preserving lives and property threatened by fire.

## II. Fire Fighter Guidelines

**Time:** 5 Minutes
**Slide:** 7
**Level:** Fire Fighter I
Lecture/Discussion

**A.** Be safe.
1. Safety should always be uppermost in your mind.

**B.** Follow orders.
1. If you follow orders, you will become a dependable member of the department.

**C.** Work as a team.
1. Firefighting requires the coordinated efforts of each department member.

**D.** Think!
1. Lives will depend on the choices you make.

**E.** Follow the golden rule.
1. Treat each person, patient, or victim as an important person.

**FIGURE 6.5** A sample lesson outline.

## III. Fire Fighter Qualifications

**Time:** 30 Minutes
**Slides:** 8–10
**Level:** Fire Fighter I
Lecture/Discussion

**A.** Age requirements
  **1.** Most career fire departments require that candidates be between the ages of 18 and 21.
**B.** Education requirements
  **1.** Most career fire departments require a minimum of a high school diploma or equivalent.
**C.** Medical requirements
  **1.** Medical evaluations are often required before training can begin.
  **2.** Medical requirements for fire fighters are specified in NFPA 1582, *Standard on Comprehensive Operational Medical Program for Fire Departments*.
**D.** Physical fitness requirements
  **1.** Physical fitness requirements are established to ensure that fire fighters have the strength and stamina needed to perform the tasks associated with firefighting and emergency operations.
**E.** Emergency medical requirements
  **1.** Many departments require fire fighters to become certified at the first responder, Emergency Medical Technician (EMT)–Basic, or higher levels.

## IV. Roles and Responsibilities of the Fire Fighter I and Fire Fighter II

**Time:** 30 Minutes
**Slides:** 11–17
**Level:** Fire Fighter I and II
Lecture/Discussion

**A.** The roles and responsibilities for Fire Fighter I include:
  **1.** Don and doff personal protective equipment properly.
  **2.** Hoist hand tools using appropriate ropes and knots.
  **3.** Understand and correctly apply appropriate communication protocols.
  **4.** Use self-contained breathing apparatus (SCBA).
  **5.** Respond on apparatus to an emergency scene.
  **6.** Force entry into a structure.
  **7.** Exit a hazardous area safely as a team.
  **8.** Set up ground ladders safely and correctly.
  **9.** Attack a passenger vehicle fire, an exterior Class A fire, and an interior structure fire.
  **10.** Conduct search and rescue in a structure.
  **11.** Perform ventilation of an involved structure.
  **12.** Overhaul a fire scene.
  **13.** Conserve property with salvage tools and equipment.
  **14.** Connect a fire department engine to a water supply.
  **15.** Extinguish incipient Class A, Class B, and Class C fires.
  **16.** Illuminate an emergency scene.
  **17.** Turn off utilities.
  **18.** Perform fire safety surveys.
  **19.** Clean and maintain equipment.
  **20.** Present fire safety information to station visitors, community groups, or schools.
**B.** Additional roles and responsibilities for Fire Fighter II include:
  **1.** Coordinate an interior attack line team.
  **2.** Extinguish an ignitable liquid fire.
  **3.** Control a flammable gas cylinder fire.

**FIGURE 6.5** A sample lesson outline *(continued)*.

4. Protect evidence of fire cause and origin.
5. Assess and disentangle victims from motor vehicle accidents.
6. Assist special rescue team operations.
7. Perform annual service tests on fire hose.
8. Test the operability of and flow from a fire hydrant.
9. Fire fighters must also be prepared to assist visitors to the fire station and use the opportunity to discuss additional fire safety information.

## V. Summary

**Time:** 5 Minutes
**Slides:** 51–53
**Level:** Fire Fighter I
Lecture/Discussion

**A.** Remember the five guidelines: Be safe, follow orders, work as a team, think, and follow the golden rule.
**B.** Fire fighter qualifications consider age, education, medical, and physical fitness, and emergency medical certifications.
**C.** The roles and responsibilities of Fire Fighter I and Fire Fighter II vary.

## Post-Lecture
## I. Wrap-Up Activities (Application Step)

**Time:** 40 Minutes
Small Group Activity/Individual Activity/Discussion

### A. Fire Fighter in Action

This activity is designed to assist the student in gaining a further understanding of the roles and responsibilities of the Fire Fighter I and II. The activity incorporates both critical thinking and the application of fire fighter knowledge.

#### Purpose

This activity allows students an opportunity to analyze a firefighting scenario and develop responses to critical thinking questions.

#### Instructor Directions

1. Direct students to read the "Fire Fighter in Action" scenario located in the Wrap-Up section at the end of Chapter 1.
2. Direct students to read and individually answer the quiz questions at the end of the scenario. Allow approximately 10 minutes for this part of the activity. Facilitate a class review and dialogue of the answers, allowing students to correct responses as needed. Use the answers noted below to assist in building this review. Allow approximately 10 minutes for this part of the activity.
3. You may also assign these as individual activities and ask students to turn in their comments on a separate piece of paper.
4. Direct students to read the "Near Miss Report." Conduct a discussion that allows for feedback on this report. Allow 10–15 minutes for this activity.

#### Answers to Multiple Choice Questions

1. A, D
2. B
3. B
4. D

### B. Technology Resources

This activity requires students to have access to the Internet. This may be accomplished through personal access, employer access, or through a local educational institution. Some community colleges, universities, or adult education centers may have classrooms with Internet capability that will allow for this activity to be completed in class. Check out local access points and encourage students to complete this activity as part of their ongoing reinforcement of firefighting knowledge and skills.

#### Purpose

To provide students an opportunity to reinforce chapter material through use of online Internet activities.

FIGURE 6.5 A sample lesson outline *(continued)*.

**Instructor Directions**

1. Use the Internet and go to **www.Fire.jbpub.com**. Follow the directions on the Web site to access the exercises for Chapter 1.
2. Review the chapter activities and take note of desired or correct student responses.
3. As time allows, conduct an in-class review of the Internet activities and provide feedback to students as needed.
4. Be sure to check the Web site before assigning these activities, as specific chapter-related activities may change from time to time.

## II. Lesson Review (Evaluation Step)

**Time:** 15 Minutes
Discussion
**Note:** Facilitate the review of this lesson's major topics using the review questions as direct questions or overhead transparencies. Answers are found throughout this lesson plan.

**A.** Name some of the physical fitness requirements established for firefighters.
**B.** What requirements do the firefighter qualifications focus on?
**C.** How do the roles and responsibilities of the Fire Fighter I and II differ?

## III. Assignments

**Time:** 5 Minutes
Lecture

**A.** Advise students to review materials for a quiz (determine date/time).
**B.** Direct students to read the next chapter in *Fundamentals of Fire Fighter Skills* as listed in your syllabus (or reading assignment sheet) to prepare for the next class session.

**FIGURE 6.5** A sample lesson outline *(continued)*.

## Theory into Practice

A set of clear learning objectives, a thorough lesson outline, and a method of ensuring that the learning objectives are met should form the backbone of every course taught in the fire service. Proper construction of learning objectives will ensure that the course meets the identified needs of the students. Without clear learning objectives, however, courses may stray from their intended purpose. To ensure clear learning objectives, use the ABCD method during the creation process.

The lesson outline is a necessity to ensure that you cover the material required to meet the learning objectives. The lesson outline should flow logically to assist in student comprehension. Using the brainstorming technique will help guide your thought process in covering all relevant topics. Arranging topics in a logical sequence requires an understanding of the learning objectives. Once it is completed, run your lesson outline by a colleague to see if he or she follows your logical sequence. If your colleague is confused, then your students are also likely to be confused.

The last step to ensure success is to create an evaluation process to confirm that the learning objectives have been met. Without an evaluation process, there is no guarantee that a student has met the goals of the class.

Instructional materials may range from handouts to overhead projectors to the hoses used during a skills practice. Often the inclusion of one instructional aid creates a need for more instructional materials. For example, if a lesson plan lists a DVD as an instructional aid, the instructional materials would need to be revised to include a DVD player and projector. Ask the following types of questions to determine what you need:

- Are additional informational resources needed to present the learning objectives to students—for example, a handout describing your department's SOPs?
- Are supplies needed to make props or demonstrations?
- Is equipment needed for the activities or skills practice?
- Is equipment needed to ensure student safety?

## Evaluation Plan

The evaluation plan is the final part of the lesson plan. Each part of the evaluation plan should be directly tied to one or more learning objectives. When writing the evaluation plan into the lesson plan, simply *describe* the evaluation plan—do not provide the actual evaluation. In other words, the lesson plan could indicate that the evaluation plan is a 50-question multiple-choice test, but it should not list the actual test questions. The test questions should be a separate document that is securely kept and only available as needed to fire service instructors. When the evaluation plan lists skills performance tests, these documents should be included with the instructional materials and handed out to students so they can prepare for skills testing. This step is covered in Part IV of this text.

## Modifying a Lesson Plan

A Fire Service Instructor II may modify lesson plans. Modifying a lesson plan occurs when a Fire Service Instructor II makes fundamental changes, such as revising the learning objectives. When these kinds of substantial changes to a lesson plan are made, the lesson plan should be completely revised, following the step-by-step process used to develop the original lesson plan. To ensure that the lesson plan is written to meet a new learning objective, follow through each step of the lesson plan development process and make the necessary changes in all sections of the lesson plan.

When modifying a lesson plan, always obtain necessary approval from the authority having jurisdiction. Even though a Fire Service Instructor II has the training to modify learning objectives, many times the change must be approved by a curriculum committee, a training officer, or the fire chief. Similarly, any lesson plan modification must comply with all agency policies and procedures. If a reference used to develop the lesson plan is updated, such as a department SOP or an NFPA standard, make sure that the reference cited in the lesson plan is current.

After modifying a lesson plan, retain a copy of the original lesson plan. This original must be kept to document the classes that were taught from that lesson plan. It can also be referred to when making future lesson plan modifications.

The fire service instructor greatly improves their ability to deliver training information to students by using a standard lesson plan format that incorporates the four step method of instruction. Consistency and accuracy of information must be relayed to varied audiences and in the event of unexpected emergency runs or other breaks that may occur during instruction, the lesson plan allows for you to pick up where you left off. Fellow instructors can use the same lesson plan and achieve similar outcomes. The lesson plan can be compared to an incident action plan as it identifies expected outcomes of a training session, resources available or needed, and provides a step-by-step measurable set of instruction material that brings a training session to a successful outcome. Existing or published lesson plans should be reviewed and modified to reflect your department procedures and practices. Utilization of fire service references and NFPA job performance requirements also provide content validity to the material being taught. Using a standard form for instruction ensures that the instructor covers many legal and ethical concerns relating to the delivery of training in the modern fire service.

# Wrap-Up

## Chief Concepts

- To provide quality instruction, use lesson plans with well-written and clearly defined learning objectives.
- A learning objective is a goal that is achieved through the attainment of a skill, knowledge, or both.
- The main components of a lesson plan are as follows:
  - Lesson title or topic
  - Level of instruction
  - Behavioral objectives, performance objectives, or learning outcomes
  - Instructional materials needed
  - Lesson outline
  - References/resources
  - Lesson summary
  - Assignment
- The four-step method of instruction is the process most commonly used for delivering fire service lesson plans. It includes these steps:
  - Preparation
  - Presentation
  - Application
  - Evaluation
- Preparing for instruction is very important. You may need to spend several hours preparing to teach a class, including reviewing the lesson plan, reserving classrooms and instructional aids, and purchasing materials.
- A Fire Service Instructor I can use a lesson plan to teach a class and may adapt the lesson plan to the local needs of the class.
- A Fire Service Instructor II can create a new lesson plan to teach a class and may modify an existing lesson plan.
- When a Fire Service Instructor II creates a lesson plan, the learning objectives must be identified. They then become the basis for the rest of the lesson plan.

## Hot Terms

**ABCD method** Process for writing lesson plan objectives that includes four components: audience, behavior, condition, and degree.

**Adapt** To make fit (as for a specific use or situation).

**Application step** The third step of the four-step method of instruction, in which the student applies the information learned during the presentation step.

**Assignment** The part of the lesson plan that provides the student with opportunities for additional application or exploration of the lesson topic, often in the form of homework that is completed outside of the classroom.

**Evaluation step** The fourth step of the four-step method of instruction, in which the student is evaluated by the instructor.

**Four-step method of instruction** The most commonly used method of instruction in the fire service. The four steps are preparation, presentation, application, and evaluation.

**Job performance requirement (JPR)** A statement that describes a specific job task, lists the items necessary to complete the task, and defines measurable or observable outcomes and evaluation areas for the specific task.

**Learning objective** A goal that is achieved through the attainment of a skill, knowledge, or both, and that can be measured or observed.

**Lesson outline** The main body of the lesson plan. A chronological listing of the information presented in the lesson plan.

**Lesson plan** A detailed guide used by an instructor for preparing and delivering instruction.

**Lesson summary** The part of the lesson plan that briefly reviews the information from the presentation and application sections.

**Lesson title or topic** The part of the lesson plan that indicates the name or main subject of the lesson plan.

**Level of instruction** The part of the lesson plan that indicates the difficulty or appropriateness of the lesson for students.

**Modify** To make basic or fundamental changes.

**Preparation step** The first step of the four-step method of instruction, in which the instructor prepares to deliver the class and provides motivation for the students.

**Prerequisite** A condition that must be met before a student is allowed to receive the instruction contained within a lesson plan—often a certification, rank, or attendance of another class.

**Presentation step** The second step of the four-step method of instruction, in which the instructor delivers the class to the students.

## Fire Service Instructor *in Action*

You are a Fire Service Instructor I who has been asked to teach an SCBA class to your department's new recruit class. The captain in charge of the training academy provides you with the lesson plan that was used during the last class. He asks you to review the lesson plan and let him know if you need anything before you teach the class in two weeks.

1. Which statement best describes your next actions?
   - **A.** Safely store the lesson plan away until the day of the class
   - **B.** Begin creating your own lesson plan and compare it to the one you were given
   - **C.** Review the lesson plan you were given and develop a timeline to prepare for the class
   - **D.** Tell the captain that a Fire Service Instructor I cannot teach this class
2. As you review the SCBA lesson plan, you notice that some of the learning objectives are no longer needed because of an equipment change. As a Fire Service Instructor I, what should you do?
   - **A.** Delete the unnecessary objectives from the lesson plan
   - **B.** Notify the captain, so a Fire Service Instructor II can modify the lesson plan
   - **C.** Teach the learning objectives anyway because they are in the lesson plan
   - **D.** Rewrite the learning objectives so they apply to the new equipment
3. The last class contained 20 recruits. The class you will teach will have 40 students. As a Fire Service Instructor I, you can adapt the lesson plan to accommodate the additional students.
   - **A.** True
   - **B.** False
4. Your department has a standard operating procedure for the use and maintenance of SCBA. Which of the following statements is true concerning the SOP and the lesson plan?
   - **A.** The lesson plan should never contradict the SOP.
   - **B.** The lesson plan may reference the SOP but you do not need to teach it.
   - **C.** The lesson plan should not include the SOP because students will learn it later.
   - **D.** The lesson plan should cover textbook material only, not SOPs.
5. As a Fire Service Instructor I preparing to teach this class, which of the following issues would you normally be responsible for?
   - **A.** Selecting the type of SCBA for your department
   - **B.** Establishing a budget for the class
   - **C.** Writing the exam questions
   - **D.** Reviewing and preparing audio/visual aids
6. Why would a less experienced fire service instructor have a more detailed lesson outline?
   - **A.** A less experienced fire service instructor should not have a more detailed lesson outline because it will be distracting.
   - **B.** A more detailed lesson plan will allow the fire service instructor to cover areas that are not part of the learning objectives.
   - **C.** The fire service instructor will have more basic knowledge about the topic, thus requiring less information in the lesson outline.
   - **D.** All lesson outlines should be the same regardless of the fire service instructor experience.
7. When reviewing a lesson plan, should you consider your personal style of presentation and adapt the plan to meet your style?
   - **A.** This is acceptable because the original fire service instructor will have incorporated his or her own personal style when developing the lesson plan.
   - **B.** This is acceptable because the original fire service instructor will have ensured that no personal style is reflected in the lesson plan.
   - **C.** This is not acceptable because the material should be about the students and not the fire service instructor.
   - **D.** This is not acceptable because fire service instructors do not have personal styles.

Fire.jbpub.com

# The Learning Environment

CHAPTER

# 7

## NFPA Standard

**Instructor I**

**4.3.2*** Review instructional materials, given the materials for a specific topic, target audience and learning environment, so that elements of the lesson plan, learning environment, and resources that need adaptation are identified. [p. 116–125]

- **(A) Requisite Knowledge.** Recognition of student limitations, methods of instruction, types of resource materials, organization of the learning environment, and policies and procedures. [p. 121–124]
- **(B) Requisite Skills.** Analysis of resources, facilities, and materials. [p. 121–124]

**4.4.2** Organize the classroom, laboratory, or outdoor learning environment, given a facility and an assignment, so that lighting, distractions, climate control or weather, noise control, seating, audiovisual equipment, teaching aids, and safety are considered. [p. 116–125]

- **(A) Requisite Knowledge.** Classroom management and safety, advantages and limitations of audiovisual equipment and teaching aids, classroom arrangement, and methods and techniques of instruction. [p. 116–125]
- **(B) Requisite Skills.** Use of instructional media and materials.

**Instructor II**

NFPA 1041 contains no Job Performance Requirements for this chapter.

## Knowledge Objectives

After studying this chapter, you will be able to:

- Describe the effect of demographics on the learning environment.
- Describe how to adapt the learning environment to suit the needs of your students.
- Describe the effect of the audience and the venue on the learning environment.

## Skills Objectives

After studying this chapter, you will be able to:

- Analyze the learning environment according to the students' needs and the learning objectives.
- Control the learning environment.
- Present to a diverse audience.

## You Are the Fire Service Instructor

The fire chief asks you to prepare and present a fire academy for the citizens of your city. The class is intended to show what a fire fighter does. If it is successful, the city plans to include funding in its annual budget to run the program on an ongoing basis.

The fire chief has suggested that five of the members of the city council act as the first citizens to go through the academy. The city council group is very diverse and consists of individuals of different nationalities, backgrounds, and genders. The five members who will participate in the academy include a lawyer, a self-made businessman, a stay-at-home mom, a dentist, and a retired military officer.

The classroom portion of the academy will be held at the fire station and the practical portions at the training tower. This class will take place over the next ten weeks, meeting every Wednesday night. You need to plan every aspect of the training.

The fire chief explains that this is the most important program that the department has ever run. The academy must be informative, challenging, and most of all, fun. The success of this program will mean additional resources for the department in years to come.

1. How can you prepare a learning environment for such a diverse group of students?
2. How will you provide a safe learning environment for these students?

## Introduction

There is an old saying: When two people are together, one is always the leader. In training, whenever two fire fighters are together, one is always the fire service instructor. The setting is not important, the time is not important, and the subject is not important—but what is important is that you use the environment to ensure that learning takes place.

The **learning environment** is not always a classroom, because learning takes place in many different locations. In many company officer training classes, the concept of tailboard chats is encouraged. Many fire officers report that a great deal is learned sitting around the kitchen table in a firehouse or standing around a charred table in a kitchen that's just been overhauled FIGURE 7.1. The place is not as important as the message delivered.

You need a solid understanding of the benefits and weaknesses of potential learning environments to select the best environment for presenting your message. For example, you would not demonstrate a hose drill in the computer room. The key point to remember is that learning occurs in many different places. Always take the opportunity to teach and share knowledge with those around you.

FIGURE 7.1 Learning can take place anywhere.

### Teaching Tip

Sometimes multiple learning environments are needed to convey your message.

### Teaching Tip

To be effective in teaching, you must know your audience, the message you are trying to convey, your best delivery method, and the ideal atmosphere in which to deliver your message. This information is not always easy to come by. You may not have knowledge of exactly with whom you are going to speak or even the context in which you will be presenting. You must be able to think on your feet and be prepared to take the presentation in the direction it needs to go in so that the learning objectives are met. Remember—a good fire service instructor always has a backup plan.

## Demographics

To deliver your message effectively, you must know who your audience is. A fire fighter is a fire fighter. There is no race, there is no male or female, and there is no ethnicity. We are no longer called "firemen"; we are called fire fighters. As fire service instructors in the new millennium, we must realize that demographic factors do not apply just to the private sector. Demographic considerations extend beyond issues of race and national origin; in the fire service, they require looking in a more holistic manner at everything a fire fighter brings to the department.

**Demographics** include the age, gender, marital status, family size, and educational background of a fire fighter. As a fire service instructor, you should look at a fire fighter's demographics and ask yourself a number of questions: Which type of skills does this fire fighter bring? Could these skills help the class to learn, or might the skills hold the fire fighter back in the learning process? Are you dealing with a highly educated individual or a fire fighter who has met the basic educational requirements? Is the student an experienced fire fighter or a rookie? Does the same class include a mix of experienced fire fighters and rookies? It is this type of demographic information that you must understand and appreciate as you approach your teaching assignment.

Another demographic consideration unique to the fire service is the different types of staffing encountered from fire department to fire department. The fire service includes career, volunteer, and paid-on-call fire fighters. The learning objectives for paid-on-call or volunteer fire fighters do not differ, but the presentation to these groups does. Because volunteer and paid-on-call fire fighters generally have full-time jobs, training accommodations must be made to fit their schedules. By gearing instruction toward the demands of the student's schedule, you allow the student to focus fully on the subject at hand and not be distracted. For example, if a class is made up of all career fire fighters, the class may be six to eight weeks long, eight hours a day. By contrast, if the same class is made up of all volunteers, the class may meet Tuesday and Thursday nights for three hours and all day on Saturdays for six months. Both classes should cover the same learning objectives to the same levels of proficiency, but the differing demographics of the participants require different schedules.

## Demographics in the Classroom

As a fire service instructor, you should be aware of how demographics affect the classroom. Always exercise caution and common sense in the classroom. Never make a comment that refers to someone's background, such as where he went to school or what her marital status is.

Gender is a hot-button issue in today's society. One change that has affected the fire service in the last few years is the abandonment of the term "fireman." Many qualified women have joined the ranks of this once male-dominated profession, so it is more appropriate to refer to students as "fire fighters." Firefighting does not, according to the fire, have a gender. While learning objectives do not change with the gender of the student, you do need to be cognizant of students' gender. If you use the term "guys" to address your students, your students need to understand that "guys" includes both men and women. Be gender neutral when developing learning objectives. The key is to be respectful of all of your students at all times.

### Offensive Language, Gestures, and Dress

The use of offensive language should be avoided—period. Some may contend that to make a point, sometimes you must swear or use slang terms; others contend that this is part of the fire service culture. This is not the case. Jokes and language of a sexual or explicit nature are unacceptable in the learning environment. As a fire service instructor, you should have a mastery of the English language. You should be able to make your point without the use of profanity. You will find that students gain a deeper respect for you when you demonstrate eloquence in the classroom.

Do not participate in improper situations. A wise fire service instructor once shared a key piece of wisdom when he said, "If you hear an off-color joke and grin, you're in." If you hear an offensive joke in your presence, show that you do not agree with what was said. It's important to take pride in your department, and one source of that pride involves showing respect to everyone.

If a student is offended by your behavior or your language, you should apologize. If you offended the student in front of the class, then it would be appropriate to apologize first to the student in private and then in front of the class. This can go a long way toward making amends. An equally important step is to have a policy in place that covers acceptable behavior and language. Your organization should have a harassment policy that addresses these issues, including remedies for offenses, reporting of incidents, and conduct. The best policy is to understand the audience to whom you are speaking. Ensure that your presentation occurs in a context that is appropriate for your audience. A wise fire service instructor used to say, "Make sure that your language would be acceptable to your grandmother."

Like words, gestures and dress can sometimes be offensive. How you dress and how you move are representations of your position **FIGURE 7.2**. Keep a neutral position at all times, which allows the student to feel free to express his or her opinion and concerns in a safe environment.

### Verbal Introductions

Fire service instructors use a variety of techniques to get the class started. One popular exercise is to have the students go around the room, introduce themselves, say why they are in the class, and explain what they hope to get out of the class. Although this practice is widely used, it has some dangers. If a student is taking the class as a mandatory requisite course, he or she may not see the benefit the class could provide.

FIGURE 7.2 How you present yourself is a reflection on the fire department.

## Ethics Tip

As a fire service instructor, should you be held to a higher standard? That is a question with which many have wrestled. If you want to be viewed as a credible professional, you must hold to a higher standard both in class and after hours. It is not uncommon for fire service instructors to hang out with students at the local pub after class. This can lead to great discussions and additional learning can take place—but it can also lead to potentially problematic issues. Toeing the line of professionalism is key. Don't let a single night of bad judgment ruin your hard-earned reputation. Respect takes a long time to develop, but it can be lost in a moment thanks to a poor choice of words.

When asked, the student may try to be funny or flip. For example, in one class several fire fighters announced that they were there for the overtime pay. The final fire fighter said, "I thought I was coming to the class for the overtime, but last night two of my fire fighters were hurt in a vehicle accident. I now want to know how to prevent it from happening to someone else." Pay close attention to the students' reasons, because their true motives may be hidden behind humor.

Another danger is that the student may come into the class with a misunderstanding of the learning objectives. Some walk into a class with a perception of what they want to learn, but not necessarily an understanding of what you need to teach. Although you need to be flexible to meet the needs of your class, the learning objectives for the class—which are laid out in the lesson plan, as discussed in Chapter 6—must be met. Listen to the needs of your students and tailor your message while meeting the learning objectives.

One benefit of student introductions is that it allows students to discover the goals of their classmates. Student introductions also allow you to evaluate students' experience levels. An added benefit is that students will know the experience levels of their classmates. This can help further the informal learning that takes place at breaks and after class, which is also called networking.

By listening as other students describe their expectations to the entire class, other students may see that they are not alone. They may find a classmate with whom they share values or circumstances. Two members of a class may have similar experiences. Once they open a dialogue after class, they may realize that they have much more in common and find someone with whom they can share information, discuss policies and procedures, and lend a sympathetic ear. The fire service is one of a few careers where members of separate (fire) companies openly and freely share "corporate intelligence." In the fire service, it is shared experience (corporate intelligence) that can save a life.

### Adapting the Class Based on Demographics

The information that you learn as a result of students' verbal introductions may require slight modifications to the lesson plan. Perhaps students will need to spend more time in one area of the class and less time in another. The information gained in this way can also help you decide the depths you need to go in the various areas of the presentation, perhaps allowing you to lengthen parts of the program and shorten others to meet the needs of the class. For example, if you are presenting a Disaster Preparedness Course in Arizona, you may not need to spend an entire hour on blizzards.

If a student requests that you cover a certain topic, you should not rewrite the entire course; however, you should keep the student's request in mind. If time and circumstances permit, you could briefly present the topic and then ask the student, "Did that meet your need?" By doing so, you ensure that students know you are available to instruct them on whatever they need to know so as to meet the learning objectives of the course. Bear in mind that as a fire service instructor, you must be prepared to either adapt your lesson plan or tell the student that the topic is outside the scope of the class.

## Teaching Tip

If you promise to cover an area or an objective, you need to ensure that you fully address any concerns or questions about that topic. This requires knowing and understanding your audience. Many fire departments have target hazards or situations that are unique to their locations. For example, many communities in the Northeast include row houses, whereas these structures are rarely found in the Midwest. When asked, "How would you handle this situation?" you should be prepared to render an opinion. By anticipating what the questions will be, you will be prepared to render an opinion because you have had time to research the question and present an intelligent response. If you do not know the answer, do not—under any circumstance—make up the answer. It is completely acceptable to say, "I do not know, but I will try to find out for you." Fire fighters respect honesty in their fire service instructors.

## Applying the JOB PERFORMANCE REQUIREMENTS (JPRs)

*Where* the student learns reflects on *how* the student learns. Setting the proper learning environment enhances and complements your instructional skills. Whether it is the classroom or a training tower, this environment needs to be safe, with limited distractions, and arranged so that all students can participate, view the demonstration, see you, and be an active part of the learning process. Experienced fire service instructors often suggest that the best way to accomplish this goal is to physically put yourself where your students will be during the session. Sit in their chairs, stand where they will view the demonstration, and identify any barriers to effective communication and learning.

### Instructor I

The Instructor I will organize the actual setting for the training session so that all participants are able to take an active role in the learning process. He or she can create a safe learning area during practical skills training and enhance student participation in classroom settings through seating arrangements and proper lighting.

### JPRs at Work

Set up your learning area so that all types of learning styles and all methods of instruction that will be used for the training can be maximized. Be able to adapt training materials to the learning environment so that all students can see, hear, and participate in the learning session.

### Instructor II

The Instructor II will create lesson plans and evaluations in a manner that will take advantage of the learning areas available to the instructor. Suggestions for classroom arrangements, student-to-instructor ratio and other delivery tips should be included in the instructor instructions for each training session.

### JPRs at Work

No direct JPRs are identified for the Instructor II for this area. Nevertheless, the Instructor II must be fully aware of where the lesson plan will be presented and how the selected instructor will present the information.

### Bridging the Gap Between Instructor I and Instructor II

If instructors can influence where the training session will take place, they consider the intended methods of instruction, the various aspects of the learning process, and the communications skills of the instructor who will conduct the session in making their recommendations for the class's location. Consider the final delivery of the lesson plan by the instructor in the design phase of this process, with the goal being to create the best possible learning environment for the student. Learning environments may include both formal and informal settings, ranging from the classroom to the apparatus tailboard. Instructors should take advantage of any opportunity to train when it arises.

### Instant Applications

1. Review your current classroom facilities and identify any barriers to effective learning or communications. Identify the methods of eliminating or reducing these barriers.
2. Review your current practical training area and identify any barriers to effective learning or communications. Identify the methods of eliminating or reducing these barriers.
3. Experiment by moving tables and chairs into different configurations to identify the best way of arranging the room for your presentation.
4. Troubleshoot the practical training area and consider alternatives to limited-view props or practical stations in the event of inclement weather or a large number of students.

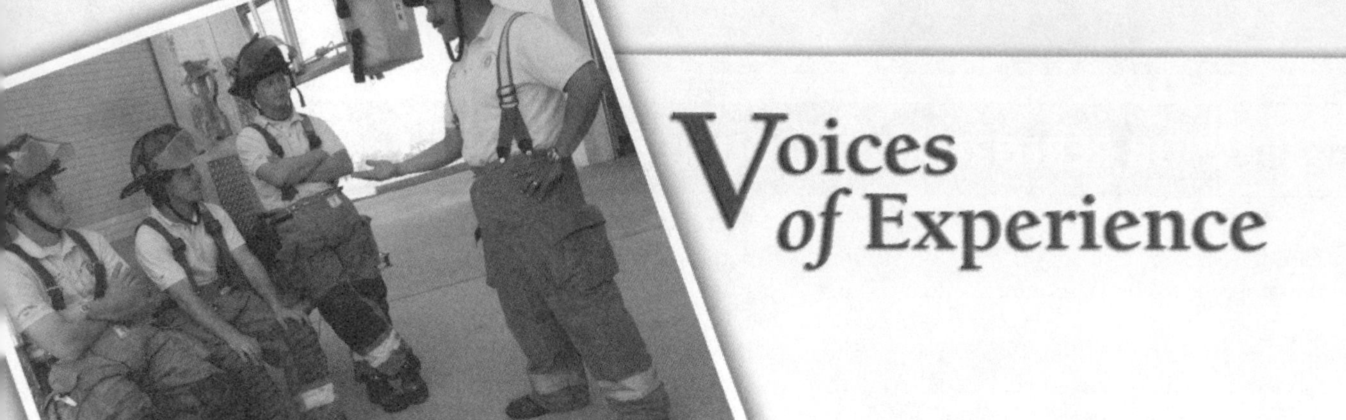

# Voices of Experience

"*I feel that the environment that we created by respecting the time constraints of our students and realizing the importance of creating a positive atmosphere for them was instrumental in the success of this program.*"

In the first Citizen's Firefighting Academy that I was assigned to coordinate and instruct for our department, we targeted public officials and civic leaders as our audience. This seemed like a good idea from the perspective of getting our message to people in the community who could become ambassadors of the fire service message within their organizations. It required the use of all of the elements of instruction to make the classroom safe and informative for a diverse group of people. We were faced each night with having only four hours to present the objectives of the class. Many of the classes required more time than was allowed and the instructors quickly saw that the structure and schedule of the class would require an adjustment to the class content and backgrounds of the students.

We had to create a learning environment for the students in the class that was safe and allowed for effective use of our students valuable time. The reality was that most of the students would not be pursuing a career in the fire service yet the importance of their participation in the program would pay long-term benefits to the department. The format of each class included lecture and practical components. This meant that we would need differing numbers of instructors to maintain the safety of the students as well as provide the instruction. It also meant that we would need a contingency plan (plan B) for each night in the event of emergency calls or other unforeseen events. The logistics became incredible. It required tremendous coordination of both on-duty and off-duty staff to assure the students received the objectives safely and effectively.

The end result of the 40 hours of classroom and practical time included a live fire component. What we gained were several excellent ambassadors for our district and the fire service. One of the students continued her training, became a paramedic, and was later hired by our district as a firefighter paramedic. I feel that the environment that we created by respecting the time constraints of our students and realizing the importance of creating a positive atmosphere for them was instrumental in the success of this program.

*Terry Vavra*
Buffalo Grove Fire Department
Buffalo Grove, Illinois

## Prerequisites

Some fire service instructors use programs that require students to take a pre-course survey. This survey might try to determine student motivation for taking the class or the student's educational level, work experience, or previous training. Its results can also be used to adapt the lesson plan based on the students' needs.

A tool used in some classes is a pre-course test or quiz, which is used to gauge the student's knowledge of the material. After reviewing the results, you can then adjust the course material accordingly so as to better meet the needs of the students. These tests or quizzes do not count toward the final grade, but they do provide an understanding of how the fire service instructor can best help each student.

Some programs require that students take prerequisite classes before attending certain advanced classes. The fact that students have completed the prerequisite classes ensures that the students taking your course have a certain base level of training and education. It also ensures that students do not get into a course that is beyond their experience level. Success in a class will be impaired if a student is already behind the knowledge level for the course. You must ensure that your students have the correct competencies before moving on to the next level.

# The Learning Environment

Learning takes place in two general settings: formal and informal. The formal environment consists of a classroom plus learning objectives, lesson plans, lesson outlines, and a final evaluation. By comparison, informal learning takes place every day in every location. Controlling the learning environment is a critical task for instructors.

## Classroom Setup

Setting up the classroom is as important to the learning process as copying handouts or creating a PowerPoint presentation. Arranging the classroom sets the tone for the learning process. The first question you need to ask is, "What type of learning do I want to accomplish?" If you plan to lecture to the students, then the room can be set up in rows facing a dry erase board **FIGURE 7.3**. **TABLE 7.1** lists various classroom setups, along with their advantages and disadvantages.

## Lighting

In addition to the arrangement of tables and chairs, having the proper lighting for the room is important. Depending on the type of presentation that you are giving, you may need to adjust the lighting. In some cases, lights may be left on during a PowerPoint presentation or during other projected presentations. In the ideal classroom, the lights will be split, allowing lights in the front of the room to be dimmed while lights in the back of the room remain on to allow students to take notes. Your podium should have a light source so that you can read your notes. If the classroom does not have split lighting, then you need to decide which type of lighting is best for your students.

**FIGURE 7.3** Your style of presentation will determine the classroom setup.

**Table 7.1 Classroom Setups**

| Method of Instruction | Learning Environment | Advantage | Disadvantage |
|---|---|---|---|
| Lecture | Classroom style | • Good choice when students are able to see the front of the room<br>• Instructor has more control in this setting<br>• This setting can be effective for demonstrating skills | • Difficult for students to see one another<br>• Depending on how close chairs and tables are, the instructor may be very limited in moving around the classroom |
| Discussion | • Small U-shaped arrangement<br>• Hollow square<br>• Conference table | • Students have the opportunity to see one another<br>• Good environment for group interaction | • Students can easily be distracted by one another<br>• Difficult for all students to see the front of the room |
| Demonstration | Large U-shaped arrangement | • Allows students to see the demonstration of skills<br>• Allows the instructor to move about the room and interact with students | • Students can easily be distracted by one another<br>• Difficult for all students to see the front of the room |

## Classroom Environment

The perfect environment for students is one in which all of their needs are met. In Abraham Maslow's hierarchy of needs, the first need relates to creature comforts and physiological needs. Warmth and shelter make you feel safe in your environment. In the teaching sense, this need would be met when the classroom has enough light, the temperature is comfortable, and the seating and work space are adequate. In the case of an outdoor classroom, it may mean a place of shelter is available where students can warm up in cold weather and cool off in hot weather.

You must also provide time for breaks so that students can attend to their personal needs. An old instructor's rule states that "The brain can learn only as long as the bottom will allow it." Some instructors (and students) may consider these breaks to be a waste; nevertheless, they are necessary both for the student and for you. In addition to enhancing students' personal comfort, the informal conversations that take place during a break can actually enhance the learning process.

### Teaching Tip

Controlling the classroom temperature is very important to the learning process. If a room is too hot or too cold, students will focus more intently on the temperature than on what is being taught. You cannot make every student happy, so you must go with the majority opinion. If possible, lean toward keeping the room cooler versus warmer. In a colder environment, students are less likely to become drowsy. Also, if the temperature is too cool, students can always put on a sweater or light jacket.

After the basic needs are met, the next highest-priority needs are those relating to safety. In a classroom environment, the term "safety" is used frequently. The first type of safety considered deals with being in a hostile environment. When performing live fire training, for example, the environment needs to be as safe as possible while allowing learning to take place. You hold students' lives in your hands, and students need reassurance that you take this responsibility seriously. The second type of safety provided in the learning environment is the ability to make a mistake. In training, whether in the confines of a classroom or on the expanse of the drill ground, students must feel confident that if they make a mistake, you will not belittle or otherwise demean them. Your task is to support students and help them learn from their mistakes. Football coach Vince Lombardi once said that a person is not measured by how many times he is knocked down, but rather by how many times he gets back up. Your students must believe that you will help them get back up.

In the next level of the hierarchy of needs, Maslow speaks of the issue of belonging. A main focus of fire training is teamwork, and the learning environment you create must be one that supports and recognizes the crucial nature of teamwork. Many of the best friends you have in the fire service are likely to be people who you met as you trained to become a fire fighter. During your training, you built a strong sense of camaraderie, which ultimately helped you in completing a very difficult task. An important part of the instruction should be providing time either before, during, or after the formal instruction to allow students to network.

The next level in Maslow's hierarchy of needs is the need for self-esteem or accomplishment in the student. You should create an environment that celebrates students' victories as they are accomplished. Status and recognition can be very powerful motivators in the learning environment. In his book *The One-Minute Manager*, Ken Blanchard talks about catching people doing things right. In a good learning environment, this idea is expanded to not only catching students doing things right, but recognizing and celebrating that accomplishment. People want to be in an environment where they can succeed.

The highest-priority need in Maslow's hierarchy is the need for self-actualization. At this level, the student is able to synthesize material by taking multiple learning situations and putting them together to solve a problem. The need for self-actualization is satisfied in the environment where you perform live fire simulations or even live fire training. Look to challenge your students to take your message to solve a problem.

### Safety Tip

Failure to provide for students' sense of safety will prevent learning from occurring.

## Other Distractions

One challenge that all fire service instructors face is the variety of distractions offered by today's communications technology. Cell phones, BlackBerries, pagers, and station alarms—all of these ubiquitous devices can cause students to be distracted from learning. Many of these distractions can be eliminated at the beginning of class by asking students to turn off or silence all electronic devices. If you have on-duty crew members in class, then make arrangements with them to turn down their pagers or radios.

Other distractions, such as people entering and leaving the classroom during a lecture, can be addressed before the class begins by posting a sign on the door stating that class is in session. Request that latecomers wait until a break to enter the classroom.

### Teaching Tip

If training is to occur with on-duty crews, consider having one student monitor a portable radio with the volume turned low.

> **Teaching Tip**
>
> Proper selection of the learning environment will determine the success or failure of the class in meeting the learning objectives.

If training is taking place outdoors, then controlling the learning environment is more difficult. Noise and distractions can disrupt the learning environment, so try to conduct your training in an isolated place. Eliminate what you can and minimize distractions as much as possible. An effective tool is to keep your groups small, so that every student is engaged in the learning process with the oversight of a fire service instructor. Also, be selective of what has to be taught outside versus what could be taught inside. Limiting instructional time outdoors is a good method of controlling distractions.

## Comfort

The learning environment encompasses much more than just the physical surroundings. It must be comfortable and enhance students' ability to achieve the learning objectives—but "comfort" here means something beyond a soft chair. That is, comfort in this sense means an environment that enhances learning. A classroom where every student can see the board and hear you enhances learning. Proper equipment is important to making learning comfortable. If you are talking about sprinklers, for example, then you should have a sprinkler as a demonstration tool. Moving the learning environment to a sprinkler control room can actually enhance the learning process even further.

## Safety

As a fire service instructor, you need to ensure both that the learning environment is not a distraction to the learning process and that it is safe for students. This seems like a contradiction when speaking about the live training environment. It is not when you follow standards such as NFPA 1403, *Standard on Live Fire Training Evolutions*, among others. Having the right number of personnel, in the right positions, doing the right things, with an emphasis on safety, makes the learning environment the safest place for fire fighters to work and learn in.

Safety also includes the absence of interruptions. Safety is the freedom to learn, to focus on the material being presented without worrying about anything else, if only for a short period of time. It is also the ability to learn in an environment that allows students to make mistakes without being subject to ridicule or damage.

Simulators can provide a safe learning environment, for example, by allowing for a structure to be burned down repeatedly. This technology enables students to experience the stress of the emergency incident while remaining in a safe learning environment. This allows for positive reinforcement, which in turn creates positive behavior. By changing scenarios slightly, behaviors can be anticipated and proper reactions can be drilled into students.

Simulators can also be used after a call in a postincident analysis. By looking at fire fighters' actions and the outcomes of those actions, students can discuss what worked and what needs improvement.

## Evaluation

The setup of the indoor classroom typically changes during testing. Specifically, the classroom should be arranged to provide each student with enough space to put both the testing booklet and the answer key on a writing space. It should also be designed to minimize the chance of wandering eyes, while simultaneously allowing the instructor to observe all students. Given that most testing is individualized, the classroom should remain quiet. If the testing is group based, you must be able to distinguish between testing noise and distractions from a group. The personal needs of students should be recognized. If a student needs to leave the room to attend to personal needs, the security of the testing process should be maintained.

## Outdoor Classrooms

Outdoor classrooms differ in more respects than just their location FIGURE 7.4. For example, outdoor classrooms are subject to the weather conditions. The physical location of your outdoor classroom can determine how you prepare for and handle the class. If you are in the South or Southwest, you may need to deal with high heat levels. If you are in the Northeast or Midwest, you may be more concerned about cold temperatures and snow. No matter where you are, the

FIGURE 7.4 Outdoor locations can present a set of challenges for the fire service instructor.

threat of rain is a priority when an outdoor classroom will be used for instruction. The safety of your students should always be your highest concern. Many fire departments use either a temperature humidity index or a wind chill index as a guide when determining whether outdoor training is feasible.

Always have a plan B. In case of inclement weather, perhaps the class can use an alternative location that is unaffected by the weather, such as an apparatus bay floor or a large unused warehouse. Plans to use these areas should be put into place before the training occurs, not during the training as an emergency measure. Rain is a fact of life. A small amount of rain may not be an issue during outdoor training, but torrential downpours and lightning are definite reasons to seek an alternative environment. Warming or cooling areas and shelter from wind, rain or other weather phenomena are necessities for your students.

Noise in the outdoor environment is also a concern during training. In most cases, you will need to project your voice so that all of the students can hear you without the use of an electronic device such as a microphone. The wind can be a deterrent in allowing your voice to carry. Normal distractions such as traffic and other ambient noises can add to the difficulty in communicating in outdoor environments.

In a live fire training scenario, communication can be difficult as well. In this environment, you need to be first cognizant of your safety. Taking off your SCBA mask to communicate not only puts you at risk, but also sends a message to the students that the use of a SCBA mask is not that important. The use of calm and controlled speech in a live fire training scenario, however, will enhance learning.

A live fire training scenario is also an environment in which testing is a constant process. You need to be close enough to the students to observe their behaviors and far enough back to observe.

It cannot be overstated that safety should be your greatest concern in the outdoor classroom. During outdoor training, the stakes become higher and the attention to detail becomes more intense. The purpose of many outdoor training sessions is to teach or evaluate students' psychomotor skills; these skills are the ones that fire fighters use to protect the lives and property of citizens. For that reason, when training and evaluating students, you need to be in the most realistic situation possible while maintaining a safe environment for students. In rope rescue, this concept is called being "redundantly redundant"—that is, there is a safety for everything.

**Teaching Tip**

The use of multiple senses as part of the learning process increases learning and long-term retention of material.

**Safety Tip**

Know your department's policies on temperature indices for training. If the temperatures reach those points, follow the policy in adapting or suspending the training activity.

**Safety Tip**

The importance of proper hydration cannot be overstated when students will participate in physical activities. Even professional athletes have suffered fatal cases of heat stroke when temperatures were not extreme.

Make sure that students' psychomotor skills are practiced and completed perfectly. A skill is either acceptable or unacceptable—there is no middle ground. If a fire fighter is asked to don a full protective ensemble and forgets to put the hood up, he or she has failed in that skill. Vince Lombardi said, "Practice doesn't make perfect; perfect practice makes perfect." Another relevant saying is, "The more we sweat in training, the less we bleed in battle."

Although it may appear that the work conducted in the outdoor classroom is more important than the instruction that occurs in the indoor classroom, both are needed to ensure that students achieve the learning objectives. Your challenge is to match the environment to the learning objectives to make sure that the student is always in the correct environment in which to learn.

### Contingency Plans

You must have a plan B when using any environment. What do you do if the projector doesn't work? What do you do if there is lightning on a live fire training day? What happens if an assistant does not arrive? All of these contingencies need to be considered and planned for. Many times the solutions are simple—for example, keeping a spare light bulb or having a rain date for outdoor training.

Another way to look at this situation is much like you would approach an emergency scene. Do you have the resources in place if something should go wrong? For example, suppose you are teaching a psychomotor drill and a trainee gets hurt. Is there an ambulance available? Who is notified? Which forms must be completed? How will the training continue if you are missing one student? On an emergency scene, we constantly plan for contingencies.

## Audience, Department Culture and Venue

### Audience

Knowing your audience is critical to your presentation's success. Depending on whether you are presenting at a conference with attendees from multiple jurisdictions or presenting

**Teaching Tip**

Always have a contingency plan. Things can and will go wrong, and having at least considered them will allow you to quickly adapt to the changing circumstances.

locally to your own department, how you present the material will change. When presenting to a diverse audience, your presentation may need to be more generic. Avoid using specifics, such as the exact procedures of any one department. Consider speaking in generalities or giving examples instead of describing mandates or specific procedures. Taking this kind of broader perspective will allow each audience member to adapt your message for their department. For example, when speaking about a procedure, give examples of several procedures used in fire departments across the country. Even when you are presenting to one department, be careful not to mistake instruction for department policy.

**Teaching Tip**

Avoid the phrase, "If I were you, I would do it this way." The best fire service instructors help lead students into making the best decisions for their own departments based on the knowledge that you provide.

Your presentation style may also need to adapt to allow for the learning environment. As your audience and your space increase in size, your presentation loses its intimacy. If you are used to presenting with a lot of student interaction, you may have to adapt your style to be a more traditional lecture because the room may not allow for personal student interaction.

## Department Culture

You must also be aware of culture when making a presentation. Here we speak not of customs associated with a particular national origin, but rather departmental culture. Every department has its own culture. It is what makes fire fighters different, yet exactly the same. It is based on the mission of each and every fire department in the world. Departmental culture specifies what we will do, what we can do, and what we won't do. You must realize that you have to put aside your personal culture when you instruct students. If you are brought in to teach, you must be able to separate facts from opinions.

For example, your misgivings about a technique do not automatically make that approach a bad idea. For example, some fire fighters might insist that positive-pressure ventilation causes the building to burn down; they may even show examples and argue that this technique introduces air into a hostile environment. These people have had unpleasant experiences with positive-pressure ventilation, but that does not mean the technique is at fault. The misapplication or misuse of any technique will end poorly. This is the information that a student needs—not personal opinions.

You must listen to the culture in which you are instructing and adapt your training accordingly. Every departmental culture says that it is in existence to save lives. Likewise, every culture says that it is in existence to protect property. But how will the members of that culture save lives and property? To what extent are they willing to risk their own lives? These are the questions that a departmental culture must answer.

As fire fighters, a major portion of our job is risk management. Risk management must be placed in context of the fire department's culture so that you can place the correct emphasis on the decision-making process. For example, although you may not agree with the decision to go offensive, you must respect and acknowledge the department's decision.

**Teaching Tip**

Culture is a strong force in education. Understand the culture of the group and work within that context to present the material.

## Venue

If you are instructing locally, the venue could be anything from the back of a fire engine to a formal classroom in a training center. Knowing where you will present the information is important so that you can tailor your presentation style to the venue. The size of the room is one factor; the seating in the room is another. The receptivity of students to your message will vary depending on whether the instruction takes place in a formal classroom with tables and chairs versus an apparatus room where students are just standing around. A formal training center with tables and chairs automatically puts students in the learning mind frame; that is, students are more focused in this venue. Because this venue represents a more formal situation, student behavior tends to be more formal. By comparison, the more casual atmosphere of the apparatus room puts students at ease and minimizes the formality of the student–fire service instructor relationship. This opens up opportunities for the discussion to go more in the direction of the students' interests.

The concept of tailboard chats is not a new one. They have been used for many years and can be an effective method of educating fire fighters. The **tailboard chat** is an informal gathering of fire fighters at the back of an engine where the company discusses various subjects. The tailboard is the equivalent to the water cooler in a business setting. It puts all members of the company at ease and on the same level when you instruct them there. For example, you might use the previous call as an opportunity to reinforce a learning objective. These tailboard chats can happen back in the station or on the scene after the call is over.

Some fire service instructors use the structure that the company was just at during a call as an example and "simulate" a fire to reinforce learning objectives. The tailboard chat may begin with you saying, "If there was a fire on the third floor of this building in the B-C corner, where would we place the engine and the truck? What would the first engine company be responsible for doing? What would the first truck company do? What if this fire was at 2:00 A.M.? At 4:00 P.M.?" This is tactics and strategy training at its best.

# Wrap-Up

## Chief Concepts

- Demographics extend beyond issues of race and national origin. In the fire service, taking demographics into account during instruction requires looking in a more holistic manner at everything a fire fighter brings to the department.
- Learning takes place in two general settings: formal (e.g. a classroom) and informal (e.g. tailboard chats).
- Knowing your audience, department culture, and the venue is critical to your presentation's success.

## Hot Terms

**Demographics** Characteristics of a given population, possibly including such information as age, race, gender, education, marital status, family structure, and location of agency.

**Learning environment** A combination of the classroom's physical and emotional elements.

**Tailboard chat** An informal gathering where fire fighters discuss various issues.

# Fire Service Instructor *in Action*

As the company officer, you are responsible for ensuring that your company is properly trained. You have conducted an assessment to determine the training needs of your crew and created appropriate learning objectives. You are now working on developing the specific lesson plan so that you can begin the process of presenting a class. You know that you have many decisions to make if you are to convey your message effectively.

**1.** A tailboard chat is a(n) __________ gathering of fire fighters where the company talks through various subjects.
   **A.** formal
   **B.** structured
   **C.** clustered
   **D.** informal

**2.** What is the most important concern in the learning environment?
   **A.** Weather
   **B.** Lighting
   **C.** Noise
   **D.** Safety

**3.** Where is the best place to conduct an SCBA class?
   **A.** In the engine room
   **B.** In the classroom
   **C.** In the environment that is most conducive to the lesson plan
   **D.** Anywhere—it doesn't matter because the material is always the same

**4.** Which of the following is *not* considered a demographic of a class?
   **A.** Age
   **B.** Course location
   **C.** Sex
   **D.** Education level

**5.** What is the best rule of thumb about using cuss words during teaching?
   **A.** Never use them.
   **B.** Do not change your natural use of them.
   **C.** Use them for emphasis only.
   **D.** Use them as often as possible.

**6.** If you inadvertently use offensive language during a presentation, what should you do?
   **A.** Act like you didn't say it.
   **B.** Make a joke about not offending anyone with it.
   **C.** Immediately apologize for it.
   **D.** Ask the female fire fighters if it offended them.

# Training Today: Multimedia Applications

# CHAPTER 8

## NFPA 1041 Standard

**Instructor I**

4.3.3* Adapt a prepared lesson plan, given course materials and an assignment, so that the needs of the student and the objectives of the lesson plan are achieved.

**(A)* Requisite Knowledge.** Elements of a lesson plan, selection of instructional aids and methods, origination of learning environment. [p. 130–145]

**(B) Requisite Skills.** Instructor preparation and organizational skills.

4.4.2 Organize the classroom, laboratory, or outdoor learning environment, given a facility and an assignment, so that lighting, distractions, climate control or weather, noise control, seating, audiovisual equipment, teaching aids, and safety are considered.

**(A) Requisite Knowledge.** Classroom management and safety, advantages and limitations of audiovisual equipment and teaching aids, classroom arrangement, and methods and techniques of instruction. [p. 130–131]

**(B) Requisite Skills.** Use of instructional media and materials. [p. 140–145]

4.4.6 Operate audiovisual equipment and demonstration devices, given a learning environment and equipment, so that the equipment functions properly. [p. 140–145]

**(A) Requisite Knowledge.** Components of audiovisual equipment. [p. 131–140]

**(B) Requisite Skills.** Use of audiovisual equipment, cleaning, and field level maintenance. [p. 140–145]

4.4.7 Utilize audiovisual materials, given prepared topical media and equipment, so that the intended objectives are clearly presented, transitions between media and other parts of the presentation are smooth, and media are returned to storage. [p.130–145]

**(A) Requisite Knowledge.** Media types, limitations, and selection criteria. [p. 130–145]

**(B) Requisite Skills.** Transition techniques within and between media. [p. 130–143]

**Instructor II**

5.3.2 Create a lesson plan, given a topic, audience characteristics, and a standard lesson plan format, so that the JPRs for the topic are achieved, and the plan includes learning objectives, a lesson outline, course materials, instructional aids, and an evaluation plan.

**(A) Requisite Knowledge.** Elements of a lesson plan, components of learning objectives, instructional methods and techniques, characteristics of adult learners, types and application of instructional media, evaluation techniques, and sources of references and materials. [p. 130–145]

**(B) Requisite Skills.** Basic research, using JPRs to develop behavioral objectives, student needs assessment, development of instructional media, outlining techniques, evaluation techniques, and resource needs analysis. [p. 130–145]

5.3.3 Modify an existing lesson plan, given a topic, audience characteristics, and a lesson plan, so that the JPRs for the topic are achieved and the plan includes learning objectives, a lesson outline, course materials, instructional aids, and an evaluation plan.

**(A) Requisite Knowledge.** Elements of a lesson plan, components of learning objectives, instructional methods and techniques, characteristics of adult learners, types and application of instructional media, evaluation techniques, and sources of references and materials. [p. 130–145]

**(B) Requisite Skills.** Basic research, using JPRs to develop behavioral objectives, student needs assessment, development of instructional media, outlining techniques, evaluation techniques, and resource needs analysis. [p. 130–145]

## Knowledge Objectives

After studying this chapter, you will be able to:

- Describe the types of multimedia tools available for the fire service instructor.
- Describe how to use multimedia tools.
- Describe how to maintain multimedia tools.
- Describe when to use multimedia tools in a presentation.

## Skills Objectives

After studying this chapter, you will be able to:

- Demonstrate how to use multimedia tools.

# You Are the Fire Service Instructor

It is the first day of class. Forty students are attending your class on responding to terrorist incidents. You spent weeks preparing for class by gathering and researching the information you feel the students need in real-world firefighting. You developed an incredible PowerPoint presentation that includes digital photographs of target hazards specific to your jurisdiction. Handouts have been duplicated based on this presentation, and refreshments are prepared and ready to be set out for the students when they arrive. Everything appears ready to go.

You practiced this PowerPoint presentation so often that you don't have to look at your notes or the slides to know what you are going to say next. When you arrive, however, you discover that the facility does not have the proper connections for your computer and the audio/video projector. Telephone calls are made between the sponsoring organization, the facility, and you, but the uncertainty begins to take you off of your game.

A facility member asks if your presentation can be transferred to his computer. Because you incorporated video and digital photographs into the presentation, the file is too large to transfer via CD-ROM.

1. Do you have a plan B in case you cannot get your presentation transferred to the other computer?
2. Should you prepare to present without your outstanding PowerPoint slide show?
3. How will you ensure that this problem never happens again?

## Introduction

A key element in most presentations is the use of a multimedia tool in some form, ranging from handouts to PowerPoint (PPT) presentations. Using multimedia tools enhances the learning process for students and can make your job easier. At the same time, multimedia tools can be a distraction if they are poorly developed or used improperly **FIGURE 8.1**. Knowing how to use multimedia tools effectively is just as important as knowing the lesson plan. When preparing to give a presentation, part of your efforts should focus on reviewing the audiovisual portions of the lesson plan.

**FIGURE 8.1** The use of too many multimedia tools can be a distraction in the classroom.

For some fire departments without access to burn buildings, technology is used to bridge the training gap in experiential learning. For this reason, an increasing number of fire departments are using presentation technologies in fire fighter training. On college campuses, most students now expect a professor to use technology to distribute subject matter to them. Whether it takes the form of lecture slides in front of the class, hard-copy handouts, or the Internet, technology makes it possible for students to get information faster and easier than ever before. Coincidentally, this trend makes technology more affordable for fire and emergency services agencies to utilize.

While technology provides you with tools to immerse your students in the subject material, it comes with additional burdens. To effectively engage students with technology, you face many questions. Which sorts of material should go on a PPT slide? Are there limits to the amount of text you should put on a PPT slide? How should you arrange the material for optimal viewing? Should handouts of the PPT slides be distributed before class, after class, or not at all? If you decide to distribute the material, should it be in the form of a printout, on a CD-ROM, or via a Web site? Transition between media types should be done smoothly with minimal interruption. Visual aids should be incorporated into your general approach to teaching in a harmonious way. There are many rules and guidelines that you can follow to ensure that you meet the goal of teaching your class effectively.

As late as the 1990s, multimedia tools used in fire fighter training included only the basics: movies, videotapes, chalkboards, the infamous easel pads, duplicated materials (handouts), audiocassettes, overhead projectors, models, cutaway demonstrators, and simulators. Good fire service instructors created their own make-shift simulators consisting of a slide projector, pictures of area buildings, a piece of glass with wooden frame, and silica sand, all on an overhead projector. When used in combination, these tools created an inexpensive fire simulator—complete with the illusion of fire and smoke. Scenarios could be changed depending on the needs of the students or at the whim of the fire service instructor.

Today, these home-made simulators have been replaced by computer, projector, and software programs that create digital combustion. With a few simple keystrokes, you can change the simulation dramatically. Enhanced through the use of digital photography, local target hazards can be imported into the simulation for a customized learning process. These multimedia tools have an enormous impact on a student's understanding of the concepts presented in the classroom.

Today's students learn differently than the students of the early 1980s. In the 1980s, technology was novel, yet cumbersome. By contrast, the new generation of fire fighters grew up with instant access to the Internet, videogame systems, and instant messaging. Their need for understanding transcends mere technology: These students have grown up with the advantage of technology in their everyday lives, so their need relates to understanding the materials and concepts presented—not the technology used to present it. Although modern-day students need to learn the same concepts as were presented to their counterpart in earlier years, the way these students learn will be enhanced by technology because it is the baseline from which the students can launch into understanding of the material.

### Theory into Practice

During the course of your career, you will find that the use of multimedia tools can greatly enhance your ability to convey your message. But use of technology can also come with some drawbacks. To minimize the potential pitfalls, consider these issues when using multimedia tools:

- Do not read from the PPT slides. All fire service instructors make this mistake at one time or another. Most such slides are outlines of the lesson plan and serve to emphasize critical points. The only time it is acceptable to read from the PPT slide is if you are reading a quote, legal definition, or legal statement.
- Select the appropriate level of lighting for the type of multimedia equipment you are using. If you want students to take notes during the presentation, the room must be dark enough to see the PPT slides, yet light enough to write.
- There is an art to transitioning from the lecture to the multimedia technology—and practice makes perfect.
- When lecturing with PPT slides being projected behind you, do not stand in front of the projection screen. It is difficult for students to see what you are talking about if you are in the way.
- Always have your multimedia equipment set up before class and the media queued. Videotapes, DVDs, or audio recordings should be set and ready to go at the touch of a button.
- Always have a backup plan. What if your computer doesn't fit the projector? Always have a plan B, and even a plan C if necessary.

## Understanding the Types of Multimedia Tools

Multimedia tools come in a variety of types and are intended for a variety of uses. For example, they include the PowerPoint and Harvard Graphics software programs, among many others; videos; DVD presentations; graphic presentations developed by a publishing company for use with its textbooks; 16-mm films; computer-based training (CBT) programs such as WebCt, Blackboard, and Angel; Internet-based training; distance learning; National Fire Academy training programs; digital audio players such as iPods; and use of personal digital assistants (PDAs). It might seem that it would be impossible to choose from such an abundance of multimedia tools, but each fills a specific niche based on the type of learning that is to take place.

Learn how to use multimedia tools before you apply them in the classroom. It may take a fair amount of preparation time to master the tools you choose to make your presentation. Sometimes use is as simple as pressing the "play" button. At other times it is as complicated as learning a whole new "language" or software program or transitioning between various types of media. It is up to you to learn how to use these tools correctly and incorporate them smoothly into your presentation skills. Don't fall into the trap of blaming a poor presentation on your inability to use the equipment, and don't depend on multimedia tools to instruct for you. You need to practice with these tools to make them an effective medium to enhance your existing instructional skills.

## Audio Systems

Depending on the class size and the arrangement of the classroom, a microphone and audio system may be needed. While this technology is not new, it is often misunderstood. Remember who your audience is. Can everyone see and hear your presentation? Do all of your students have the same hearing abilities? Just as some people don't like to wear bifocals, some do not want to admit that their hearing just isn't what it used to be.

When deciding on which kind of audio system you will use, keep in mind your presentation style. Do you move around a lot, or do you stay at the podium with your notes? Think about these issues when you must choose between a wireless microphone or wired microphone setup.

A wireless microphone can be attached to your clothing and the transmitter hidden from view. This looks more natural and allows your hands to remain free for writing. Wireless microphones offer you the most freedom to move around the classroom, but they have some problems, too. Wireless systems may have an issue with frequency usage, such that you might hear voices broadcast that are not supposed to be part of your program. In addition, wireless microphones have limitations in terms of how far the microphone can be from the remote receiver.

The greatest limitation for wired systems relates to the cable. Is it long enough for you to move around effectively? Some wired microphones are permanently attached to the podium, in which case you lose the freedom to move around the classroom. Another disadvantage of wired systems is the potential for a mishap if your feet become entangled in the wire.

## Slide Projectors

Slide projectors have been around for a long time. Many fire fighters can remember the days when the expert was "the guy who drove 50 miles and had the slides." Today, we don't use slide projectors quite as much, but this somewhat-dated technology still offers some distinct advantages. For example, slide projectors are simpler and easier to operate than an overhead projector. The 35-mm slides are kept in a carousel that is placed on top of the slide projector. Because the 35-mm slides are already in their assigned order, you simply push the advance button or reverse button to change the slide.

In general, 35-mm slides deliver a much sharper picture than overhead transparencies, though you need to be aware of the possibility of the keystone effect FIGURE 8.2. Projection screens tend to hang straight down, but slide projectors (especially portable ones) often must be tilted to aim the image of the 35-mm slide over the heads of the audience members. Whenever a slide projector sits at an angle other than directly perpendicular to the screen, some distortion will occur. Typically, the top of the image will appear wider than the bottom (hence the term "keystoning"). Many of the latest slide projectors provide a degree of keystone correction. And remember—sometimes the simplest solution is the best solution. Some projection screens can be tilted to help alleviate the keystoning effect.

Another downside of the slide projector is the fact that the 35-mm slide carousel takes up more room than transparencies. The carousel is larger and heavier than a three-ring binder. If you have more than one carousel to carry, they may become difficult to handle. Also, if you drop your carousel and the ring holding the slides in place falls off, your slides will scatter all over the place. Unless you numbered your slides, putting them back in the proper order is difficult and time-consuming. In addition, 35-mm slides, like all projected multimedia tools, are subject to constraints related to lighting and classroom arrangement.

The largest drawback to the slide projector is the fact that 35-mm slides are difficult to produce. Adding text or captions to these slides must be done professionally or by using expensive equipment. Today, only a dwindling number of companies produce 35-mm slides. Finding a slide projector may also be difficult. Many organizations no longer have them or, if they do, the bulb does not work.

**FIGURE 8.2** **A.** A slide with the keystoning effect. **B.** A slide projected correctly.

## Presentation Programs

Presentation programs include a variety of software programs developed by software vendors. Companies such as Microsoft (the maker of PowerPoint) and Macromedia, for example, create software programs to assist customers in creating portable presentations for business, education, and government. Each presentation program has its own unique features and benefits that you can use to enhance your instruction. Some are easier to use than others; some allow for greater flexibility. For the purposes of this chapter, we refer to presentation programs with the acronym PPT, reflecting the fact that most fire departments use the PowerPoint program FIGURE 8.3. If you use another software program, the basic style rules are still relevant.

The ability to embed video and audio clips into your presentation is one advantage of using PPT, and it adds a new layer of sophistication and professionalism to your presentation. A disadvantage is the potential for over-reliance on technology as part of your instruction. If your computer doesn't work or problems crop up between the computer and projector, then you cannot use the PPT presentation. If your instruction relies solely on your PPT slides, then you will have

## Applying the JOB PERFORMANCE REQUIREMENTS (JPRs)

After you decide on the learning environment, the next step is to identify how the lesson plan will be presented. The more of the student's senses that you can appeal to during the learning process, the more effective the training will be. Using visual aids is one way to improve a lecture presentation by appealing to the senses of sight and sound, but remember to match the equipment to your instruction style. Today's students desire the inclusion of multimedia and visual aids in presentations and as much hands-on practice as you can deliver. Part of your continuing education responsibilities should be to keep up-to-date on the latest delivery methods and technology available for use in the classroom.

### Instructor I

When delivering the lesson plan in a classroom setting, a multimedia tool will probably be used to enhance the lecture. The Instructor I must be knowledgeable in the use of the media and also how to adapt to classroom setting limitations. All media supplement the objectives being presented, and all of them have pros and cons for their usage. Take the time to practice their use well before your presentation and adapt them to the learning environment.

### Instructor II

The learning environment is a big part in the selection and use of media in your presentation. Today's training applications require a mix of multiple types of media and visual aids. Media should be designed to enhance the learning, not to replace traditional learning methods. For example, a DVD needs to be presented and discussed and made relevant by the instructor using the media.

### JPRs at Work

The Instructor I will organize the learning environment and operate all audiovisual equipment to present the content of the lesson plan so that all learning objectives are effectively covered. Multimedia tools include a wide range of projectable aids, handouts, Internet-based products, and DVDs or videos. The Instructor I must be skilled in using all types of media equipment available.

### JPRs at Work

NFPA does not identify any JPRs that relate to this chapter, although the Instructor II will have a great amount of influence in choosing which multimedia equipment is used when he or she develops a lesson plan. If choosing a mass-produced curriculum or designing your own, consider the skills of the instructor presenting the material and prepare backup media in the event of equipment failure.

### Bridging the Gap Between Instructor I and Instructor II

The Instructor I presents objectives using a lesson plan and media selected by the Instructor II. In some situations, the Instructor II may be conveying a message that the Instructor I does not completely understand. It is sometimes difficult to teach another person's original thought and work. The Instructor II must be clear in the construction of the message and the media used to present the objectives. Work together to ensure clarity in the message being presented.

### Instant Applications

1. Experiment with a PowerPoint presentation. Which color schemes work best for dark classrooms? Which work best for bright classrooms? Learn how to change a PowerPoint presentation quickly, so that you can instantly adapt to existing conditions.
2. Modify a PowerPoint presentation by adding text, photos, or other illustrations to clarify the message.
3. Select a learning platform such as an online or distance-learning course. Discover what the benefits of that platform are for your students.

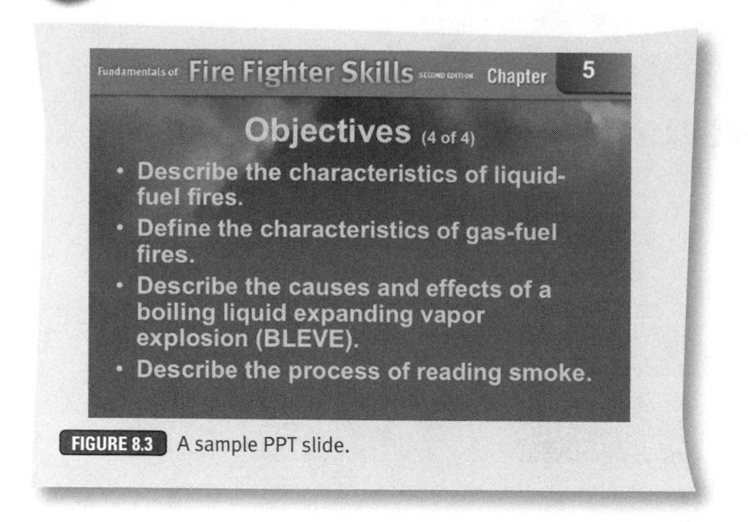


FIGURE 8.3 A sample PPT slide.

FIGURE 8.4 A digital projector can enhance your presentation.

to reschedule your presentation. Another disadvantage is the overload factor. Some PPT creators tend to include too much information, too many sounds, and too many visual effects on their slides. Keep your presentation simple and concise for greatest effect.

## Visual Projectors

You may find it useful to refer to the textbook to illustrate your message. Instead of requiring your students to get out their books and turn to the right page, you can use a visual projector to show the material to the entire class. This multimedia tool can project almost any one-dimensional printed material onto a screen. The visual projector has backlighting, overhead lighting, and color capabilities that can be tied in with a computer. It combines the flexibility of the overhead projector with some of the capabilities of a high-tech device such as an LCD projector. Major disadvantages of visual projectors include their costs and unwieldy size.

### Teaching Tip

Block print should be used whenever possible when written material that will be projected. Script is difficult to read and should not be used.

## Data Projectors

Liquid crystal display (LCD) projectors and digital light processor (DLP) projectors (also known more simply as data projectors) are used with video players, DVD players, computers, and visual projectors to project documents, images, and motion pictures so that a group of students can easily view the material simultaneously. Data projectors can present material on a screen ranging from 2 feet wide to 30 feet wide FIGURE 8.4. Important features to consider when selecting a data projector are the ability to show videos or DVDs directly, audio capabilities, screen resolution, and lumens, the unit used to measure the brightness of light emitted by the project. Many data projectors have become more affordable in recent years. In addition, the size of the projector is fairly compact, so these devices can be moved from facility to facility with little effort. The various pros and cons should be examined before purchasing a data projector. As with a slide projector or overhead projector, having an extra bulb available is a good idea. However, bulbs for LCD and DLP projectors are very expensive.

You should also consider the issues that can arise in the classroom when you use a data projector. First, connections can be an issue, because the right kinds of cables must be available. Also, make sure that the cables are long enough, because short cables will hinder your setup and use of the projector.

You must learn the proper start-up procedures for your data projector as well. Generally, the data projector must be on before a computer is turned on. Learn how to get the computer to output to the data projector. Some computers will automatically recognize the data projector device, but often you must press a control-function key to activate the connection. Some of your presentations may not be compatible with the data projector, and video clips that play on your computer screen may not project properly with such a device. Given these caveats, always test your presentation before class and have a plan B.

## Videotapes

Videotapes have been around for more than 20 years. The technology standard is the VHS format. Most VHS videotapes consist of a magnetic tape encased in a plastic cassette, similar to an audio cassette tape, only larger.

Video cassette recorders (VCRs) and video cassette players (used to play VHS tapes) are very affordable. The VCR requires another component for viewing the images—either a data projector or a television. If you are using a television to view the videotape, keep in mind the size of the class and lighting. The larger the class, the larger the screen must be. A 27-inch or larger screen is adequate for most classrooms. Also think about the audio component. Most televisions have adequate sound quality for classroom presentations. In larger classrooms where data projectors are used, however, a more sophisticated audio system may be required.

Videotapes are copyrighted material. As a consequence, they should not be copied unless you obtain permission to do so from the video production company. An alternative to using copyrighted material is to produce your own videos. The price for video cameras continues to decline, allowing some fire departments the opportunity to develop their own training materials. Videos can be produced through a variety of sources, including your local television/cable provider. You can take advantage of software programs such as Final Cut Express, Final Cut Pro, and Adobe's Premiere Elements to produce quality videos for training purposes. But remember: Students don't want to watch your home movies. Your videotapes must be of professional quality before they can be screened in the classroom. Amateur videos are no different than the overused war story.

## DVDs

Like CD-ROMs, DVDs are available as either blank or preprogrammed media. Information ranging from video clips to audio clips to PPT presentations can be stored on a DVD. DVD players and burners are often standard hardware on new personal computers. With a DVD burner and digital media editing software, you can develop your own digital presentations and ensure that the end result is a professional product.

In addition to a DVD player (either as a separate device or in a computer), DVDs require either a television or data projector for viewing. As with use of videotapes, you must remember the audio component when showing DVDs. Depending on the size of the classroom, you may need additional external speakers.

## Digital Cameras

To capture students' attention and keep them engaged in your presentation, you may want to include local examples in your multimedia presentations. If you are giving a PPT presentation on the implications of special occupancies for preincident planning, you might want to include some local examples, such as the assisted care center on Main Street or the group home on Center Street. A digital camera allows you to easily capture an electronic image, upload it onto your computer, and insert the image into your PPT presentation.

Digital cameras are available at every price point. You can purchase anything from a simple point-and-shoot digital model to a professional digital camera with multiple lenses. The resolution of these cameras is constantly changing and improving. As of this printing, it is advisable to purchase a digital camera of 4 megapixels or higher.

The number of megapixels matters because it directly affects the image quality. To see how this works, take a newspaper from the recycling bin and look at a photo very closely with a magnifying glass. The photo is made up of tiny dots of color, called pixels. In printing, pixels (dots) are measured in units of number of dots per inch (dpi). The more pixels or dots per inch, the sharper the resolution.

A printing press would never consider printing an image at less than 360 dpi, but a data projector is only capable of projecting at a resolution of 72 dpi. What does this mean for you, the fire service instructor? You can use smaller-size image files in your PPT presentation than if you were going to actually print out the image. A pixel-heavy image file will just take up time and space in your PPT presentation. Save these images' files at a lower resolution, thereby creating smaller files and saving space and time in your presentation.

## Flash

Flash refers to both the Adobe Flash Player and a multimedia authoring program used to create content for Web applications, games, and movies. The Flash Player, which was developed and is distributed by Adobe Systems, is an application that is readily available in most Web browsers. It features support for vector and raster graphics, a scripting language, and bidirectional streaming of audio and video. Since its introduction in 1996, Flash has become a popular method of adding animation and interactivity to Web pages FIGURE 8.5. Several software programs and devices are able to create or display Flash images. Flash files (traditionally called "flash movies") carry the file extension .swf. These files may be either

FIGURE 8.5 Flash is a popular way to add animation.

accessed as objects on a Web page or played in a stand-alone Flash Player. Your local community college may offer classes or workshops on this program.

## Scanners

While the scanner can be a helpful tool, its use does raise copyright concerns. For example, suppose you find a picture of a house fire in your local newspaper. This picture may perfectly illustrate your message about the dangers of leaving food unattended on the stove. Excited to add this image to your PPT presentation, you head over to the office scanner with your laptop, scan the newspaper picture, save the image file onto your laptop, and drop the image into your PPT presentation. If you haven't contacted the newspaper and received written permission to use the newspaper's photo in your presentation, you are committing a copyright violation.

## Presentation-Ready Programs

Prepackaged software programs are a valuable resource when you are planning your instruction. For instance, Jones and Bartlett Publishers provides instructors with presentation-ready programs for use on computers at the local level. Thanks to these programs, there is no need to spend valuable preparation time developing your own presentation; instead, you can simply use the professionally developed materials. Of course, you must still familiarize yourself with the presentation before class, rather than seeing it for the first time in front of the class. Professionally developed presentations are intended to be dynamic instructional tools, used to enhance students' learning experience, but not to replace the experienced instructor.

## Distance Learning

Distance learning does not require computers, modems, DVD players, or any other fancy technology. Instead, it simply consists of instruction from a locale "a distance away" from a training center or college/university. Such a program can be as simple as a package of printed materials sent through the mail or as complex as Flash animation presented during a video conference. Today, the majority of distance learning appears to be accomplished through a combination of reading assignments, conferencing, video presentation, testing with proctors, and written assignments. Student evaluations are processed through an instructor.

## Computer-Based Training

Computer-based training (CBT) software provides a virtual course environment with tools for course preparation, delivery, and management. With this software, you have the tools to rapidly prepare course materials and efficiently manage day-to-day teaching tasks. Many of CBT programs offer features that allow you to facilitate collaborative learning, personalize content based on students' unique needs, and positively affect learning outcomes. WebCt, Blackboard, and Angel Learning Systems are all examples of CBT software.

CBT is not limited to students in colleges and universities; it has also changed the way in which private industry and public safety agencies train their staff members. On the surface, it allows for self-paced learning, easy access to information, and a student evaluation system. Several users in several different locations can use the same material at the same time, establishing uniform information transfer.

CBT or e-learning can be a powerful way of teaching. CBT can inform, illustrate, test, and demonstrate complex concepts simply. For example, students can view Flash animations with a voice-over that explains how a fire pump works internally. They can also click on parts of the fire pump image to see labels or learn more details. If real-world experience is desired, some systems have built-in simulators to enable students to experiment with key concepts while using different parameters. This allows students to benefit from real-world learning while ensuring that they can make mistakes without endangering personnel or equipment.

### Ethics Tip

The fire station is a natural collection of diverse individuals. Some may be good at one thing, while others are good at something else. At times, some personnel may want to take advantage of this fact by having one person take the online test and give the correct answers to the rest of the class. This is cheating. Not only is it unprofessional, but it can be life-threatening. If a fire fighter does not know which fire extinguisher to use on a grease fire because a buddy helped him with the answers on a test, the grease fire isn't going to help the fire fighter out with the right answer. If asked to cheat, don't.

With CBT or e-learning, both the instructor and students receive almost instantaneous feedback on students' performance. When a student takes a test, the score is reported through a gradebook function along with any questions missed, plus the correct answers. The fire service instructor can also take advantage of this information to evaluate the student's understanding of the material. Evaluating the reports from students' tests enables you to determine whether information needs to be added or removed from the course and what needs to be provided to students to increase comprehension. In addition, a gradebook is set up automatically to track the progress of each student through an entire course of delivery.

The popularity of e-learning has exploded in the past few years. In fact, many colleges and universities have doubled or tripled their course offerings via the Web. Some institutions even offer entire associate or bachelor's degree programs over the Internet. Recognizing this fact, many students now expect courses to be offered in this manner. Colleges, universities, fire departments, and other training organizations are identifying how to integrate e-learning into the instructional experience. Of course, no technology can ever replace the wisdom and experience of a seasoned fire service instructor. E-learning is not a threat to instructors, but rather a tool that can be used to address challenges posed by live training. For example, in the future, virtual reality simulators could be used instead of live burns.

Hybrid training represents an alternative to totally Internet/computer-based training. It takes the best parts of the classroom environment and partners them with computer-based elements so that the student has greater flexibility in completing assignments. Texts, assignments, and exercises can be completed online, while discussion and hands-on learning can be completed in the classroom. This creative alternative is useful in attempting to meet the needs of both the student and the fire service instructor. It can be difficult for any fire fighter—not just volunteers—to be in a classroom at the same time, five days a week.

For fire service instructors who plan to instruct via e-learning systems, consider the following issue: Do you have the time to commit to this method of instruction? When instruction is presented in the classroom, all students meet at a set time period, ask questions, and get immediate responses. With e-learning, students have the same expectations, yet the course is available on their schedule, not yours. As a consequence, you may need to check e-mail or discussion boards throughout the day and night to quickly give your students feedback unless you first establish how and when you will communicate with them.

**Safety Tip**

Multimedia tools inevitably require a power supply. It is all too easy for both students and instructor to trip on power cords snaking around the classroom.

Designing e-learning courses can be difficult and time-consuming. Understanding the language of e-learning is only the beginning. All the rules of keeping the course interactive, challenging, and interesting to students still apply. Can you adequately address this issue with the time you have available? According to Elearningguild.com, it can take hundreds of hours to produce one hour of material for an e-learning system. If you are following a traditional course length of 16 hours for continuing education or 45 hours for college or university level, this represents a considerable time investment. The advantage is that, once the course is developed, it can be easily updated for new challenges, new information, or improved operational capabilities.

In the past, the focus has been placed on the instructor developing the e-learning course. This may not be the best use of a fire service instructor's talents, however. In some cases it might be wiser to work with one of the professional outsourcing avenues to develop an e-learning solution that is right for you.

## Internet-Based Training Provided by a Vendor

A myriad of vendors provide internet-based training. With this kind of training system, the student simply logs onto the vendor's Web site, signs up for a course, and pays a fee to the vendor. The vendor then provides a password to the student so that the student can view the desired course of study.

Today, vendors provide educational materials for fire, EMS, and homeland security personnel, among many others. In many cases, this type of training has approval from state regulatory agencies for continuing education units. It is important to realize that these programs are not free—someone must cover the vendor's development costs. Additionally, these courses do not provide much flexibility for instructors. Courses come either "as is" or as a stand-alone component. If you are considering the use of this type of program, be prepared to have minimum interaction with the students who take the course and to accept certificates for the students from the vendor upon successful completion of the training program.

## National Fire Academy CD and Virtual Academy

Since 1975, nearly 1.5 million students have received training through the National Fire Academy (NFA) on many different types of training platforms. NFA is continually evaluating its training methods for fire fighters. This organization offers courses via residential delivery on its campus in Emmitsburg, Maryland, but the scope of this program is obviously too small to meet the needs of the fire service community across the United States. To fill this gap, NFA offers coursework via off-campus delivery, thereby reaching students through a distance-learning system. This system includes regional delivery, direct delivery, and the National Emergency Training Center (NETC) virtual academy. The last of these options—the NETC virtual academy—leverages technology through the combination of CD distribution, simulations, and Internet-based resources.

NFA courses can be delivered right to the student's door via CD-ROMs that are mailed directly to the student. The student views the content and takes a test evaluating his or her understanding of the material presented. Upon successful completion of the test, a certificate is mailed to the student. Each course is available to the firefighting student at no cost.

In the NETC virtual academy, students register online, view course material via the Internet, and take a course examination. Upon successful completion of the course, a certificate is provided to the student. The NETC system incorporates the use of slides, audio/video presentations, audio presentation, and quizzes with interactive participation. Through interactive participation, for example, students can assess their comprehension of the materials before they participate in testing. In this active learning environment, students gain confidence while undergoing training in an atmosphere where any mistakes will not cause physical injury.

In addition to the NETC virtual academy, the NFA participates in an academic outreach program. If fire service personnel cannot attend traditional college courses owing to their work hours or location, they can earn a degree in fire science through independent study as part of a distance-learning program offered by seven colleges and universities across the United States. For more information on this or any other program offered by NFA, visit the organization's Web site (http://www.nfaonline.dhs.gov/).

## Audiocassettes

Audiocassettes have been around since the early 1960s, though the first models were somewhat unwieldy. Currently, audiocassette players are small enough for someone to wear while jogging. Because they are so portable, these devices can be very useful for both students and fire service instructors. For example, an audio recording of a telephone call to a dispatch center can be used to illustrate how communications are conducted. This technology allows the student to hear the conversation and judge the material on the audiocassette.

Audiocassettes are especially useful in table-top exercises, lending an air of reality to the training by giving the impression of a receiver on the radio. Entire presentations can be prepared on audiocassette for students to listen to away from the classroom, adding another dimension to the learning program. Audiocassettes are relatively easy to produce and similarly easy to reproduce for students to use outside of the classroom.

FIGURE 8.6 Digital audio players allow students to listen to lectures and view images outside the classroom.

## Overhead Projectors

Low-tech solutions are sometimes just as effective as the newest technology. Overhead projectors, for example, have been used in training for more than 40 years and were once considered the industry standard for quality educational enhancement. The overhead projector displays materials that have been developed on transparencies and then manually placed on the projector for viewing on a projection screen. This low-tech device is often a good choice as your plan B in case your high-tech devices fail.

### Safety Tip

Given the number of forms of media being used in presentations, instructors might be tempted to violate the local fire codes by exceeding the rating of the power cords or the outlet. Practice what you preach: Don't use multiple power strips or lightweight extension cords.

## Digital Audio Players and Portable Media Devices

Digital audio players, such as the iPod, are creating a new horizon for digital recording. In the past, the digital audio player was used to download music and music videos. Through the creative work of many fire service instructors, digital audio players are becoming a mainstream technology for students at most universities. Before long, digital audio player usage within the fire service will be commonplace. These devices are so compact that students can carry them easily and download lectures or pictures that can be reviewed while traveling to class FIGURE 8.6.

Some digital audio players allow students to use an audio record feature, similar to tape recorders. Several media outlets are actively engaged in pushing the technological edge by offering downloads to these devices featuring such notable fire service figures as Chiefs Billy Goldfeder and Rick Lasky.

Students may also find that digital audio players allow them to complete class assignments. In some courses, students may be able to use desktop audio/video editing software to create and edit projects for course assignments. The results can be uploaded to the training center's computer system for presentation, or content playback can occur immediately via classroom audiovisual equipment. Digital audio player devices can store not only images and audio, but also entire presentations created in PowerPoint, word processing documents, or other software.

As an instructor, you can benefit greatly from these devices as well. The possibilities for their use are endless: oral examinations, classroom performance and practice, group discussions, field interviews, interviews with content experts, classroom lectures, and discussions or presentations. The possibility of a podcast replacing a webcast allows you to make the materials for a class more widely available, too.

As with other technologies, we must wait to see what benefits this technology will ultimately yield. As more digital systems become available and the costs for them are driven lower by greater demand, fire service instruction may be transformed from a system of learning confined to a specific locale to one where information can be obtained "on the fly."

Another portable media device, the personal digital assistant (PDA), allows students to effectively carry a computer in the palm of their hand. PDAs are making an entrance into educational media as well. While known by various names, PDAs can include technology such as cellular telephone capabilities; computer operating systems such as Linux, Windows, and others; e-mail; web browsing; and much more. Their use can bring new dimensions to educational media, but it can also introduce potential problems. For example, cheating, use of calculator functions during examinations, and answering of telephone calls in the middle of coursework are all new challenges associated with PDAs. For information transfer, however, PDAs combine the benefits of digital audio players and the flexibility of mobile communications technology.

As technology adapts, improves, and advances, new opportunities will be found to augment educational media. Satellite communications, cell phones with media downloads, and text messaging—all are new frontiers for you to provide quality training to fire service personnel.

## Satellite Programming

Satellite television is available worldwide. Vendors such as Dish Network and Direct TV market such services to residential customers to replace their landline cable systems. Many fire stations are equipped with systems that rival local and national commercial broadcast companies by utilizing a 3-meter base station. Two of the most popular satellite broadcasters for fire service personnel are the United States Fire Administration (USFA) and the Fire and Emergency Training Network (FETN). The USFA provides interactive Internet and satellite distance learning, making a wide variety of programs readily available to any community in the United States. Through the Preparedness Network (PREPnet), all programming is available to any community with access to the Internet, or a C-band or Ku-band analog satellite dish. Vendors such as FETN, whose programs are produced by Trinity Learning Corporation, may require you to have an additional satellite dish to take advantage of their programming.

## Virtual Reality

Virtual reality devices allow students to interact with a computer-simulated environment. By displaying images either on a computer screen or via a headset, the goal is to immerse the student in an environment that can be controlled by the student and the programmer. This sensory experience is designed to force feedback from the student as he or she interacts with the environment. While this technology is, in essence, a simulation, the visual, auditory, and tactile sensation create an artificial world wherein the student can make mistakes and cause injury.

The virtual environment can be designed to mimic a real-world scenario such as a burning grocery store or it can be set up to resemble a video game. This technology is developing daily by quantum leaps, and it may become the standard means of training in the near future. Unfortunately, the currently available technology does not achieve the realistic look and feel instructors want for their students. In the future, as faster processors and more cost-effective systems emerge, instructors may be able to apply the best possible technology for the benefit of their students.

## Simulations

Like virtual reality systems, simulators allow students to safely interact with hazards in a controlled environment. This practical application was developed by the military to train soldiers for tactical operations during peacetime. Students today have the benefit of growing up with simplified simulators. With the advent of popular video games such as Halo and Doom, many students are accustomed to being immersed in an imaginary world. Today, gaming has taken on new dimensions, as newer systems mimic the movements of the operator to accomplish a mission in the game. Simulations—no matter how simple or complex—allow students to be challenged and to learn from their mistakes.

Simulations have been developed both commercially and non-commercially. Commercially available simulators include gForce Technologies, ATC Flight Simulator Company, Fire Studio by Digital Combustion, and Advanced Digital Management Simulator (ADMS) by Environmental Tectonics Corporation. ADMS is an interactive virtual reality-based team training system that provides emergency responders with the opportunity to develop skills in command, control, coordination, and communications. Incidents faced in ADMS simulation include aircraft accidents, terrorist acts, hazardous materials spills, fires, and natural disasters, with the programs being designed to test and validate students' understanding of emergency operations.

While not the same as live fire training, after a simulation, the student walks away with concepts that can be applied and institutionalized long before he or she faces a live fire event. As with an actual live fire, replication of exact circumstances is difficult at best, yet with the simulator the same conditions can be repeated until the outcomes are acceptable. Simulators are no substitute for live fire training, of course. Even so, given that there are fewer fires to respond to and that regulations governing live fire training are becoming stricter, simulators offer an alternative to live fire training or no training at all.

## Models

Many fire service instructors find the use of models helpful when teaching basic principles prior to going into the field. Models are built by the fire service instructor. To be effective, they should be built to scale or all pieces in the entire model should be on the same scale. Such models are often incorporated into the table-top exercises used for incident command training. For example, a model city may be built for the various types of scenarios that could be encountered within a jurisdiction.

Models can be used to teach basic principles before students go out into the field. A good example of this application would be a trench rescue scenario that uses a scale model to demonstrate demarcation of zones, placement of apparatus, shoring techniques, and incident management. Once the students have mastered the basic principles using the models, then they can progress to an actual trench. One disadvantage when using models is that you have to build your own models, which may prove difficult for all students to see in a large classroom.

FIGURE 8.7 Paper easels can be a valuable addition to your presentation.

## Chalkboards and Erasable Boards

Chalkboards can be used to illustrate points the instructors believe are important. Chalkboards have been around since the 1800s and have several advantages: They are very easy to use, most classrooms provide one, they are relatively inexpensive, and they can be reused.

One downside is the time spent with your back to the classroom, writing information on the board. You can alleviate this problem by arriving early and writing your points on the board prior to class.

Another issue related to chalkboards is not so easy to overcome—namely, how do you contain the large amount of chalk dust? With students who are allergic to chalk dust or have asthma, creating a healthy learning environment can be a challenge. In these cases, you should look into the availability of erasable boards. Erasable boards utilize dry-erase markers, thereby avoiding the dust from chalk, and are used in the same way as the chalkboard. Most classrooms are outfitted with either a chalkboard or, as in most modern facilities, an erasable board.

## Pads and Easels

This simple technology comprises a large drawing pad that is used to emphasize material. Pads and easels are relatively inexpensive and can be reused. Similar to the chalkboard, you illustrate points on paper, which can then be torn off and posted on walls or retrieved for use in future class presentations FIGURE 8.7. The pad can be developed before, during, or after a presentation, becoming a flexible learning device. The greatest advantage to the pad and easel is its flexible use in small-group settings. Members of each group can write their thoughts and ideas in response to the activity and post them for others to see, review, and discuss. Each group can write its responses to activities and post them for later discussion. The pads can also be used for later discussion to review information or to refer back to while clarifying a point or concept.

When using pads/easels (and chalkboard/erasable boards), legibility is very important. If you cannot write so that students can read the material, then you should have the material preprinted. A simple tip to help improve legibility is to use lined easel pads or to lightly draw lines with pencil onto the paper. Then, use the pencil to do any writing before class. During the presentation, the pencil marks are not visible but can quickly be traced to give a nice, neat visual.

## Handouts

Remember the old days when the teacher handed out those smelly blue mimeographed pages? Today, handouts play just as vital a role in student comprehension and retention. While the technology for handouts has not changed much since those early school days, handouts remain an effective way to enhance lecture material and provide additional information. Happily, they no longer require typing on a mimeo pad: The copy machine has made it cost-effective to produce a variety of materials for students to take with them when the class is over.

**Teaching Tip**

Include your name and contact information on each student handout. This will allow students to ask questions at a later date, or allow others to contact you for more information regarding similar training. Handouts allow students to review the information on their own time and gain a more comprehensive understanding of the material.

# How to Use Multimedia Tools

## General Guidelines for Slide Design

When using PowerPoint software or a similar program, instructors need to follow several simple rules to make it easier for students to visualize their message and avoid distractions.

# Voices of Experience

As a fire service instructor, one of the expressions I remember hearing most often is this: "Those that can, do; those that can't, teach." I don't believe this is true. There are many great instructors out there who both do and teach. It takes a special talent for someone to translate information so that other individuals can understand it and then incorporate that information into how they operate on the fire ground.

This was proven to me early in my career by Lieutenant Jim Notary of the North Washington Fire Protection District in Denver, Colorado. He was our academy coordinator. He had the technical skills and a background in firefighting, yet he was able to make me understand that the principles he was teaching could save my life, my buddies, or my teams. That was in the 1980s. At that time, Lieutenant Notary (now Fire Chief Notary) incorporated every trick in the book—from hands-on training to audiotapes, videotapes, movies, simulators, and more. Yet, he did not depend on those "tools" to get us through. He worked with a team of instructors—each skilled in one area or another, and as a group, an outstanding team. Each instructor was well prepared, knowledgeable, and skilled.

*"Lieutenant Notary also taught us not to depend on technology, even on the fire ground."*

Lieutenant Notary also taught us not to depend on technology, even on the fire ground. We learned that it is the fire fighter's ability to be creative, innovate, adapt, and overcome that will put the fire out. A fire will never be extinguished by a PowerPoint presentation. And as instructors, we need to remember that it is the creativity and innovation of the instructor that will help the student understand the most complex of concepts. Fire Chemistry, while a difficult subject, is no more difficult than Firefighting Tactics and Strategy. The instructor simply needs to find the common denominator between the student and the concept. When found, the student's light will go on. And when that light goes on, the student and the team he or she operates with will be safer and better fire fighters because of it.

*Kurt Larson*
Northwest Florida Fire Academy
Pensacola, Florida

When you begin developing a presentation with such software, you will be asked if you wish to utilize a "design template," "blank presentation," or "auto content wizard." If you have not created a presentation in the past, the easiest way to get started is to use the wizard or select a design template. The software will then guide you through the process of making choices about background color, font, graphics, and other aspects of the presentation. After you have completed your design, the resulting presentation should look and feel professional.

### Font

The words you choose impart to your audience the message you are trying to convey, but the appearance of your message can distract or maintain students' attention, thereby affecting how well they heed that message. Fonts are a critical factor when you are developing a presentation for others to read and understand. Some fonts are better choices than others at getting your message across. For instance, sans serif fonts are better choices than serif fonts for enhancing readability when the material will be displayed on screen. Serif fonts have small embellishments or lines at the base of each letter. These embellishments make it easier to follow a line of text on the printed page, but they are a distraction on a screen. Select a sans serif font (e.g., Helvetica or Arial) instead of a serif font (e.g., Times New Roman) for your presentation.

Font size is crucial. You will find many rules for determining the proper font size for presentation settings. A general rule of thumb is to use a font size of at least 28 points for body text and 38 points for heading text.

Some people are colorblind, so avoid using color combinations that might cause these individuals difficulty, such as red text on a green background. Normally you will use light text on a dark background in projected presentations, but pay attention to the strength of the image projected by the projector.

The shades of color you see on your monitor when you are developing your presentation are not necessarily the same ones you will see when projecting that presentation. Before showing your presentation to students, project it through the equipment you will use during class. You may find you need to change colors or fonts.

Keeping the right amount of blank space on a slide is important. If you find yourself wanting to reduce the font size so that you can cram more text onto the screen, consider redesigning the slide so that you have less text on it. Limit the number of words on your slides to either 7 × 7 (seven lines, no more than seven words each) or 5 × 5 (five lines, no more than five words each). Always remember the needs of your audience: Is your focus on your message or how much you can pack onto one slide? Add more slides for greater impact in your PPT presentation. Not only does this make the text easier to read, it helps to keep students' interest throughout the program.

### Image Selection

Illustrate your PPT slides with images to engage the audience. Such images appeal to the senses. Remember the old saying, "A picture is worth a thousand words"? This is especially true for fire service personnel, because fire fighters tend to be hands-on learners. If students can relate their actions to visual images, the presentation is likely to have a greater impact.

Images can also be used to support classroom discussions. They help to move students through stages of understanding, allowing them to organize their thoughts and responses based on patterns in the image. For example, in a safety presentation, a slide might contain an image of an intersection with a car slammed into a power pole, fluids on the ground, and people milling around. Ask the students to work together to identify safety patterns and decide what they should be aware of in response to this situation. You might then present a slide that shows improved patterns. The discussion would continue, supported at each stage by a slide that exhibits the patterns identified at that stage.

Appropriate images can often be obtained from the clip art files provided with many commercial presentation programs. If you want to add or create your own images, use a digital camera, scanner, or graphics program. Cameras and scanners generally come with their own software, allowing you to save your images in the format that you prefer. If you want to create your own graphics, you will need graphics software (also known as a drawing program). Commercial programs such as Illustrator, Photoshop, Paint, and Paint It are readily available. Each has its own language and setup, so you will need to spend some time becoming familiar with the software before you attempt to edit or create images with such a program.

### Sound

Sounds can either be helpful or a hindrance. Files containing sounds for use with computers carry a variety of file extensions. Be sure your computer can accommodate the particular file extension (i.e., the type of file) so that it will perform properly. For example, files with a .mov extension are intended for use with Quicktime media programs, whereas .voc files are intended for use with Soundblaster programs. One kind of file may not work with the other program.

Depending on the source from which you obtained a sound clip, some of the material may be copyrighted. Find out whether you need to obtain permission for its use before

#### Teaching Tip

If you are using an overhead projector and transparencies, don't use dark backgrounds. They will wash out, diminishing the impact of your presentation.

#### Safety Tip

Inevitably, your preferred medium will crash at the worst possible time. Always back up your material in at least one additional location such as a CD, thumb drive, or other storage device to prevent a catastrophic loss.

you include the sound clip in your presentation by contacting the source of the clip.

### Video and Animation

Locally produced video can add to the realism of the presentation and help hold students' interest. Students can absorb only as much as their attention can absorb. If your images aren't interesting, you have lost your students.

Like sound clips, video takes up a large amount of storage space on a computer. Essentially, every frame of video is comparable to a .jpeg photo with sound. Usually, video is captured at a rate of 24 to 30 frames per second. If you have a 5-minute video, that equates to approximately 7200 pictures (24 frames × 60 seconds × 5 minutes). That 5-minute video will consume 1.08 gigabytes of storage space. Most training videos are 15 to 30 minutes long, so they require considerable storage space. If you plan on editing this video, you will need to approximately double the size estimate. More information on storage space considerations can be obtained from the Internet by contacting the software developer's Web site.

Animation is similar to video in that it consists of a series of still images put together to create motion. Animation is a very powerful, captivating tool, but one that is difficult and time-consuming to produce. It took a team of animators more than three years to develop the Disney film *Snow White*. The old adage "Time is money" applies here, so consider using animations that can be freely copied from various sources and incorporated seamlessly into your presentation.

Animations have also been used in reconstructions of fire scenes. For example, some of the best re-create the Worcester, Massachusetts, warehouse fire that occurred in December 1999. Animation files generally have a .gif extension.

## Suggestions for Uses of Slides

If you plan to use slides to illustrate or otherwise support a lecture, remember that lecture notes displayed on your slides play a different role in the lecture than handwritten lecture notes that only you can see. Don't try to make them play the same role, or you may find students reading your slides instead of listening to you.

You can use slides in the lecture to list the major points of your presentation. Several major points might stay on the screen as you develop each of them in turn, providing a way for students to place each point in the larger context.

Also use slides to list important terms. Again, one slide with several terms may remain on the screen for some time, allowing you to refer to each term as you introduce it.

In addition, slides can be used to create prompts for group work, being projected for all groups to see as the students do their work. This helps students move from one stage to the next when you manually change the slide or through the use of a timer built into the slide presentation. Slides can also contain breakouts that indicate how each group will handle different aspects of a scenario.

At times, you may find that you need to use audio or video as part of your presentation. This becomes possible with a PowerPoint presentation, thereby eliminating the need for videotape machines, audiocassette players, or other expensive A/V equipment. Streamlining the technology employed in the class will not only keep you on track, but also enhance the flow of your presentation. This can be a simple process. The audio or video can be started when you click on a slide, for example, or it can be started when you click on an **icon** (a pictorial representation of an object used to represent documents, file folders, and software) on the slide. To work correctly, the files need to be in the correct location. If you copy the file to a different medium (CD-ROM), the program may not locate the audio/video file when replayed.

Remember, more is not always better. Sometimes a simple, clean slide is better than one with bells and whistles such as sound and animation. Multimedia tools should *enhance* your students' understanding of the material; they are not designed to make you look cool with your techie toys.

When inserting video onto a slide, keep in mind the visual area of the slide. Several complications can arise when incorporating video into a presentation. If the image is too big, the picture appears out of focus. If it is too small, it is just distracting. If the computer processor is not fast enough to show the video, the video playback will appear jerky or not play at all, or the picture may appear out of focus or with **ghosting** (faint shadows to the right of each letter or number). Check your video on the equipment you will be using for the presentation before you choose not to use a videotape machine or DVD player. Don't give up quality for convenience.

## How to Maintain Technology

With any technology, maintaining its quality is an important consideration. Some vendors charge an additional fee to repair or replace damaged materials. When purchasing multimedia tools, get the maintenance contract in writing. Although this contract may seem to be just an additional step or an extra cost, it can pay dividends when you need it. Remember—fire fighters can and do break things.

Some multimedia products require less maintenance than others. For instance, a chalkboard will not get a computer virus—but it also cannot transition from one slide to the next at the touch of a button. Be aware of the trade-offs between high technology and good old-fashioned methods. You need to decide based on a number of factors (e.g., cost, ease in transportation, audience) what is appropriate for your event or presentation.

## Antivirus Protection

Like everything else in the computer world, computer viruses have evolved over the years. Massive use of e-mail and the Internet in the 1990s may have triggered the enormous growth of viruses. Remember the Michelangelo virus? In retrospect, it was a minor problem for computer users. The appearance of the Melissa virus in 1999 changed public perceptions, however; it is credited with causing the first worldwide computer virus epidemic. Today viruses can have a lethal impact on the privacy of information on the computer.

To fight this threat, vendors have developed sophisticated software programs that require frequent updating as new viruses arrive daily. Each update provides software with the ability to counteract or isolate the virus so that the computer remains clean. While many vendors exist that offer both fee-based protection and **freeware** (copyrighted software provided free by the author, who maintains the copyright), the best-known computer security companies—for example, Norton, McAfee, PC-cillin, Panda, and AVG—offer computer operators some reassurance that their information and computers will not be hacked.

Unfortunately, viruses are only part of the problem. Computer users may also encounter worms, which are self-replicating programs. Worms use a network to send copies of themselves to other systems and can do so without any user intervention. Unlike viruses, worms do not need to be attached to an existing program. They harm networks by consuming bandwidth, whereas viruses always infect or corrupt files on a targeted computer.

To protect against worms, viruses, and other potentially harmful programs, today's computer users are well advised to include **firewalls** on their systems. Similar to the term with which fire fighters are familiar, a computer firewall acts just like a building firewall by preventing fire or harm from getting from one side to the other; that is, it acts as a virtual wall that separates a trusted environment from an untrusted environment by controlling and regulating the traffic between the two. Firewalls do have a drawback, in that they can slow down or complicate communication between computers.

## Servers

Many fire departments use computer servers if the department has multiple stations. In essence, the server allows the creation of an **intranet** within the fire department FIGURE 8.8. In this type of network, only those belonging to the organization can access the network. The typical server is a computer system that operates continually, controls the network, and receives requests for services from other computers on the network. Several computers can communicate with the main computer (server), which stores data and programs, so that the other computers do not have to. The server is centralized to the fire department and can allow the department to handle report writing, data storage, simultaneous users, and much more in one convenient location.

FIGURE 8.8 The typical server is a computer system that operates continually, controls the network, and receives requests for services from other computers on the network.

Many servers are dedicated to this role, but some may also be used simultaneously for other purposes. Servers today are physically similar to most other computers, although their hardware configurations may be optimized to accommodate their various roles. The biggest advantage of such a system is that the server allows you to place a program on it for use by multiple computers, thereby saving the cost of purchasing multiple copies of some software.

## When to Use Multimedia Presentations

Multimedia presentations allow students to see, hear, and (in some cases) touch, which enhances learning. Such applications should be used to simplify complex theories or hypotheses. Multimedia resources can be used to reinforce materials through exercises, pictures, or words that relate to the information you are presenting. They can summarize previously integrated material for use in the current program so that you do not waste a lot of time covering prerequisites. With today's learning styles, multimedia presentations help keep students interested as well.

Despite all of these advantages, there are some drawbacks to using multimedia in training. In particular, the failure of the media to perform as designed can put instructors at a loss. Such a failure can be as simple as the bulb on the slide projector burning out or as complex as the computer presentation not interfacing properly with the projector. You must be versatile enough to be able to overcome these glitches and not let failure impede what you are trying to present. If it does happen, then you must have a plan B.

Perhaps the biggest issue facing instructors who want to use multimedia is determining when it is appropriate. Multimedia resources are not always needed. Remember—the technology always needs to be appropriate for the audience and the content to be presented.

The classroom has been known as a place where networking and information combine to create the learning experience. It allows the students to learn from one another as well as from the instructor and the materials presented. At its best, this environment tends to encourage students to learn for themselves. Learning is like the old saying, "You can lead a horse to water but you cannot make it drink." The same is true for students: You can present all the material in the world but, if students do not want to learn, they won't. Multimedia tools try to break down the resistance to learning by engaging the student.

Multimedia should be used whenever its application makes it easier for students to grasp the material. It does not matter if the application means bringing the emergency incident into the classroom, demonstrating a nonthreatening environment, exemplifying safety concerns, or utilizing simulation or distance education. The only reason to use multimedia tools is to help students learn.

**Teaching Tip**

Give credit where credit is due. If materials came from a journal or other publication, remember to credit the authors. Regardless of how you found the material or how long ago the information was developed, give credit to its author. This will assist you in presenting credible and defendable information. The fire service has had an unhappy reputation of borrowing information and calling it research; do your part to dispel this image.

## Selecting the Right Applications

Selecting the right media application is critical. When used as an instructional aid, the right tool allows you to clarify the information you present to students. Essentially, these features organize the way in which students receive the material. Diagrams, flowcharts, and graphics impart information that can be retained by students who are visual learners. The lecture or audio cues assist auditory learners. Used in combination, these methods engage students by involving many of their senses.

Emphasis alone will not guarantee that every student retains the information, but it makes it possible to clarify difficult concepts. Through visual representations, students can travel to places that are unreachable in real life. For example, imagine the advantage of being able to follow a drop of blood through the body to understand the circulatory system. In this case, words are not an adequate medium to get the point across, yet the desired effect is easily achieved through simulation.

## Troubleshooting Commonly Encountered Multimedia Problems

There are so many things that can go wrong when using multimedia applications. Instructors need to know and practice a few alternatives. TABLE 8.1 can assist with some of the more common problems. The simplest things that can go wrong can often be easily fixed by taking a deep breath and thinking about the problem and potential solutions. Take the time to work with your equipment and know its limitations. Be prepared for the possibility that something may go wrong. Have a plan B, a plan C, or even a plan D.

Sometimes solutions to problems are not as easy as referring to a simple chart. That is, sometime you need to be able to diagnose the problem and verbalize it. This step is necessary to use the *help* files contained within most software programs. To find these files, look at the menu bar at the top of your screen. It typically lists options such as File, Edit, and View; at the very end of that list is Help. You can either double-click on the word Help or press the Alt and H keys simultaneously to open this script. Then select the type of help you need.

**Table 8.1 Common Troubleshooting Procedures**

Try each "solution" in order. There is no need to continue down the list if the first step solves your problem. If your problem remains unsolved, you may need more expert assistance.

| Symptom | Solution |
|---|---|
| Presentation projector does not turn on/off | 1. Point the remote directly at the projector and try again.<br>2. The projector may be unplugged.<br>3. The remote may need new batteries. |
| No audio from podium | 1. Make sure the audio system is turned on.<br>2. Make sure the microphone is turned on. |
| No audio from wireless microphone | 1. Make sure the audio system is turned on.<br>2. Make sure the body pack is turned on.<br>3. The microphone may need new batteries. |
| VCR, audiocassette, or DVD will not play | 1. Check to make sure the VCR, DVD, or other player is plugged in.<br>2. Eject and reload the media and try again. |
| No audio from VCR, DVD, or computer | 1. Check the volume controls to make sure they are not turned down.<br>2. Make sure the audio system is turned on.<br>3. Make sure connections are made to the audio output on the device and the input on the audio system. |
| No light from visual presenter | 1. Check that the light has not been switched off on the base unit. |
| No image from computer/laptop | 1. Check that the correct input source has been selected.<br>2. Make sure the computer has not been turned off.<br>3. Press the Alt and F5 keys simultaneously. |
| Projector light will not come on | 1. If you hear a fan, make sure the lens cover is off.<br>2. If you hear a fan and the lens cover is off, you may need to replace the bulb. |
| Computer will not read media | 1. Take the disk out and confirm that it was inserted properly.<br>2. Make sure the computer has the proper software needed for the media to work properly. |

**Teaching Tip**

To determine whether your multimedia presentation is appropriate and useful, it needs to meet the following criteria:

- Be appropriate for the audience and topic. Is the presentation geared toward the specific group of students, is it easy to grasp the material, and is the material right for the subject matter?
- Cover the material completely and accurately.
- Be easy to use and dependable.
- Be current and up-to-date. You may need to identify trends and potential future challenges.
- Promote expected behavior and outcomes, such as demonstrating the use of PPE or apparatus positioning.

# Wrap-Up

## Chief Concepts

- A key element in most presentations is the use of multimedia tools.
- Multimedia tools come in a variety of types and uses, ranging from audio systems to Flash animation to simulations.
- As with any technology, maintaining the quality of multimedia tools is important.
- Multimedia tools should be used to simplify complex theories or to assist instructors in reinforcing their message through exercises or images.
- Technology is no replacement for quality training.

## Hot Terms

**Firewalls** Software that acts as a virtual wall, separating a trusted environment from an untrusted environment by controlling and regulating the traffic between the two.

**Freeware** Copyrighted software that is provided free by the author, who maintains the copyright.

**Ghosting** When viewing text, faint shadows that appear to the right of each letter or number.

**Icon** A small, pictorial, on-screen representation of an object used to represent documents, file folders, and software.

**Intranet** A network in which only those belonging to the organization can access the network.

# Fire Service Instructor *in Action*

Things seem to be going well in your career. You have been promoted to Battalion Chief of Training. You are finally getting to apply your skills in making great presentations. Your first big assignment is to teach a refresher course on strategy and tactics. Given that only a few new topics, such as compressed-air foam, must be included in the class, you can pretty much use the previous curriculum. But you also know that while the content is still current, the presentation is antiquated.

You decide to update the presentation from 35-mm slides to a PowerPoint slide show. You also want to embed some photos and video files into your presentation. You have some great footage from a recent house fire that provides a good example of how to read the smoke. You also know that dispatch has the sound files to go with it.

**1.** If you decide to use PowerPoint slides, which other equipment will you need to give the presentation?
- **A.** An overhead projector
- **B.** A slide projector
- **C.** A data projector
- **D.** An iPod

**2.** Which file format could be associated with the video clips?
- **A.** .mov
- **B.** .tif
- **C.** .wav
- **D.** .pdf

**3.** Which of the following statements is true?
- **A.** Video files are essential for a good class.
- **B.** Including sound files will hamper learning.
- **C.** Only multimedia tools that support learning should be included.
- **D.** Both A and C are true.

**4.** If you decided to turn this into a distance-learning course, what does that mean?
- **A.** It will be a computer-based course.
- **B.** The course will be provided "a distance away" from the training center/college/university.
- **C.** The course requires college credit to be assigned to it.
- **D.** The course cannot be computer based.

**5.** Which unit is used to measure the brightness of the light emitted by a data projector?
- **A.** Megapixels
- **B.** Pixels
- **C.** Megahertz
- **D.** Lumens

**6.** ________ are an effective means of providing students with materials to take home.
- **A.** Movies
- **B.** Handouts
- **C.** Cassettes
- **D.** Slides

Fire.jbpub.com

# Safety During the Learning Process

# CHAPTER 9

## NFPA Standards

**Instructor I**

**4.4.2** Organize the classroom, laboratory, or outdoor learning environment, given a facility and an assignment, so that lighting, distractions, climate control or weather, noise control, seating, audiovisual equipment, teaching aids, and safety are considered. [p. 150–160]

**(A) Requisite Knowledge.** Classroom management and safety, advantages and limitations of audiovisual equipment and teaching aids, classroom arrangement, and methods and techniques of instruction. [p. 150–160]

**(B) Requisite Skills.** Use of instructional media and materials.

**4.4.5** Adjust to differences in learning styles, abilities, and behaviors, given the instructional environment, so that lesson objectives are accomplished, disruptive behavior is addressed, and a safe learning environment is maintained. [p.150–160]

**(A)* Requisite Knowledge.** Motivation techniques, learning styles, types of learning disabilities and methods for dealing with them, and methods of dealing with disruptive and unsafe behavior.

**(B) Requisite Skills.** Basic coaching and motivational techniques, and adaptation of lesson plans or materials to specific instructional situations.

**Instructor II**

**5.4.3*** Supervise other instructors and students during training, given a training scenario with increased hazard exposure, so that applicable safety standards and practices are followed, and instructional goals are met. [p. 150–160]

**(A) Requisite Knowledge.** Safety rules, regulations, and practices; the incident command system used by the agency; and leadership techniques. [p. 150–160]

**(B) Requisite Skills.** Implementation of an incident management system used by the agency.

## Knowledge Objectives

After studying this chapter, you will be able to:

- Discuss the relationship between training and fire fighter safety.
- Describe the 16 fire fighter life-safety initiatives.
- Describe how to ensure safety in the classroom.
- Describe how to promote and teach safety by example.
- Describe your responsibility to student safety during training.
- Describe the laws and standards pertaining to safety in live fire training.
- Discuss how to develop safety as part of your department's culture.

## Skills Objectives

After studying this chapter, you will be able to:

- Demonstrate how to lead by example.
- Demonstrate safety in the classroom.
- Influence safety through training.

## You Are the Fire Service Instructor

You are a newly appointed training officer. Your chief has asked you to conduct live fire training on a recently acquired residential structure. The fire officers and fire fighters are enthusiastic about this training opportunity and are inquiring about when they will be able to go to this house. You want to provide your personnel with realistic and challenging training, but you begin to wonder if the risks outweigh the benefits for this type of training. Your responsibility is to prepare your students for every type of incident, and you know this must be done in a safe manner. You have some questions that need to be answered before the first drill begins.

1. How will you keep your students safe during this type of drill?
2. Are there any standards or laws that apply to live burn training?
3. How can the training be conducted safely while still maintaining a degree of realism?

## Introduction

Each year in the United States, countless fire fighters are injured during training. On average, 10 fire fighters are killed each year during training-related activities. This trend has accelerated over the past few years and is a growing concern among fire service leaders.

As a fire service instructor, you must take measures to stop the preventable injuries and deaths that occur during training. Of course, for training to be effective, it has to be realistic and, therefore, a certain level of danger will exist in all training events. Live fire training tops the list of high-risk training events, but this same high risk can be found in many other training activities.

You must always place students' safety ahead of any performance expectations for the training session. This mandate includes everything from the assignment of safety officers to rehabilitation protocols to increasing instructor–student ratios to ensuring that enough trained eyes are watching out for the safety of all students. You have the standards, technology, and information needed to protect your personnel from training tragedies. You also need to apply relevant checklists, standards, and control measures to your training sessions correctly to keep your students safe. Fire fighters want meaningful drills that allow them to apply their skills, knowledge, and abilities. It is your responsibility to provide this training while keeping your students safe FIGURE 9.1.

FIGURE 9.1 It is your responsibility to provide meaningful drills that allow fire fighters to apply their skills, knowledge, and abilities while keeping your students safe.

## The Sixteen Fire Fighter Life-Safety Initiatives

In 2004, fire service experts and leaders convened to discuss solutions to slow the number of line-of-duty deaths experienced by the fire service. Their work culminated in 16 initiatives that were given to the fire service with the goal of reducing line-of-duty deaths by 50 percent over a 10-year period TABLE 9.1. These initiatives should provide both guidance and motivation to you while you are developing and conducting any training. The initiatives highlight the importance of training as a strategy that will help reduce the number of fire fighter line-of-duty deaths. As a fire service instructor, you will be specifically charged with carrying out initiative 5 every time you instruct a course.

## Leading by Example

As a fire service instructor, you will use many methods of delivery to teach. One of the most powerful methods is leading by example. Fire fighters, regardless of their rank, years on the job, or past experiences, will look to you and make mental notes on how the job *should* be done FIGURE 9.2. You

**Table 9.1 16 Firefighter Life Safety Initiatives**

| | |
|---|---|
| 1. | Define and advocate the need for a cultural change within the fire service relating to safety; incorporating leadership, management, supervision, accountability and personal responsibility. |
| 2. | Enhance the personal and organizational accountability for health and safety throughout the fire service. |
| 3. | Focus greater attention on the integration of risk management with incident management at all levels, including strategic, tactical, and planning responsibilities. |
| 4. | All firefighters must be empowered to stop unsafe practices. |
| 5. | Develop and implement national standards for training, qualifications, and certification (including regular recertification) that are equally applicable to all firefighters based on the duties they are expected to perform. |
| 6. | Develop and implement national medical and physical fitness standards that are equally applicable to all firefighters, based on the duties they are expected to perform. |
| 7. | Create a national research agenda and data collection system that relates to the initiatives. |
| 8. | Utilize available technology wherever it can produce higher levels of health and safety. |
| 9. | Thoroughly investigate all firefighter fatalities, injuries, and near misses. |
| 10. | Grant programs should support the implementation of safe practices and/or mandate safe practices as an eligibility requirement. |
| 11. | National standards for emergency response policies and procedures should be developed and championed. |
| 12. | National protocols for response to violent incidents should be developed and championed. |
| 13. | Firefighters and their families must have access to counseling and psychological support. |
| 14. | Public education must receive more resources and be championed as a critical fire and life safety program. |
| 15. | Advocacy must be strengthened for the enforcement of codes and the installation of home fire sprinklers. |
| 16. | Safety must be a primary consideration in the design of apparatus and equipment. |

**FIGURE 9.2** As the fire service instructor, you will be demonstrating how the job should be done.

## Teaching Tip

Prior to class, survey the classroom or teaching environment and correct any safety concerns that may exist.

play an extremely influential role in determining the performance of your students. Given this responsibility, you must assure that your practices and principles clearly demonstrate the safety values employed in the fire service. If you do not correct an unsafe practice during training, your students will mistakenly believe that you condone the error. You must set a positive tone and embrace safety as the number one priority of every lesson plan. In the classroom and on the drill ground, use the controlled environment of the training arena to point out where things go wrong in real applications.

## Safety in the Classroom

Safety during training begins with safety in the classroom. First, consider the location of the classroom. While you may not have complete control over this decision, you can still affect some things. Look at the general classroom environment. Are there adequate fire exits or other unsafe conditions such as remodeling in the area or icy steps leading to the classroom? If the classroom is in an engine bay, are there excessive exhaust fumes? Evaluate the lighting conditions in the classroom. Not all students have great vision, and either low or high lighting may create eye strain. Excessive noise can be a problem as well, not only because it interferes with learning, but also because it can cause unnecessary damage to students' hearing.

Next, review the physical arrangements of the classroom. Unsecured extension cords running throughout the classroom may trip either you or the students. Prior to each class, reroute or tape down extension cords. Also, be aware that fire stations may be the recipients of hand-me-downs from other city agencies. Often, chairs and tables that have outlived their usefulness in other areas are pressed into service in a fire department classroom and could be prone to collapsing. Be sure that each item is in good condition to prevent injuries to students.

At the beginning of class, be sure to raise safety awareness with the students. Go over the locations of the nearest fire exit and fire extinguisher locations. If you are teaching a class to an on-duty crew or a group of volunteer fire fighters, you should address what to do if an alarm sounds. For example, if you are teaching a class of on-call volunteer fire fighters, have them sit near an exit so that if they need to respond during class they can leave without interrupting the class. Explain the emergency plan should a fire alarm activate or a severe weather event or natural disaster occur. It is important to emulate the behavior you teach to the general public and to remind your students that safety is paramount.

### Hands-on Training Safety

Hands-on training includes all training activities conducted outside of the classroom environment. Whether it consists of a simple tailboard chat on nozzles or a full-blown live fire training exercise, hands-on training is often the most powerful method of instruction for a student. Unfortunately, hands-on training also produces hazards that must be considered and mitigated. Slippery surfaces, sharp or jagged metal, and products of combustion need to be accounted for. Risk management practices need to be applied to limit, reduce, or remove risk exposure to students. Environmental factors have to be considered as well, such as wind, rain, and storms. Mother Nature can pose a significant risk to your students.

## Personal Protective Equipment

Determine the appropriate level of personal protective equipment (PPE) after reviewing the lesson plan. Lesson plans often specify the level of PPE required. However, local standard operating procedures (SOPs) may also exist that specify the level of PPE for training. A general statement of the PPE requirements for the training session needs to be part of your presentation. As an instructor, you must always wear the PPE required for the drill being conducted. The appropriate level of PPE depends on both the potential hazards and your department's SOPs.

The best practice is to always use the same PPE in training as in real-world incidents. In some departments, it may be acceptable to conduct ladder drills using helmets, gloves, and boots as opposed to full fire gear and SCBA due to the skill being taught or reinforced. This modification to the PPE policy may make the drill safer by reducing the potential for heat stress. Another option may be to begin the training session in full levels of PPE and then to reduce the PPE required after signs of heat stress appear or when the skill has been performed properly. When these adaptations are made, you must make very clear to your students that the modification is only for the training session and should never be employed when responding to actual emergencies. If any level of hazard exists in the session, then the highest level of PPE available should be used.

## Rehabilitation Practices and Hands-on Training

NFPA 1584, *Standard on the Rehabilitation Process for Members During Emergency Operations and Training Exercises*, requires that rehabilitation protocols be established and practiced at training events that involve strenuous activities or prolonged operations. As a fire service instructor, you must be aware of the content of this standard and apply it to these types of training sessions. These practices further reinforce the importance of following appropriate rehabilitation practices on the fire ground FIGURE 9.3. Training should mirror as closely as possible the actual fire-ground operation. Medical care and monitoring, hydration, and cooling/warming materials are just a few of the rehabilitation considerations that this standard requires to be available during training events.

FIGURE 9.3 Rehabilitation is an important part of training.

### Safety Policies and Procedures for Training

As a fire service instructor, you must have a firm grasp of the safety policies mandated by your fire department. The safety policies required on the fire ground must be reinforced and practiced during training. Indeed, one of the more common causes of injuries during training is failure to follow established safety practices. You have a responsibility to assure that all policies of the department are followed during training. Policies such as incident command, fire-ground accountability, rehabilitation, and full PPE are followed on the emergency scene, and adherence to those policies must not be set aside in training. Enforce those policies in the same manner a safety officer or incident commander would at an emergency incident.

You must also be informed of and follow the manufacturer's recommendations for using equipment. Read the accompanying owner's manuals for equipment such as chainsaws and heavy hydraulic equipment. Often the limitations of these tools are assumed, and an injury occurs when the tool is used in a manner counter to its intended use. When new apparatus is delivered to your department, you should be present for the initial training by the apparatus builder. Essential information presented by the manufacturer needs to be recorded, practiced, and reinforced whenever that piece of equipment is used.

### Influencing Safety Through Training

As a fire service instructor, you are vitally important in setting a good example of safe fire-ground practices. Constantly remind yourself that you are the role model of a safe and effective fire fighter. Everything from your fitness level to the way you wear your uniform should exemplify your values

## Applying the JOB PERFORMANCE REQUIREMENTS (JPRs)

One of the primary goals of training is to improve students' knowledge and skills in a subject so as to increase their ability to do their jobs safely. Opportunities exist throughout the entire learning process to educate and practice specific safety skills. During classroom training sessions, objectives relating to safety can be added to almost every subject area. Every practical session requires diligence in each component of the training to ensure that students and instructors alike remain safe during the session.

### Instructor I

The Instructor I understands the application of safety policies and practices those policies in every training session. Lessons learned through experience and case studies help in providing valuable references to the proper safety precautions taken during training. In each lesson presented, the instructor emphasizes topical safety points.

### JPRs at Work

Organize and adjust the learning environment to ensure the safety of students and instructors at all times. This includes applying NFPA standards such as NFPA 1403, *Standard on Live Fire Training Evolutions*, and following established procedures for conducting hands-on training.

### Instructor II

The Instructor II is responsible for the inclusion of safety-related behaviors in each lesson plan that is developed. Classroom sessions may not present obvious physical hazards, but the instructor must incorporate information on the safe applications of the objectives in the field. In practical sessions, safety rules for each evolution must be reviewed prior to the start of each session.

### JPRs at Work

Supervise other instructors and ensure the application of safety policies and standards during the training session. Understand how standards and safety policies apply to various types of learning.

### Bridging the Gap Between Instructor I and Instructor II

The Instructor I must learn how to apply safety rules to the training process. When developing skills sheets and lesson plans, the Instructor II must be clear when describing issues relating to safety performance. This will allow for easier application by the Instructor I. The Instructor I should seek opportunities to learn about the many standards that apply to training and identify ways to implement safety information in every session.

### Instant Applications

1. Review local policies relating to safety in training.
2. Read NFPA 1403, *Standard on Live Fire Training Evolutions*, and review the sample checklists used for conducting live fire training exercises.
3. Research fire fighter line-of-duty deaths in training accidents and review the recommendations of the investigators regarding ways to prevent similar occurrences.

### Theory into Practice

Creating a safety-conscious student must be a primary goal of virtually all training in the fire service. While there are inherent dangers in their occupation, fire fighters should constantly strive to reduce those risks through good awareness. When considering the appropriate level of protective gear to use during training, consider the exercise's safety implications for students.

First, is PPE necessary to ensure students' safety during the training activity? If so, full PPE must be used and any inconveniences mitigated. For example, because full PPE is needed for live fire training, be sure to provide rehabilitation. Second, consider whether *not* using full PPE will make the training safer. For example, perhaps hoods need not be worn during ladder practice because of the potential heat stress in July. Third, consider whether *not* wearing full PPE will create a false perception that real scenes can be safely mitigated without its use. Fourth, ask yourself if you are considering reduced PPE owing to peer pressure or if this decision is *truly* in the best interest of the student.

Each of these considerations should be evaluated prior to making a determination of the level of PPE to be used. Safety cannot be an afterthought or an unwanted formality: Instead, it must be the first and foremost consideration to ensure the lives and safety of all.

related to safety and adherence to rules. The example that you set sends a far more powerful message to your students than the theories and concepts you present in a lecture or drill.

Be aware of how influential a role the fire service instructor plays, especially with new fire fighters. New fire fighters are unaware of what is expected of them—that is, what is considered right versus what is considered wrong. With these students, whatever you do or say will be construed as the right thing. If you use improper techniques or fail to follow safety protocols while teaching, students will perceive that as the right way to do things. This faulty knowledge becomes dangerous when students leave your protection and perform on the emergency fire ground.

Your primary responsibility during a drill or training session is to ensure the safety of all students attending the drill. The actions of a fire fighter on the fire ground are the result of training. If you allow students to use unsafe practices in training, they will repeat those same actions on the fire ground, because decisions and actions on the fire ground are based on previous training.

You have many responsibilities related to the safety of your students during training, but one that stands above all others is the responsibility to have students practice safe principles and practices during training so that they remain safe on the fire ground. As part of your duties as an instructor, you must correct poor skills or bad decisions as they occur. If you witness something being performed incorrectly or a safety procedure not being followed, but you do not correct the

### Teaching Tip

Review safety policies that pertain to the drill you are practicing. Reinforce the safety policies during the drill. Also, inform students about where they can find the local safety SOPs for your department.

student's error, then you are reinforcing the incorrect behavior. This constant monitoring and correction is important to do with new recruits but can be more difficult with experienced personnel. Correcting unsafe or poor behavior demonstrated by a veteran fire fighter can be very challenging, but it is a challenge you cannot ignore. Correcting an unsafe skill should be the same whether the fire fighter is a raw recruit or a seasoned veteran.

## Planning Safe Training

You should begin to address safety at the earliest stages of course development and preparation, while formulating or modifying lesson plans. Analyze and critique your training to assure that safety measures are correctly addressed and emphasized. It is too late to address safety issues after an injury has occurred. Preventing injury becomes much easier when the hazards are identified and addressed early in the planning process.

An important part of the planning process is determining how many fire service instructors are needed to conduct the planned training. Two factors should be considered in making this decision: the type or risk level of the training and the skill level of the students.

Training can be classified based on risk as either high risk, medium risk, or low risk. An example of high-risk training would be live fire evolutions in an acquired structure or swiftwater rescue training. Medium-risk training would include fire extinguisher training or ladder evolutions. Low-risk training would include a classroom lecture or replacing a sprinkler head in a sprinkler lab.

The skill level of those being taught should also be considered. High-risk training of new recruits should generate different concerns than high-risk training involving veteran fire fighters.

Other areas of planning include assembling the equipment that may be needed. Sometimes outdated equipment or equipment that has been replaced by upgrades within the department is used when conducting drills. Using old or outdated equipment has long been an accepted practice in

### Teaching Tip

After development of a lesson plan, have an experienced fire service instructor—for instance, a safety officer—review the lesson plan to identify any safety concerns. The department safety officer or safety committee may be able to assist in this task.

the fire service, but it has often led to injuries that could have been prevented. If the tool or apparatus does not meet the accepted performance standards, then it should not be used in training. You should also question whether using an outdated power saw, for example, is beneficial to students given that this type of saw is not used on current apparatus.

You are also responsible for inspecting the equipment being used in the training to assure that it meets the required standards prior to the drill. Don't assume that the equipment is in safe and operable condition. If your department does not have the equipment available to make the training safe, then it needs to find alternative equipment by borrowing items from a neighboring fire department or your state fire training academy. Acquisition of the appropriate equipment should be arranged prior to the drill.

As a fire service instructor, you are responsible for protecting students from physical and emotional harm both in the classroom and on the drill ground. Most of this chapter focuses on means to protect students from physical injury. Nevertheless, instructors cannot overlook the psychological dangers students encounter during training. These hazardous scenarios also must be prevented.

## Hidden Hazards During Training

During training, students are asked to perform various tasks, recall policies, or demonstrate the ability to perform. When placed in an environment of peers, most students realize that their performance is being evaluated—not only by the instructor but by their fellow classmates. Failure to accomplish the task or answer the question correctly can label the student as not competent or someone who cannot be trusted on the fire ground. This situation is particularly dangerous because the typical personality of fire fighters is driven by the need to succeed, regardless of the conditions or situation.

This scenario sets the stage for disaster if a fire fighter is placed in an overwhelming situation, such as being asked to conduct a simulated rescue of another fire fighter or a rescue mannequin who may be twice the candidate's size or weight. If the fire fighter is asked to perform this task while fellow classmates are watching, it forces the student to attempt the impossible. Often, the student makes a valiant attempt, only to suffer an injury in the process. You have a responsibility to construct the learning environment so that situations such as these do not arise.

Do not minimize your levels of performance or drop your benchmarks so that every student can achieve success; this is not realistic. At the same time, for you to stand idly by as a student is placed into a scenario where he or she cannot perform successfully without being exposed to physical harm is unacceptable.

## Overcoming Obstacles

Money may be a minor obstacle in the training process, because instructors often face demands to train students while using limited resources. In the past, the fire service has been creative in developing methods and practices to accomplish drills or tasks with minimal cost. For example, when an acquired structure is not available for training on overhaul, it is easy to construct a wooden simulator from scrap 2-inch by 4-inch (2 × 4) lumber and cover it with discarded gypsum board that can be obtained from a lumber supply shop, often at very little expense **FIGURE 9.4**.

When foam is needed for a fire stream training session and expired or surplus foam is not available, you can make

### Ethics Tip

What message are you sending to your fire fighters? Look at the equipment being used during training. Is it the best and most up-to-date equipment in the department? New fire fighters learn the culture of the organization from their first day on the department—and the equipment used for training is one method that communicates that culture. Is your department—and are you personally—communicating a culture that emphasizes safety?

When some fire departments replace front-line equipment, they send the equipment that was replaced to the training division. For example, a truck company's chainsaw that does not have a chain brake might be replaced with the newest equipment, and the old saw goes to train the new recruits. Does this practice reinforce a culture of safety for those students who may have never even used a chainsaw before? Similarly, new recruits need to build their confidence until they can determine margins of safety for themselves. Using a new ladder, for example, allows recruits to build confidence in how much safety is designed into fire service ladders.

As fire service instructors, we have an obligation to the future of the fire service. Specifically, we have an obligation to develop fire fighters who have internalized our concern for safety. Teach your students through words and actions that as fire fighters, we all have an obligation to see that everybody goes home.

**FIGURE 9.4** If an acquired structure is not available for training on overhaul, it is easy to construct a wooden simulator of scrap 2 × 4 lumber and cover it with discarded gypsum board.

*Source*: Courtesy of Forest Reeder

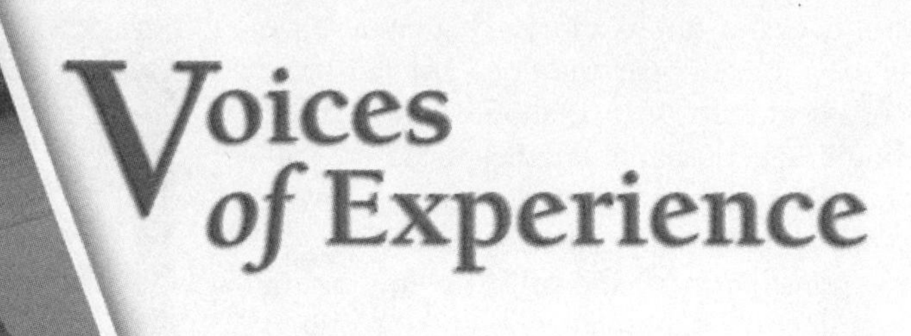


On a sunny, warm October morning, I took a day off to attend to some personal projects. The training staff was conducting a rapid intervention team (RIT) drill for personnel who missed the previous training drill. The drill consisted of a "downed fire fighter" scenario and was the last part of a four-part training program. The fire service instructors conducting the drill were well versed in how it would be run and had conducted the same drill numerous times before.

Then I received a page. One of my instructors was injured and being transported to the hospital. My stomach flipped, and I began to worry about what happened and how serious the injury was. I met the ambulance at the emergency room to find that one of our veteran instructors was on the cot, with a broken leg. He was conscious and alert, and could explain to me that he made a simple mistake that caused the injury.

*"The injury ultimately was career-ending."*

The instructor was setting up for the drill and used an 18-foot straight ladder to access the first-floor flat roof of our training tower. When he ascended the ladder, no one was heeling the ladder and the bottom kicked out. The instructor rode the ladder down. His injuries included a shattered leg that was broken in at least seven places, along with a broken arm. Due to complications, the possibility of an amputation loomed for several weeks, but, thankfully, did not occur. The injury ultimately was career-ending.

An investigation was launched and recommendations were made, the most obvious being that all ladders must be heeled by a fire fighter when someone is climbing them. The class being taught was well planned and included lesson plans, skills sheets, and checklists. However, the one component that was overlooked was the safety of the instructors. In this case, the students were not even on the drill site at the time of the injury. The injury was totally preventable.

This incident reinforces the responsibility of the instructor-in-charge to pay attention to every detail, including your instructor staff. Don't let your guard down, anticipate potential hazards and correct them, and assure that all personnel involved are aware of and practice the safety protocols, especially your instructor staff.

*Bryant Krizik*
Orland Fire Protection District
Orland Park, Illinois

## Theory into Practice

Analysis of your audience will help you select the best method of instruction for your training session. Some training sessions will require you to demonstrate the proper techniques in a step-by-step fashion, pausing to allow for questions and to provide clarification. This method may be appropriate for entry-level or new-skills training. Other sessions will require you to review the objectives or desired outcomes of the training session along with key safety behaviors. This approach may be applied to training sessions for experienced fire fighters or when doing evaluations of skill levels.

a homemade solution using dish soap liquid and water. Of course, if you are using something other than equipment or products that would normally be used in the evolution, make sure that they will not produce a safety hazard.

One item that cannot be minimized is personnel: An adequate number of personnel must be present to conduct the training safely. This is particularly true when conducting live fire training. Live fire training is the pinnacle of all firefighting training, but it is also one of the most hazardous drills. The hazard becomes even more pronounced when the drill occurs with fewer personnel than is acceptable for a safe environment.

## Live Fire Training

NFPA 1403, *Standard on Live Fire Training Evolutions*, describes the positions that must be staffed when conducting live fire training. It also discusses the prerequisite training that must be completed before students can participate in live fire exercises. This standard must be consulted before the live burn exercise and followed to the letter when conducting live fire drills. You have a responsibility to assure that all positions required are staffed at your training sessions. Do not attempt the drill with anything less than a full complement of staff.

Positions that must be filled during live burn activities include the assignment of a designated **safety officer**, who has received training and background in the responsibilities of this position during live fire training. This safety officer is responsible for the intervention and control of any aspect of the operation that he or she determines to present a potential or actual danger at all live fire training scenarios. An **instructor-in-charge** is required to plan and coordinate all training activities.

The proper student-to-instructor ratio when teaching **practical skills** is recommended to be 5:1. In some training sessions where multiple fire fighters are training, you will need a large pool of fire service instructors to assist you and maintain the 5:1 ratio of instructors to crew members. This 5:1 ratio must be maintained through the evolutions and is set so that the typical span of control is not exceeded.

Each fire department must set its own guidelines regarding which fire service instructors are qualified to deliver live

## Safety Tip

When using hands-on teaching methods, be sure to have an adequate number of fire service instructors present to ensure students' safety as well as learning.

fire training. It may be appropriate to assign these positions at other high-risk training sessions as well.

In many fire-ground incidents, if a particular position is not filled, then the incident commander is responsible for completing all of the functions of that position. Likewise, in training, the instructor-in-charge must perform the duties of the safety officer and assistant fire service instructors if these positions are not filled. If company officers are available to assist with training, the responsibilities associated with these positions must be covered in depth before training.

You must pay attention to detail when planning and conducting drills where students may face a risk of injury. When conducting practical drills, many variables come into play and many things can go wrong, resulting in an injury to a student. One way in which you can help minimize that possibility is to pay special attention to every detail of the drill. Mentally walking through the evolution during the planning session and drawing upon your own experiences or case reports of actual incidents can help develop your awareness of what might potentially go wrong. This review will allow you to develop backup plans or, in the best case, to eliminate the hazard altogether.

When the drill includes large numbers of people who you may not know (e.g., in disaster drills where many outside agencies are invited to participate), your preplanning skills will be put to the test. Asking for help from other fire service instructors or agencies that have conducted similar exercises will help you in covering all of the bases during the planning phase. As the instructor-in-charge, run through each aspect of the session, always asking, "What might go wrong here?" Aspects of the training to be considered need to include the weather, topography, equipment being used, and the steps of the task.

## Safety Tip

When using assistant fire service instructors, ensure that they have the appropriate level of knowledge and skills to fulfill their roles based on your department standards.

Additional fire service instructors should be employed at training sessions where the following conditions are present:

- Extreme temperatures are expected.
- Large groups of people are participating.
- Training will take place over a long duration.
- Training involves complex evolutions and procedures where additional instruction and safety oversight may be valuable.

## Teaching Tip

Establish a file of newspaper and magazine articles pertaining to fire training accidents. Refer to this file as you prepare your lesson plans to assure that the mistakes that led to a prior tragedy are not repeated in your training session.

## Developing a Safety Culture

Fire service instructors are in the position to select personnel to assist with their training programs. Often company officers are used to help deliver the session or watch over a skill being evaluated. As an instructor, you have the responsibility to know the people you are asking to help conduct this training. What are their qualifications? What are their values to fire service safety? Despite your best efforts to keep your training safe, you cannot be everywhere, at all times. You must rely on other fire service instructors to help supervise the drill.

Fire service instructors who are not competent or confident with the subject being presented should not be placed in that situation. The motivation of your subordinate fire service instructors must be assessed as well. An instructor who has a propensity to show off or try to prove that he or she is superior to students is an accident waiting to happen. Often the focus of such an instructor is not on the students or their safety. Know your staff, their attitudes toward safety, the way they operate on the fire ground, their motivations for being fire service instructors, and their teaching methods. All support personnel for the drill, including those who simply assist in prop construction, should be included in this assessment, because this information is essential to conduct training safely.

## Ethics Tip

Is it ethical for a fire service instructor to teach a topic without the practical experience to support it? Can a fire fighter teach leadership skills to company officers? Can those without Aircraft Rescue Fire Fighting (ARFF) experience teach a class solely through extensive research? Can a fire officer effectively teach other fire officers solely based on his or her own personal experiences without receiving any relevant education in the subject? You will likely have to come to grips with these questions at some point during your tenure teaching other fire fighters.

As a fire fighter, you must have confidence in yourself—you would not be successful in the fire service without it. Nevertheless, that confidence can be detrimental to you and others if you are not aware of your limitations. As a fire service instructor, you have an obligation to your students, your department, and your profession. Teaching based solely on experience without research is nothing more than a history lesson given from a single perspective. Teaching based solely on research without experience is a regurgitation of other people's opinions without grounding in reality.

If you are asked to teach a course in an area where you have little background, you owe it to everyone to decline the request. If you are asked to teach a course in which you don't have enough interest to research the topic for a broader understanding, you owe it to everyone to decline the request. You have an ethical responsibility to provide your students with high-quality training. Don't fall into the trap of teaching a class in which you aren't interested or experienced, hoping that your students will not notice. They will.

## Anticipating Problems

As part of the preparation for instruction, you need to assess potential problems and rectify them before they result in an injury. You need to train your eye to pick out these hazards on first glance. It may be easy for a company to throw up a ladder and leave it unheeled; you should be able to foresee the potential consequences of that action and take steps to make sure that hazard is addressed. Most experienced fire service instructors have this quality already. The more fire-ground experiences you have, the better you will become at recognizing the small details that might escalate into a full-scale disaster.

Your fire-ground experiences translate to the drill ground, and you should apply your firefighting experiences to every evolution you instruct. Tell your fire fighter students what can go wrong, what to watch out for, and how to be prepared for changing conditions throughout the training evolution. If a smoke generator will be used for the training session, let students know how it compares to smoke at a real incident, if the situation may be significantly different from the training experience.

Often training tragedies are not the result of one catastrophic event. Instead, they represent the culmination of several small safety violations that combine to result in a fire training tragedy. The instructor-in-charge may identify that an attack line of a smaller size is appropriate for the amount of fuel being used if it is advanced without delay. But hose lines can get kinked, water may be slow to the nozzle from the pump, and conditions may deteriorate and lead to a delayed attack that can bring the fuel load above the capacity of the slowly advancing attack line. Collectively, these elements are a prescription for disaster.

Changes in wind or weather, crew fatigue, or student motivation may also affect the evolution. Your guard must stay up until the evolution is complete. You might imagine that if

## Safety Tip

Review the causes of accidents for the type of training you will be conducting to ensure you have taken all appropriate preventive measures.

this one thing goes wrong you can quickly correct it, but the escalation of events often occurs so rapidly that you cannot gain control of the one small mistake before an injury takes place. If you see a potential hazard, correct it immediately.

## Accident and Injury Investigation

Despite all of your careful preparation, sometimes an accident might happen. In such circumstances, it is important to conduct a comprehensive investigation of the incident. A key factor in any investigation is the attitude held by those conducting the investigation. The purpose of such an investigation should be to find the reasons for and contributing factors to the accident or injury—not to assign blame. If the main focus is on finding the cause of the accident, then individuals will be more honest and open to this important process.

The investigation should include all of the events, equipment, and personnel who were present when the incident occurred. This information will help to determine the cause of the injury and enable fire service personnel to put control measures in place to prevent future occurrences. Keep all injury investigation reports on file and periodically review them to determine if any trends are emerging that might cue you to change a policy or procedure that contributes to injuries. When the injuries are compared over a period of time, trends in their incidence often become apparent. You should see that modifications are made to the training program based on this information.

## Student Responsibilities for Safety During Training

Every student participating in training has a stake in ensuring his or her own personal safety. Each student is responsible for his or her actions as well as the actions of other crew members, training team members, and fellow students when it comes to addressing hazardous conditions. Students should be knowledgeable about the training to take place and should meet all of the prerequisites specified before participating in the drill. This can be accomplished through formal training or by individual study. Students should also adhere to all PPE requirements specified for the evolution and wear all SCBA or other assigned safety devices. Observance of all safety instructions and safety rules throughout the training session will assist in maintaining the overall safety profile of the training session. The student should also be honest in knowing his or her strengths and weaknesses. If a student is not prepared for the skill or evolution, he or she should notify you so that the class is not put at risk.

## Legal Considerations

Fire service instructors are held accountable for planning, lack of planning, adherence to standards and codes, and decision making during the training process. Chapter 2 described the many types of liability that arise during training. As a fire service instructor, you have a responsibility to stay informed about all laws and standards governing fire fighter training and safety. These laws and standards change over time, so you must keep up to date with the most current editions. Fire service instructor networks, Web sites, and electronic bulletin boards are all popular methods of staying current with these ever-changing rules.

In many cases, the standard of care or best practices relating to training safety may be tied to your department's SOPs. As mentioned in earlier chapters, those SOPs must be referenced within the scope of training along with the application to the individual job description.

Another legal test that may be applied to training is the "reasonable person" concept. This standard tests your decisions against the decisions made by others with similar training and background, asking what your peers would do when faced with the same situation.

Many of the NFPA standards in the 1400 series provide guidance in the application of different types of training exercises. For example, NFPA 1451, *Standard for a Fire Service Vehicle Operations Training Program*, provides information about the development and delivery of a vehicle training program. Within that standard and others such as NFPA 1500, *Standard on Fire Department Occupational Safety and Health Program*, it is required that training on vehicle driving take place before the fire fighter is allowed to respond to emergencies as a driver. If the department does not adopt that policy as a standard practice, the failure to do so might be held against the department in a lawsuit.

Most of the safety protocols that govern fire fighter training are embedded in laws that describe broader topics, such as Occupational Safety and Health Administration (OSHA) rules on respiratory protection (or SCBA) and hazardous materials. The NFPA standard specific to conducting live fire training safely is NFPA 1403, *Standard on Live Fire Training Evolutions*, which addresses conducting live fire training in acquired structures, gas-fired live fire training structures, non-gas-fired live fire training structures, exterior prop fires, and Class B fuel fires. Many of the requirements for these exercises must also be applied to skills training.

**Safety Tip**

Strictly adhere to all recognized standards and laws for the type of training you are conducting.

NFPA 1403 was developed as a result of fire fighter fatalities and injuries in live fire training accidents. This standard is almost entirely devoted to practices that must be followed to conduct live fire training safely. Although NFPA standards are not laws, you can be held accountable to them just the same. For example, a training officer was recently held criminally liable for a fire fighter fatality that occurred during live fire training because he did not follow the guidelines in NFPA 1403 when conducting the drill.

Laws and standards are intended to keep fire fighter training within a framework of safe practices. Some critics might suggest that the requirements place so many restrictions on the training that it is no longer realistic or relevant. This is simply not true. First and foremost, fire service instructors have a responsibility to keep the training safe and to follow procedures (TABLE 9.2). The notion that you have to discard safety requirements to make the training realistic is contrary to the entire idea behind conducting training sessions. The standards, rules, and laws established for safe live fire activities must be adhered to during skills training as well.

**Table 9.2 General Training Safety Procedures**

| Training Location | Procedures |
| --- | --- |
| Classroom Training | Make sure all applicable department SOG's are referenced within your lesson planning.<br>Ensure all participants are aware of facility safety procedures; fire alarm, smoking, restricted access areas, etc.<br>Identify trip, slip/fall, and other facility hazards. |
| Practical Training | Specify all learning objectives and performances. Use incident action plans as appropriate for high-risk training such as Live fire evolutions.<br>State all required PPE for instructors and students.<br>Identify all department procedures for skills being performed (use the same Incident Command System as required during actual operations).<br>Identify and demonstrate any hazard control activities that have been taken or are in place.<br>Review all applicable safety procedures (Emergency Procedures; MAYDAY, Back-up assignments, communications, rehab, EMS standby or aid, accountability and rapid intervention teams).<br>Check all communications equipment.<br>Follow all applicable standards for training session (NFPA 1403 for live fire training). |
| Keys to Safe and Successful Training | Identify companies/members who are in or out of service.<br>Identify and document lead instructor.<br>Notify supervising officers of all training events.<br>Brief all participants on all training expectations.<br>Monitor above practical training safety factors throughout (act as Instructor/Safety Officer/Commander).<br>Immediately STOP all unsafe acts or conditions; Don't try to train through them.<br>Debrief all participants on how training expectations were met and identify opportunities to improve individual and company performance.<br>Return all equipment and personnel to service; refueled, recharged, rehabbed and ready for the next response. |

# Wrap-Up

## Chief Concepts

- A fire service instructor is a role model for students. For this reason, he or she should always reinforce safety policies as part of the teaching process.
- Safety during training begins with safety in the classroom.
- In hands-on training, risk management practices must be applied to limit, reduce, or remove risk.
- Knowledge of the department's safety policies and the equipment manufacturer's recommendations is the key to keeping students safe during training.
- NFPA 1403, *Standard on Live Fire Training Evolutions*, is the nationally accepted standard for all live fire burn training.
- Fire service instructors are responsible for both the physical and emotional well-being of students while in training.
- Fire service instructors are held accountable for their planning and adherence to standards and SOPs during the training session.

## Hot Terms

**Instructor-in-charge** An individual who is qualified as an instructor and designated by the authority having jurisdiction to be in charge of fire fighter training.

**Practical skills** Tasks or jobs that require the physical performance of a fire fighter to achieve.

**Safety officer** An individual who is appointed by the authority having jurisdiction and is qualified to maintain a safe working environment at all training evolutions.

## Fire Service Instructor *in Action*

A local resident stops by the fire station and offers you the use of his home for a live burn. The house is scheduled for demolition, and your department hasn't had any live fire training for a while. You see this event as a great opportunity to do some real fire fighter training.

The owner states that the home needs too much work to make it habitable. There are some weak spots in the roof and lots of overgrown brush along the sides. You know the area well. The water supply in the hydrant system is plagued with poor pressure and poor maintenance. The area is fairly well populated, including a school on the next block. The owner wants to move on his plans to demolish the house and gives you a couple of days to plan the training. You have an old training facility nearby, which was abandoned that could be used for the classroom portion of the training.

**1.** Why should you note the nearest fire exit location in the classroom?
- **A.** It is required by NFPA 1500.
- **B.** Students must know where to go in the event of a fire.
- **C.** It is required by NFPA 1403.
- **D.** Fire service instructors should practice what they preach.

**2.** You should ensure students' safety in the classroom by first considering which of the following issues?
- **A.** The location where the class is to be held
- **B.** The temperature of the classroom
- **C.** The presence of a designated safety officer at all classes
- **D.** The appropriate number of classroom instructors as specified in NFPA 1403

**3.** Which NFPA standard covers live fire training?
- **A.** NFPA 1500
- **B.** NFPA 1403
- **C.** NFPA 1503
- **D.** There is no NFPA standard for live fire training

**4.** What is the recommended instructor-to-student ratio for high-risk training?
- **A.** 1:1
- **B.** 2:1
- **C.** 5:1
- **D.** 10:1

**5.** If you decide to conduct the live fire training and a student is killed, which of the following is *not* true?
- **A.** Your actions will be compared to the actions of other instructors.
- **B.** Your actions will be compared to national standards and laws.
- **C.** Your actions will be compared to departmental policies.
- **D.** Your actions will be compared to the actions of the average person on the street.

# Evaluation and Testing

PART

# IV

# Evaluating the Learning Process

# CHAPTER 10

## NFPA 1041 Standard

**Instructor I**

**4.5** **Evaluation and Testing.** [p. 166–187]

**4.5.1*** **Definition of Duty.** The administration and grading of student evaluation instruments. [p. 166–187]

**4.5.2** Administer oral, written, and performance tests, given the lesson plan, evaluation instruments, and the evaluation procedures of the agency, so that the testing is conducted according to procedures and the security of the materials is maintained. [p. 183–187]

(A) **Requisite Knowledge.** Test administration, agency policies, laws affecting records and disclosure of training information, purposes of evaluation and testing, and performance skills evaluation. [p. 183–187]

(B) **Requisite Skills.** Use of skills checklists and oral questioning techniques. [p. 181, 184]

**4.5.3** Grade student oral, written, or performance tests, given class answer sheets or skills checklists and appropriate answer keys, so the examinations are accurately graded and properly secured. [p. 166–187]

(A) **Requisite Knowledge.** Grading and maintaining confidentiality of scores. [p. 173–183]

(B) **Requisite Skills.** None required.

**4.5.4** Report test results, given a set of test answer sheets or skills checklists, a report form, and policies and procedures for reporting, so that the results are accurately recorded, the forms are forwarded according to procedure, and unusual circumstances are reported. [p. 183–185]

(A) **Requisite Knowledge.** Reporting procedures and the interpretation of test results. [p. 183–185]

(B) **Requisite Skills.** Communication skills and basic coaching. [p. 183–185]

**4.5.5*** Provide evaluation feedback to students, given evaluation data, so that the feedback is timely; specific enough for the student to make efforts to modify behavior; and objective, clear, and relevant; also include suggestions based on the data. [p. 186–187]

(A) **Requisite Knowledge.** Reporting procedures and the interpretation of test results. [p. 186–187]

(B) **Requisite Skills.** Communication skills and basic coaching. [p. 186–187]

**Instructor II**

**5.5** **Evaluation and Testing.** [p. 166–187]

**5.5.1** **Definition of Duty.** The development of student evaluation instruments to support instruction and the evaluation of test results. [p. 166–187]

**5.5.2** Develop student evaluation instruments, given learning objectives, audience characteristics, and training goals, so that the evaluation instrument determines if the student has achieved the learning objectives; the instrument evaluates performance in an objective, reliable, and verifiable manner; and the evaluation instrument is bias-free to any audience or group. [p. 166-185]

(A) **Requisite Knowledge.** Evaluation methods, development of forms, effective instructional methods, and techniques. [p. 166-185]

(B) **Requisite Skills.** Evaluation item construction and assembly of evaluation instruments. [p. 166-185]

**5.5.4** Analyze student evaluation instruments, given test data, objectives and agency policies, so that validity is determined and necessary changes are accomplished. [p. 167–168]

(A) **Requisite Knowledge.** Test validity, reliability, and item analysis. [p. 167–168]

(B) **Requisite Skills.** Item analysis techniques. [p. 167–168]

## Knowledge Objectives

After studying this chapter, you will be able to:

- Describe how to develop student evaluation instruments.
- Describe standard testing procedures.
- Describe how to analyze a student evaluation instrument.
- Explain the role of testing in the systems approach to training process.
- Describe the types of written examinations.
- Describe how to administer testing.
- Explain the legal considerations for testing.
- Describe the process of providing feedback to students.

## Skills Objectives

After studing this chapter, you will be able to:

- Demonstrate how to prepare an effective exam for student evaluation.
- Demonstrate how to grade student evaluation instruments.
- Demonstrate how evaluations are proctored and results are recorded.
- Demonstrate the methods for providing feedback on evaluation performance to students.

## You Are the Fire Service Instructor

As a fire service instructor, you believe in the four-step method of skill training: prepare, present, apply, and evaluate. Your testing materials, both in the classroom and on the job, reflect a complete and fair assessment of the required knowledge and skills for your course.

You have been a fire service instructor for many years and are aware that each new class includes students who do not like to be in a classroom. Other students enjoy the challenge of learning something new but experience difficulty when they have to perform on-the-job skills or take written tests.

You receive a call from Fire Officer Kelly. One of your new students is Fire Fighter Piepper, a capable candidate who has had a problem with meeting some of the requirements of his probationary period. You assure Fire Officer Kelly that you understand the situation and will keep him apprised of the progress that Fire Fighter Piepper is making. In addition, you promise to meet with Fire Fighter Piepper and give him any individual help that is required.

1. What will you do as part of each step of the four-step process to ensure you are prepared for your first day of class?
2. Which evaluation and testing methods should you use for new recruits learning basic skills?
3. What will you do to help ensure Fire Fighter Piepper is successful?

## Introduction

Testing plays a vital role in training and educating fire service personnel. A sound testing program allows you to know whether students are progressing satisfactorily, whether they have learned and mastered course learning objectives, and whether your instruction is effective. Without a sound testing program, there is no way to assess student progress or determine which learning objectives have been mastered. Likewise, it is impossible to determine the effectiveness and quality of training programs without evaluation.

Fire departments across the country continue to encounter problems with testing programs. Consider these common problems in testing:

- Lack of standardized test specifications and test format examples
- Confusing procedures and guidelines for test development, review, and approval
- Lack of consistency and standardization in the application of testing technology
- Failure to perform formal test-item and test analysis routinely
- Inadequate fire service instructor training in testing technology

This chapter provides the necessary information and training to help overcome these and other common weaknesses and, thus, lend strength to your testing program.

## Development of Tests

Tests should be developed for three purposes:

- To measure student attainment of learning objectives
- To determine weaknesses and gaps in the training program
- To enhance and improve training programs by positively influencing the revision of training materials and the improvement of instructor performance

The three basic types of tests are written, oral, and performance.

## Written Tests

**Written tests** can be made up of eight types of test items. These eight types of written test items are presented in this chapter:

- Multiple choice
- Matching
- Arrangement
- Identification
- Completion
- True/false
- Short-answer essay
- Long-answer essay

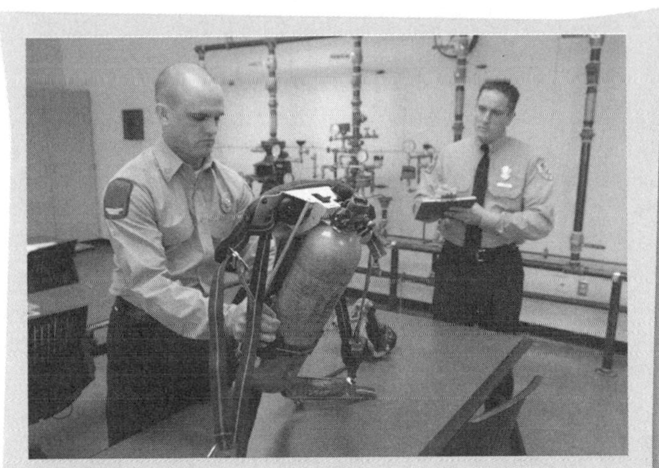

FIGURE 10.1 Oral tests assess mastery of knowledge, skills, and abilities before a performance test.

## Ethics Tip

The multiple-choice exam is typically the most common form of written test because many feel it is the easiest to create and the easiest to grade. Given that many different learning styles exist, some students are inevitably better able to demonstrate competency in a different format than other students. If the primary purpose of the testing process is to ensure competence, to what extent should you consider using multiple testing methods to ensure competency?

## Oral Tests

**Oral tests**, in which the answers are spoken either in response to an essay-type question (oral content) or in conjunction with a presentation or demonstration (oral presentation), are not used extensively in the fire service. Nevertheless, this type of test has a place in technical training. Oral tests are given in a structured and standardized manner to determine the student's verbal response while assessing his or her mastery of knowledge, skills, and abilities considered to be important on the job. Such tests primarily focus on safety-related issues. This kind of test allows students to clarify answers and instructors to clarify questions. Oral tests are effective in ascertaining student knowledge and understanding that are not conveniently measured through other types of tests.

Following is an oral test example:

> Given specific situations, name the types of personal protective equipment necessary, and explain how to don each item and how to remove these pieces. Focus on the most important safety-related steps.

Oral tests used in conjunction with performance tests should focus on critical performance elements and key safety factors FIGURE 10.1. The overhead or directed questioning technique seems to work best when a group of students are preparing to take a performance test. Specific oral test questions can be used when a student is taking a performance test. Oral tests used in conjunction with performance tests are usually not graded, but rather are designed to reinforce key safety factors for students and the fire service instructor.

## Teaching Tip

It is important to outline the required/expected responses to oral test questions to ensure that proper feedback is given to the student.

## Performance Tests

**Performance tests** (also known as **skills evaluation**) test and measure a student's ability to perform a task. This category of tests includes laboratory exercises, scenarios, and job performance requirements (JPRs). Any performance test should be developed in accordance with task analysis information and reviewed by **subject-matter experts (SMEs)** to ensure its technical accuracy.

Following is an example of a performance test item:

> Perform a daily inspection of self-contained breathing apparatus (SCBA).

## Standard Testing Procedures

Within a training department, the test development activities should adhere to a common set of procedures. These procedures should be included in standard operating procedures (SOPs) or guideline formats. Taken as a whole, this valuable document should offer test development concepts, rules, suggestions, and format examples. The procedures and guidelines specified in this manual should be used by fire service instructors throughout the test-item development, test construction, test administration, and test-item analysis processes.

## Test-Item Validity

When developing a test, it is crucial to make sure that each item actually measures what it is intended to measure. All too often, tests contain items that are totally unrelated to the learning objectives of the course. The term "valid" is used to describe how well a test item measures what the test-item developer intended it to measure. Taking steps to ensure validity will prevent your test items from measuring unrelated information.

Four forms of test-item validity are distinguished:

- Face validity
- Technical-content validity
- Job-content validity/criterion-referenced validity
- Currency of the information

Although this chapter deals primarily with face validity and technical-content validity, you will encounter methods in the chapter to help ensure that your test items have job-content validity/criterion-referenced validity as well.

### Face Validity

**Face validity** is the lowest level of validity. It occurs when a test item is derived from an area of technical information by an experienced subject-matter expert SME who can attest to its technical accuracy and can provide backup evidence to prove its correctness. Each level of validity requires documentation and evidence.

### Technical-Content Validity

**Technical-content validity** occurs when a test item is developed by a (SME) and is documented in current, job-relevant technical resources and training materials.

### Job-Content/Criterion-Referenced Validity

**Job-content/criterion-referenced validity** is obtained through the use of a technical committee of SMEs who determine that the knowledge being measured is required on the job. In the case of criterion-referenced validity, professional standards such as the NFPA Professional Qualification Standards, which are based on job and task analyses, may be used to establish the value of the test items. This level of validity should be carefully documented with each test item so that anyone can trace the validity information to the specific part of the NFPA Professional Qualification Standards and the reference material used to develop the question.

### Currency of Information

A new area that you need to consider in test development is the currency of the information. When conducting evaluations, including the most current information that a student should know and use is as important as the information being relevant to the job.

**Teaching Tip**

Ensuring that all test items are valid is critical for successful student evaluation.

## Test-Item and Test Analyses

Test analysis occurs after a test has been administered. Three questions are usually answered in post-test analysis:

- How difficult were the test items?
- Did the test items discriminate (differentiate) between students with high scores and those with low scores?
- Was the test reliable? (That is, were the results consistent?)

If a test is reliable and the test items meet acceptable criteria for difficulty and discrimination, the evaluation instrument is usually considered to be acceptable. This process, which is known as **quantitative analysis**, uses statistics to determine the acceptability of a test. If a test is not reliable and the test items do not meet acceptable criteria for difficulty and discrimination, a careful review of the test items should be conducted and adjustments made to the test items as necessary. Known as **qualitative analysis**, this type of in-depth research study organizes data into patterns to help determine which test items are acceptable.

The purpose of test analysis is to determine whether test items are functioning as desired and to eliminate, correct, or modify those test items that are failing in this regard. Sometimes it takes a cycle or two before real improvements are achieved in a particular set of test items. The best way to achieve acceptable test analysis results is to ensure the validity of each test item as it is developed and to follow a standard set of test development procedures as the test is constructed and administered. See Appendix A for additional information about analyzing evaluation instruments.

## The Role of Testing in the Systems Approach to Training Process

The **Systems Approach to Training (SAT) process** was developed by the U.S. Military during the early 1970s and 1980s, though many improvements have been made since the initial development of this training approach. The effectiveness and efficiency of SAT have been well documented in training journals, academic studies, and actual practice by leading businesses and industries in the United States and around the world.

Dr. Robert F. Mager—often referred to as "the father of the performance objective/learning objective"—played a key role as a leading researcher during the development of performance-based or criterion-reference instruction. Mager introduced the idea of learning objectives that have three distinct parts:

- The learning objective is task-based, using verbs as part of the learning objective that implies doing something. Sometimes these verbs are referred to as the action part of the learning objective.
- The learning objective deals with the condition(s) under which the action is to be performed.
- The learning objective contains a standard or measure of competence.

Mager was a strong believer in the use of learning objectives and outcomes-based learning. Following is an example of a learning objective that meets his criteria:

> Given a self-contained breathing apparatus (SCBA), the fire fighter will don the SCBA and place it in operation within 1 minute.

All forms of testing and training should be based on learning objectives. In technical training, whether the tests are written, oral, or performance-based, they must be based on the learning objective.

Within any training program, testing serves three purposes:

- Measure student achievement of learning objectives
- Determine weaknesses and gaps in the training program
- Enhance and improve training programs by positively influencing the revision of training materials and improving instructor performance

In a performance-based training program, the goal is for all students to master all learning objectives. This mastery is

## Applying the JOB PERFORMANCE REQUIREMENTS (JPRs)

For the learning process to be complete and reach its maximum potential, it needs to be evaluated. Evaluation can come in many forms—from written evaluations to hands-on evaluations to ongoing evaluation during the delivery of training. Numerous legal and ethical considerations exist during the evaluation phase of the learning process. As the fire service instructor, you must be aware of these considerations at all times.

### Instructor I

As part of the delivery of training, the Instructor I will be responsible for administering written evaluation and practical skills sessions.

### JPRs at Work

Administer evaluations of student learning and grade evaluations according to the type of evaluation used. Record evaluation scores and provide feedback to students about their performance.

### Instructor II

The Instructor II develops evaluation instruments to measure the student's ability to meet the performance objectives.

### JPRs at Work

Develop and analyze evaluation instruments to ensure that the student has achieved the learning objectives and that the evaluation is a valid measure of student performance.

### Bridging the Gap Between Instructor I and Instructor II

The Instructor I will administer the evaluations developed by the Instructor II while observing the legal and ethical principles of test administration. The Instructor I must learn to identify the desired outcomes of evaluations so as to improve his or her instructional skills. Both levels of instructors must ensure that the evaluation measures the stated objectives for the lesson plan and that the evaluation correctly measures the learning process.

### Instant Applications

1. Review a recent written examination and compare the student responses to the answer key. Identify any test questions that were either too easy (every student answered the question correctly) or that had poor success (there is a high percentage of failures).
2. Review your local authority policy and practice posting test scores and student completion records.
3. Develop a strategy to improve student performance based on a poor evaluation result.

evaluated or measured by the use of performance tests and safety-related oral tests. If this is the case, then what purpose does the written test serve in a performance-based training program?

Written tests are used to take a "snapshot" of students' knowledge at predetermined points throughout a course of instruction. For example, a pre-course exam may be used to determine a student's level of knowledge upon entering a course. A mid-course test or formative exam can determine a midpoint level of knowledge, and a final exam or summative exam assesses the student's knowledge at the end of the course. These snapshots provide information on how well the student is progressing. If that progress is not satisfactory at any point, additional instruction can be provided to bring the student up to the required knowledge level. In this way, testing allows the instructor to provide needed assistance before the student's lack of knowledge becomes critical. Lack of knowledge mastery is often a primary reason for poor or nonperformance of a learning objective and a task on the fire ground.

## Test-Item Development

The fire service is committed to safety and productivity through improved training programs and courses. Testing, which is an important part of any approach to training, encompasses two activities:

- The preparation of test questions using uniform specifications
- A quantitative/qualitative analysis to ensure that the test questions function properly as measurement devices for training

The development of objective (based on facts) test items requires the application of specific technical procedures to ensure that tests are valid and reliable. **Validity** means the test items (questions) measure the knowledge that they are designed to measure, and **reliability** means the test items measure that knowledge in a consistent manner. The fire service is fortunate that it has the NFPA's Professional Qualification Standards to use as a basis for criterion-referenced training and testing. When developing your own tests, you should carefully consider using the technical references in these standards as part of your validity evidence and documentation.

To develop a comprehensive test, it is important to have a large group (bank) of test items available. The test-item bank can be organized on a course basis, by standard reference, by job title, by duty, or in any other way that seems sensible and logical. This type of organization permits a test developer or fire service instructor to alternate test questions or produce different versions of a test for use between classes or among students within a class. This balancing of test versions is important given today's sometimes contentious legal environment and must be applied on a procedural basis to reduce the possibility of any discrimination occurring as a result of the testing program.

To begin building statistically sound tests, SMEs first write test items. These experts should be technically competent and actually working in the job for which the test items are being developed. The rationale for requiring this level of competence is readily apparent: Changes are occurring at a more rapid rate than ever in the fire service, and it is essential that test items reflect current practice. This initial activity provides the first level of validity—namely, face validity.

From this point, test items are analyzed on the basis of their individual performance in the test using test-item analysis techniques and data collected from responses of students taking the test. Once data are collected from approximately 30 uses of a test item, the first test-item analysis can be completed. The test-item analysis data provide specific information needed to establish a quality test-item bank.

## Development of Test Items According to Specifications

Test specifications permit fire service instructors and course designers to develop a uniform testing program that relies on valid test items as a basis for constructing reliable tests. Writing and developing test items is difficult for fire service instructors. In particular, it is difficult to write a test item the first time that is both content valid and reliable; instead, most valid test items and reliable tests evolve over time as a result of refinement through use. To facilitate this process, test-item writers and test developers must follow basic rules or specifications to create a valid and reliable test. This section presents and briefly explains the test specifications, which are intended as a guide for developing test items and tests with initial technical-content validity and testing reliability.

Two terms used to describe the quality of test items/tests are *validity*—the ability of a test item/test to measure what it is intended to measure—and *reliability*—the characteristic of a test that measures what it is intended to measure on a consistent basis. Put simply, a valid and reliable test is one that measures what it is supposed to measure each time it is used. A test item has content validity when it is developed from a body of relevant technical information by a SME who is knowledgeable about the technical requirements of the job.

Written tests are designed to measure knowledge and acquisition of information, but they do have some limitations. One major limitation is that testing knowledge and information in this way does not ensure that a student can *perform* a given task or job activity; it simply indicates whether the student has the knowledge *required* for performance. Skill development (the ability to perform tasks or task steps) occurs through actual performance, such as during evolutions on the training ground. Skill development may be documented through skills evaluations that take place during and at the end of training.

A simple analogy using golf explains the difference between testing for knowledge and information and actual performance of a task. For instance, a person can read numerous books, watch films, view videotapes, attend golf tournaments, and perform other information-gaining activities about golf. After completing these activities, the person could probably garner an acceptable score on a written test about golf. Even with all this information and a perfect score on the written test, however, the person probably could not actually play a game of golf and score less than 100 after 18 holes **FIGURE 10.2**.

FIGURE 10.2 Obtaining knowledge and applying knowledge are separate skills.

**Teaching Tip**

There are three justifications for testing:

- To measure student attoinment
- To determine weaknesses and gaps
- To improve instructor performance and training materials

This concept can be applied to the many tasks necessary to score well in any course. The basic skills associated with any task must be practiced and kept sharp for the individual to be a competent performer in the real world. Of course, these same conditions apply to being a public service professional. Whether the job is in firefighting, policing, administration, supervision, or training, the person's success reflects his or her ability to apply the information needed to achieve levels of professional competence. A balance of must-know and need-to-know information and competent performance is the foundation of performance-based learning.

**Ethics Tip**

Some of the areas covered by national standards may not apply to the students whom you are teaching. For example, fire fighter standards require a basic understanding of wildland firefighting principles, but you might be teaching fire fighters in a large, urban department with no wildland interface. Similarly, you might be teaching a very small, rural organization that does not have any standpipes in its jurisdiction. Is it acceptable to omit material from certification courses when the material will not be used by the students?

Professional qualifications are nationally recognized minimum standards for various levels of fire personnel. If you do not teach your students to the standard, you will have failed as a fire service instructor and jeopardized the very theory underlying certification—namely, ensuring a standardized level of competency. Although you may feel relatively confident that some areas of the material will not be used by some or all of your students, you will be doing them a disservice if you do not cover all of the learning objectives for the course. Imagine a student who later joins a larger department and does not understand the functioning of a standpipe system, or a fire fighter who attends the National Fire Academy but does not know the term "backfire" or "back burn." Professional standards are just what they say. Don't undermine your credibility by not teaching all of the material in the NFPA standard.

The following specifications, examples, and test development criteria will help ensure that fire service instructors, SME, and curriculum developers prepare written test items, construct written tests, and performance tests that measure what they should measure and function reliably over a long period of time.

## Development of Written Tests

The written test-item type to be developed should reflect two factors:

- The knowledge requirement to be evaluated
- The content format of the resource material

As you research and identify passages of relevant information in the resource material, ask yourself which type of written test item the material might support. For example, a passage full of terms and definitions might support developing a matching question, whereas a passage containing a scenario or other complex situation might support multiple-choice questions. You may want to note this observation somewhere in the resource material as you go through the review and research phase of course development. One technique that works well is to use a different-colored highlighter for each type of test item. When you return to write test items, the analysis for the type of test item will already be complete.

As a fire service instructor for my fire department, I often have the opportunity to work with new recruits as they participate in training to acclimatize them to our department. This training ranges from technical training, such as ropes and water rescue, to firefighting basics, such as forcible entry and ventilation. It is always a challenge to ensure that the fire fighters I am training are learning what they need to know in the street to perform in a safe, efficient manner. It is this evaluation of the learning process that occupies much of my time during instruction.

One particular training evolution comes to mind as especially notable. I had instructed the class to ladder a home, place a roof ladder, and perform vertical ventilation using a chainsaw. As the fire service instructor, I ascended the roof first to supervise the students. One new recruit was visibly nervous about being evaluated. The fire fighter and his partner ascended the roof and began an attempt to ventilate. The chainsaw quickly died. Once the chainsaw was restarted, they began again, struggling through the use of the chainsaw while standing on the ladder very high up on a roof.

*"It is this evaluation of the learning process that occupies much of my time during instruction."*

At the conclusion of recruit training, the fire fighter was assigned to my ladder company. Recalling the lack of confidence he had displayed during training, I worked to revisit the training evolution with him. We discussed his performance and his perceived strengths and weaknesses. We reviewed the ventilation procedure numerous times, using the tools and operating guidelines pertaining to the operation. The true test came when we arrived at a working house fire that required immediate roof ventilation. He quickly ascended the roof with the chainsaw and began to cut a large opening without difficulty, successfully ventilating the residence.

When we discussed the operation later, he expressed to me how important our continued communication and training was. He explained that it increased not only his technical skills, but also his confidence in performance of the skill.

The evaluation of the learning environment is an intuitive and continual process. Evaluation can take place in a formal training session or on the rear bumper of an engine. Final exams aren't always on paper—sometimes they are on a roof, with heavy smoke pushing from a residence and a chainsaw in your hand.

*David Pennington*
Springfield Fire Department
Springfield, Missouri

## Complete a Test-Item Development and Documentation Form

Test-item development and documentation forms serve many purposes other than the most obvious—recording the test item. This kind of form is also useful for the following tasks:

- Connect planning/research with test-item development efforts
- Record pertinent course-related information
- Record the source of test-item technical content
- Provide a record of format and validity approval
- Record the learning objective and test-item number for banking

By documenting the required information on the form, you are using it to accomplish all of these purposes.

Collectively, the information you place in the blanks will serve to link the identifiers of the program or course and the reference with the test item. Each blank is important to proper identification, so fill in each one carefully with the appropriate information. A continuation sheet may be used when developing a test item that is too long to fit on a single form.

The **reference blank** is where the current job-relevant source of the test-item content is identified. In this blank, supply enough information to pinpoint the specific location within the resource material where you found the information you are using to develop the test item. Entries might appear as follows:

> Starting at the top of page one of the test-item development and documentation form, complete it as follows:
>
> - **Program or Course Title.** In this blank, put the title of the program or course for which you are writing a test item. An example would be "Fire Officer I."
> - **Learning Objective Title.** In this blank, put the learning objective you are working on in preparing test items. An example would be "Fire-Ground Operations."

### Components of a Test Item

Test items are made up of components that collectively create a testing tool. The part of the question that asks the information is known as the stem **FIGURE 10.3**. The choices are made up of a correct answer and distracters. An important part of every question is the reference material that documents the standard on which the question is based and the source of the information in a textbook.

## Selection-Type Objective Test Items

## Multiple-Choice Test Items

The multiple-choice test item is the most widely used test item in objective testing. It also contributes to high test validity and reliability estimates. The multiple-choice test item is the favorite among test-item developers because it comes nearer to incorporating the important qualities of a good test—content validity, reliability, objectivity, adequacy,

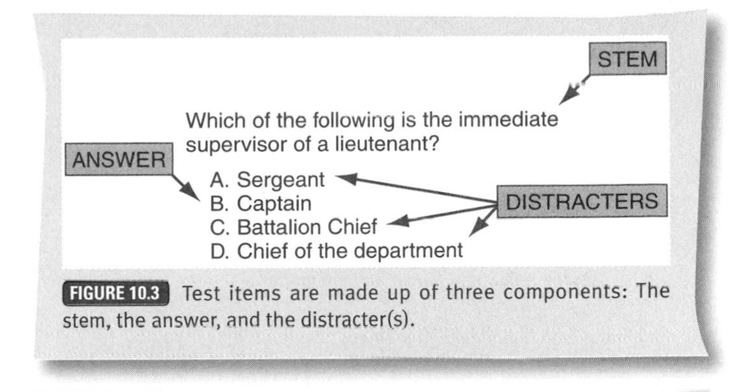


**FIGURE 10.3** Test items are made up of three components: The stem, the answer, and the distracter(s).

Fire department synthetic rope, new or used, should be kept

**A.** Warm, dry, and in a sunlit place.
**B.** Cool, moist, and in a sunlit place.
**C.** Neatly coiled and stored out of sunlight.
**D.** Damp and in a dark place.

**FIGURE 10.4** An example of a properly formatted multiple-choice test item.

> **Teaching Tip**
>
> While the multiple-choice question is the most popular type of test item, it is essential to choose the testing method that best ensures competency of the learning objective, rather than simply falling back on habit when developing evaluation tools.

practicality, and utility—than any other type of test item. In addition, the ease of grading of the test items and the ability to provide immediate feedback to students make multiple-choice items extremely flexible for the fire service instructor.

The most important advantage of the multiple-choice test item is that it can be used to measure higher mental functions, such as reasoning, judgment, evaluation, and attitudes, in addition to simply knowledge of facts. The multiple-choice test item is so versatile that it can be used to measure almost any cognitive information. Its major disadvantage is the fact that several well-written test items may be needed to properly cover the learning objective.

### Format Example for a Multiple-Choice Test Item

**FIGURE 10.4** provides an example of a properly formatted multiple-choice test item.

### Suggestions for Preparing Multiple-Choice Test Items

- Select one and only one correct response for each item. The remaining distracters should be plausible but wrong, so that they serve as distracters for the correct response.
- Construct the same number of responses for each test item. Four responses are preferable. More than four

response choices may be used and sometimes are necessary, but they do not improve the validity of the test.

- Responses having several words should be placed on one or two lines with space separation from other responses. If responses consist of numbers or one word, arrange two or more of them on the same line, positioned in a consistent manner.
- Provide a line or parentheses on the right or left margin of the page where students can write their responses. This placement permits ease of test taking for the student and encourages use of a grading key for more rapid and accurate test grading. Disregard this step if you will be using a machine-scannable answer sheet.
- Change the position of the correct response from test item to test item so that no definite response pattern exists.
- Prepare clear instructions for students.

## Matching Test Items

The matching test item is actually another form of multiple-choice testing. This type of question has limited uses, but functions well for measuring such things as knowledge of technical terms and functions of equipment. Care should be taken while preparing matching test items, because all information contained in the question must be factual. There should be no more than four items to match with five choices for providing matches. This type of test item is relatively easy to develop, but requires more space on the test.

The matching test item is particularly useful when a question requires multiple responses or when no logical distracters exist. Such a test item is very useful for assessing technical vocabulary development, identifying tools, and performing similar kinds of low-level cognitive information measurement.

### Format Example for a Matching Test Item

FIGURE 10.5 provides an example of a properly formatted matching test item.

Directions: Match the terms listed in column A with the definition provided in column B.

| Column A | Column B |
|---|---|
| 1. Oxidation | A. Rapid oxidation with heat and light |
| 2. Conduction | B. Oxygen combining with other substances |
| 3. Convection | C. Heat energy carried by electromagnetic waves |
| 4. Radiant heat | D. Heat transfer by circulation through a medium |
| | E. Heat transfer by direct contact |

FIGURE 10.5 An example of a properly formatted matching test item.

### Suggestions for Preparing Matching Test Items

- Make sure that similar subject matter is used in both columns of the matching test item. Do not mix numbers with words, plurals with singulars, or verbs with nouns.
- Use up to four items in column A and up to five choices for the match in column B. This prevents students form earning credit by simply applying the process of elimination.
- Include only one correct match for each of the items to be matched.
- Arrange statements and responses in random order.
- Check for determiners or subtle clues to the answers.
- Provide a line or parentheses beside each response statement where students can indicate the answer. Arrange the parentheses or lines in a column for ease of marking and grading. This step is not required if you are using a machine-scannable answer sheet.
- Prepare directions to students carefully. Analyze students' ability to follow the instructions to ensure that the instructions function as intended.
- Attempt to keep each matching test item on one page.
- When preparing the draft of questions, follow the format guide to expedite typing, reproduction, response by the student, and rapid grading.

## Arrangement Test Items

The arrangement test item is efficient for measuring the application of procedures for disassembly and assembly of parts, start-up or shutdown, emergency responses, or other situations where knowing a step-by-step procedure is critical. These test items lose strength and efficiency steadily as time lags between the paper/pencil solution and the actual performance of the procedure. For this reason, you should plan to administer an arrangement test in close proximity to an opportunity to perform the actual procedure, disassembly, or assembly.

The major disadvantage of arrangement test items is inconsistent grading. A student may miss the order of one step and, therefore, miss the order of all steps that follow, even though the remaining steps may be in the right sequence after the error. Consider giving some credit for steps that are properly sequenced. The test item should be developed in agreement with actual procedures used on the job if procedures are required. Arrangement test items are typically not effective if the procedure is much longer than ten steps, however. In cases where it is important to test more complicated procedures, it is better to group test items so that they focus on key points in the procedure. Knowledge of emergency procedures, fire drills, triage, or other related procedures where knowing exactly what to do under critical conditions can be measured with high reliability using an arrangement test item.

### Format Example for an Arrangement Test Item

FIGURE 10.6 provides an example of a properly formatted arrangement test item.

**Directions:** (*Caution—This is a two-part question worth 3 points.*)

The six steps of the basic method that should be used in fire or explosion investigations appear out of order below. **First,** number the steps in the proper order by placing numbers (1–6) in the blanks beside the steps. **Next,** select the answer that matches yours from choices A–D.

______ Conduct the investigation.

______ Collect and preserve evidence.

______ Receive the assignment.

______ Analyze the incident.

______ Prepare for the investigation.

______ Report findings.

**A.** 1, 4, 5, 2, 3, 6 **B.** 2, 4, 5, 1, 3, 6

**C.** 6, 3, 4, 5, 1, 2 **D.** 3, 4, 1, 5, 2, 6

Answer: D

FIGURE 10.6 A properly formatted arrangement test item.

### Suggestions for Preparing Arrangement Test Items

- Make sure that arrangement test items developed for procedures are based on current procedures. Such test items should relate to emergency performances or to things that must be performed in a certain order. They are useful when you want the performance to occur without references or verbal input.
- Have one or more SMEs review the test items for technical accuracy.
- Make sure the steps of the procedure are given out of order and that steps with clues to the correct order are separated.
- State exactly what credit will be given for correct sequencing of all steps.
- Include cautionary statements with the procedure to be rearranged. Example: Order the RIT team to enter a structure. (Caution: Report order to the incident commander.)

## Identification Test Items

The identification test item is essentially a selection-type test using a matching technique. Its major advantage over the matching test item is that it evaluates the ability of the student to relate words to drawings, sketches, pictures, or graphs. These test items do require reasoning and judgment, but basically focus on the ability to recall. Identification test items can also be considered as arrangement test items for assembly- and disassembly-type tasks.

Identification test items provide excellent content validity and contribute to high test reliability. Unfortunately, these items are sometimes difficult to develop and may be costly to produce. With today's computer publishing capabilities, digital cameras, and other technologies, however, identification questions can often be rather easily developed using your own fire equipment and apparatus features as the basis for the identification test item. Using this technique ensures that the test item is job related and makes it easier to transfer required knowledge from the instructional setting to the job.

### Format Example for an Identification Test Item

See FIGURE 10.7 for an example of a properly formatted identification test item.

### Suggestions for Preparing Identification Test Items

- Select a graphic, art, or other object for inclusion in the question.
- Place the object into a drawing or illustration software.
- Identify the parts or items you want identified. (Caution: Don't identify more than seven or eight things to be labeled, or even fewer items if the object will be cluttered by arrows or letters.)
- Develop the desired alternatives A–D such that only one answer gives the correctly identified objects.
- The identification test item may give a point value of more than 1 for the correct answer. A rule of thumb is to award one point for every two correct responses.
- Have a SME review the technical accuracy of the question and its answer alternatives. Be prepared to provide your validity evidence and documentation.
- Try out the test item several times with the target population to make sure it measures what it is intended to measure.

## Completion Test Items

A completion test item is not easy to develop, and it is difficult to achieve a respectable reliability estimate during the first application of such a question in a test. Instead, reliable test items of this type are developed through several uses and refinement steps. Test-item development skills are essential

**Directions: (*Caution - This is a two-part question worth 5 points.*)**
**First**, label the tools depicted below by placing the number in the blanks provided.
**Then**, choose the answer below (A–D) that matches yours.

____ McLeod
____ Pulaski
____ Back pack pump
____ Shovel
____ Drip torch

Answers
A. 3, 5, 4, 2, 1
B. 2, 4, 1, 3, 5
C. 5, 3, 1, 2, 4
D. 4, 3, 1, 2, 5

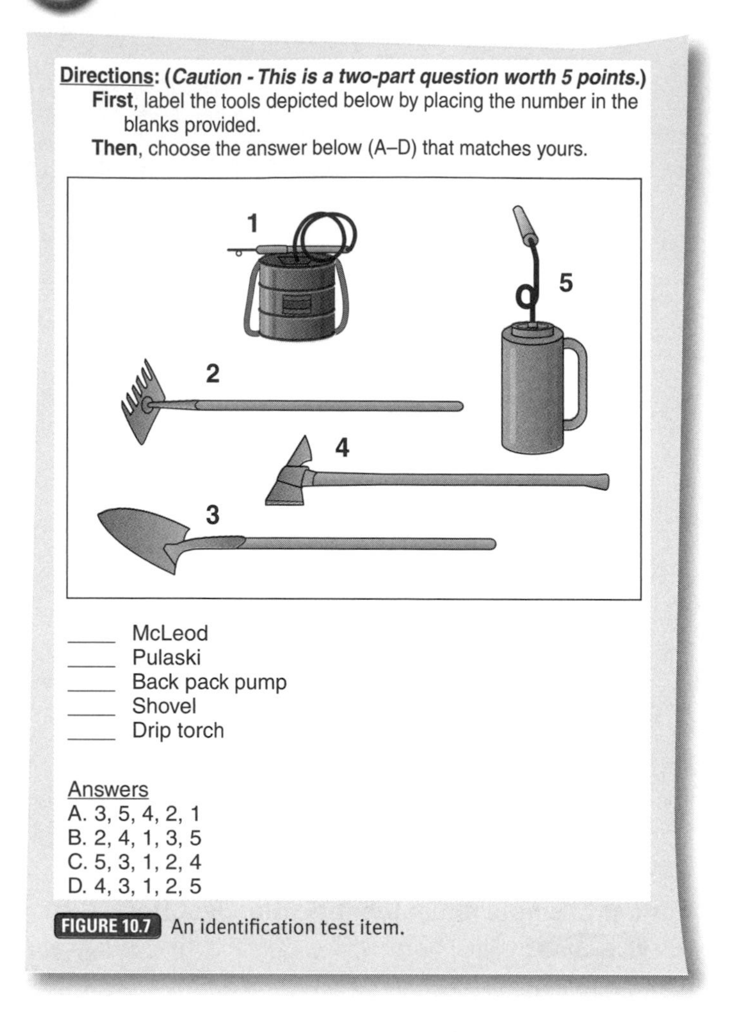

**FIGURE 10.7** An identification test item.

Foams in use currently are of the mechanical type and must be __________ and __________ before they can be used.

**FIGURE 10.8** Supply-completion test item.

for the test-item writers and SMEs who review the technical content of completion test items. The best practice for developing and using this type of test item is to follow specific procedures and to develop two test items for every one item that is expected to be included in the final group of test items.

Completion test items are of two types: supply and selection. The supply type requires the student being tested to supply the response that completes the test-item statement. The selection type requires the student being tested to select from a list of responses for completing the test item. The supply type tests primarily the recall capability of the student, whereas the selection type focuses on recognition and analysis.

### Format Example for a Supply-Completion Test Item

**FIGURE 10.8** provides an example of a properly formatted supply-completion test item. This type of test item requires the instructor to score the test item manually and to have a list of synonyms as possible correct answers.

Foams in use currently are of the mechanical type and must be __________ and __________ before they can be used.

**A.** proportioned, blended **B.** stirred, aerated
**C.** mixed, proportioned **D.** proportioned, aerated

Answer: D

**FIGURE 10.9** Selection-completion test item.

### Format Example for a Selection-Completion Test Item

**FIGURE 10.9** provides an example of a properly formatted selection-completion test item.

### Suggestions for Developing Completion Test Items

- Sources of completion test items include system descriptions, equipment operation manuals, technical specifications, procedures, and safety practices. Multiple-choice questions also provide a source of completion items, but care should be taken to avoid using that multiple-choice item again on the test (so as to reduce the number of give-away answers).
- Make the statements as brief and factual as possible. Do not use misleading or contradictory words or terms.
- Avoid use of responses that match verb tense for one blank. It is important to have several singular and several plural responses or to use the same verb tense when developing selection-type items.
- Have two or more SMEs review each test item for technical accuracy and quality of response selections.
- Selection-type questions should have two to three more possible responses than there are blanks.
- Avoid the use of more than two blanks per statement *except* when asking for major linking parts (e.g., The four parts of a pump body are the __________, __________, __________, and __________.).
- Be aware that supply-completion test items require manual scoring and may require a list of synonyms as possible correct answers. The quality of students' handwriting can also be a problem. It is not unusual for persons taking a test to scribble something in a blank and hope for credit if they don't know the answer or have doubts about their response. The supply-completion test item is also considered subjective, because instructors can withhold or provide a point based on whether they like or dislike the student taking the test (the halo effect).

## True/False Test Items

True/false test items have been controversial for years. Test-item developers have experimented with numerous ways to make true/false, right/wrong, and yes/no responses more valid and reliable. The major problem with two-answer selection

**Directions:** Circle either "T" for true or "F" for false.

As a fuel's surface-to-mass ratio decreases, its ignitability increases.

Answer: F

FIGURE 10.10 Typical true/false test item.

**Directions:** Read the following statements regarding diversity programs and select your answer from choices A–D.

**Statement 1:** A diversity program creates an environmental and cultural change in an organization.
**Statement 2:** Achieving a culturally diverse organization is ethically and managerially a worthwhile accomplishment.
**Statement 3:** A diversity program is a single effort intended to integrate an organization to meet government mandates.

**A.** Statement 1 is true; statements 2 and 3 are false.
**B.** Statements 1 and 2 are true; statement 3 is false.
**C.** Statement 1 is false; statements 2 and 3 are true.
**D.** All three statements are true.

Answer: B

FIGURE 10.11 A complex true/false test item.

**Teaching Tip**

Using multiple forms of test questions can increase validity of the test. Use the question formats that best ensure the learning objective is met.

test items such as the true/false question is the guessing factor: A student who marks a response without reading the question has a 50 percent possibility of giving the correct response. Application of two or three rules for taking a true/false test could get the student a higher score just by guessing alone.

### Format Example for a Typical True/False Test Item

FIGURE 10.10 shows an example of a typical true/false test item.

### Format Example for a Complex True/False Test Item

A more complex true/false test item can be developed that uses a multiple-choice approach to the test item. FIGURE 10.11 shows an example of a complex true/false test item.

### Suggestions for Developing True/False Test Items

- True/false test items do not cover the subject matter in depth because of the limited application possible in most technical information or courses. Use of true/false test items should be limited to factual information at the basic level of an entry-level course such as that taken by Fire Fighter I personnel. There is little discernible value for the true/false test item in technical instruction when student knowledge and performance measurement is required for safe and efficient operation of equipment and emergency task performances. For this reason, true/false questions should be avoided in operations and maintenance programs.
- The complex true/false test item is much better suited to technical courses and information. These items should be developed from a specific area of knowledge. Statements should be brief and technical in nature. This type of question requires higher-level cognitive skills for success.
- If used in a multiple-choice format, true/false test items can be objectively scored.

## Essay Test Items

Modern test developers do not consider the **essay test** to be an objective form of testing. This position is based on research that examined the major weaknesses of this testing method. Weaknesses that affect the objectivity of essay tests include the following issues:

- Essay tests tend to focus on limited points in a course of study. Because of the length of student response and time available for the fire service instructor to grade tests, essay tests cannot be comprehensive in coverage of the subject matter.
- The length of the student response tends to cloud the actual knowledge required by the question with extraneous knowledge closely related but not directed to the question. Students can "beat around the bush" without ever really addressing the question.
- The order in which test papers are graded affects the grading. The papers graded first tend to receive lower grades than those graded last, especially when tests are long and are graded one exam at a time. Grading one question at a time for all students tends to improve the consistency of grading on an essay test.
- Handwriting affects grades. The better the student's handwriting, the higher the grade tends to be, even when the quality of the response is varied. Typewritten papers get higher grades than handwritten papers, even when answers are identical.
- Grading varies widely even among fire service instructors who are recognized SMEs. Research studies have identified variance ranges as great as three letter grades and 25 points. Such variations in grading occurred even when fire service instructors were provided with grading keys and a set of scoring rules for the questions.
- The "halo effect" influences the grading practices of some fire service instructors. The more they know about a student (either favorable or unfavorable), the more the grade is affected. The personality of the student tends to have a unique effect on the grade received on an essay test.

- Writing essay questions requires a lot of time on the part of fire service instructors. Initial essay questions must be refined over several uses before they can be expected to function in a reasonably reliable manner.
- Students require much more time to prepare for and write answers to essay questions. Long tests tend to be fatiguing to students and can sometimes encourage lengthy responses containing extraneous information.
- Poorly prepared essay test items reinforce negative learning. In other words, the answer the fire service instructor receives tends to support the wrong interpretation of facts and conditions even among students considered to be outstanding.

Research has shown that a fire service instructor can develop an objective test in approximately the same amount of time that it takes to develop an essay test. However, the objective test will achieve higher content validity, comprehensiveness, discrimination, and reliability than is possible with an essay test.

Although use of essay test items has many disadvantages, these types of questions can be improved to minimize some of their inherent weaknesses. The essay test item requires the student to recall facts from acquired knowledge and to present these facts logically and in writing. Writing answers provides the student with an opportunity to express ideas and attitudes, interpret operational situations, and apply knowledge gained to an individual solution. A good technique for answering essay questions can be developed by a fire service instructor and taught to students. Outlining a response to an essay question and marking the key points so that they may be emphasized in the answer is a valuable approach to improve performance.

### Formatting Essay Test Items

It is important to word the essay question properly and to develop a preliminary outline of the key points expected in the response to the question. This exercise helps the instructor determine whether the question will actually elicit the intended responses. The outline also serves as a guide for grading student responses and assigning relative point values to the key points in an answer before the test is given.

A hierarchy of words can be used to develop essay test items that will range from easy to difficult. For instance, test items that require the student to "describe," "explain," "discuss," "contrast," and "evaluate" follow an ascending order of difficulty. That is, asking the student to "describe an event in plant operations" is much easier than asking the student to "evaluate the same event and contrast that event with the implications for plant safety." In the essay test, each of these types of questions should be asked so that the relative difficulty of the test is maintained. The easier questions (describe, explain) should be presented earlier in the test.

### Suggestions for Preparing Essay Test Items

- Review the learning objectives. Test items should be developed by considering the learning objectives and using them as the basis for testing.
- Make a list of major points that are contained in the body of the course material to be covered by the test. This list should focus on major points, rather than trivial details. Check the list to confirm that it is comprehensive in its coverage of the subject matter, keeping in mind that an essay test contains only a few test items and requires more time both for the student to answer and for you to grade.
- Arrange the list of major points in an order similar to that used when presenting the information or assigning reading material for outside study.
- Identify the major points that you want students to emphasize. These major points provide a less difficult essay question for students to begin with on a test. The use of a relatively easy question for the first item will get students ready for the more difficult questions that follow.
- Prepare a question or two requiring students to describe and explain a key point or major concept in the subject matter. Again, this approach takes students to the next level of difficulty in an essay exam.
- Prepare the most difficult essay items last and place them toward the end of the exam. These questions usually require students to contrast, compare, and evaluate a situation or a complex set of facts. It is a good testing practice to use the earlier questions in an exam to form the basis for these contrasts, comparisons, and evaluations. Placing difficult questions at the end of an exam has the added value of permitting well-prepared and knowledgeable students to pull the facts together and express personal views in terms of the facts. It is not uncommon for students to expand their knowledge well beyond that expected by the fire service instructor and to offer plausible solutions that may not have been presented previously. These unexpected results may be rewarded with extra credit.
- Take the test. Answer the questions in terms of the course material presented, assigned reading, and lab and shop activities. Once you have prepared your answers, prepare a detailed outline of each essay test-item response. Determine the amount of credit to be given for each key point in the outline. Determine, in advance, where extra credit will be awarded for answers that clearly excel, and determine the number of points to be deducted for major omissions in test-item answers. These outlines then become the scoring guide and key to help ensure consistent grading from question to question and from one exam to another.
- When essay exam questions are completed, review them once again. While reviewing the exam, keep in mind the exact response behavior desired from students and those specific abilities and knowledge that students should possess relative to course materials covered by the exam.

### Grading the Essay Test Item

A crucial activity to make an essay exam more objective is the use of a scoring key with associated criteria for consistent grading. Following are a few suggestions that can make the grading process more reliable:

- Fold exam cover sheets back or cover names before beginning any grading activity. This permits the application of the grading criteria as anonymously as possible.

- Grade one question at a time on all exam papers before moving to the next question. This tends to keep the focus on the grading key and the response at hand.
- If the exam has an effect on a student's career or promotion, it is good professional practice to have another qualified person give a second grade to the exam. An average of the two grades provides more objectivity.
- Provide some flexibility in the grading key or outline for giving additional credit for answers that produce results beyond those expected when the test was developed. Penalties for omission of key points must also be deducted from the test score. Make sure, however, that the omission penalties are uniformly applied to all questions and exams.
- The *whole method*, which can improve grading of exams, begins with a preliminary reading of the exam papers. During the reading process, a judgment is made of each exam in terms of four groupings: excellent, good, average, and below average. Each exam is then sorted into one of these preliminary groups. Once the grouping is completed, grading is accomplished in the same manner as mentioned previously. A shift from group to group of one or more exam papers is not uncommon when the scoring is completed.

Remember—be as objective as possible when applying the grading criteria. Essay tests, even under the best of conditions, have low validity and reliability. Do everything possible to avoid a further decline in value because of poor testing practices.

### Prepare Essay Test Item Grading Key

It is important to develop a preliminary outline of key points expected in the response to the essay question FIGURE 10.12. While helping to determine whether the question will actually elicit the intended response, this exercise also facilitates assigning relative point values to the key points in an answer before the test is given.

In outline format, list the key points that are expected in the answer to the essay test item. Assign a weighted point value to each key point in the outline, where the weight reflects the point's degree of importance to the overall answer. Key points of a critical nature or those pertaining to safety-related activities and information should be given higher weighted values. Because of the importance of grading essay test items, the methods for increasing objectivity should be considered.

## Performance Testing

Performance testing is the single most important method for determining the competency of actual task performances. For this reason, the major emphasis in performance-based instruction is the validity and reliability of the performance tests. The focus on student performance is the critical difference between traditional approaches to instruction and the use of performance-based instructional procedures. Performance must be evaluated in terms of an outside criterion derived from a job and task analysis. The justification for training program knowledge and skill development activities centers on the concept that the activities are derived from and will be applied to training for tasks performed on the job. This focus prevents the inclusion in programs of "nice to know" information and irrelevant skill development activities.

**Question**

State four precautions for handling or working with caustic soda, including the appropriate first-aid actions, in the event contact with caustic occurs.

| **Grading Key** | **Point Value** |
|---|---|
| 1. Hot water must be used when dissolving caustic soda. Eyes, face, neck, and hands should be protected. | 1 |
| 2. Wherever caustic soda is stored, unloaded, handled, or used, abundant water should be available for emergency use in dissolving or diluting and flushing away spilled caustic. | 1 |
| 3. General first aid is of prime importance in case of caustic coming in contact with the eyes or skin. Prolonged application of water to the affected area at the first instant of exposure to caustic is recommended. | 2 |
| 4. a. If even minute quantities of caustic soda (in either solid or solution form) enter the eyes, the eyes should be irrigated immediately and copiously with water for a minimum of 15 minutes. | 2 |
| b. The eyelids should be held apart during the irrigation to contact of water with all the tissues of the surface of the eyes and lids. | 1 |
| c. If a physician is not immediately available, the eye irrigation should be continued for a second period of 15 minutes. | 2 |
| d. No oils or oily ointments should be applied unless ordered by the physician. | 1 |
| **Total Point Value** | **10** |

FIGURE 10.12 Example of an essay test item grading key.

**Theory into Practice**

Putting together a valid testing instrument may be one of the most challenging parts of the entire learning process. To achieve success, the test-development process requires creating good questions such that a student who knows the material is able to provide the correct response, whereas a student who does not is not able to provide the correct response. The test also should allow those students who know the material the best chance to get more correct answers than those who know much less.

To accomplish this goal, take each learning objective and create questions that demonstrate competency for that learning objective. Use various styles of questions to discern the level of comprehension. For example, writing a scenario and having students apply the material concepts is much more difficult than creating questions where students can recite rote information from the text.

The final demonstration of job knowledge and skills application in fire and emergency medical training programs is the completion of specific on-the-job tasks.

For example, skills are developed based on job related standards. During lecture the students learn the material and during performance testing they are required to perform the skill. The instructor uses a skills checklist during the application step of instruction to record the performance of the student as they perform the skill **FIGURE 10.13**. The skills checklist matches the standard and the material taught in the course. This is what makes the skills checklist valid. When administered over a length of time, it consistently measures student performance and becomes reliable.

## Test-Generation Strategies and Tactics

Tests can be put together in many ways. In particular, you can do it yourself or use a computer or Web-based test-item bank.

### Instructor-Made Tests

The strategy of doing it yourself is generally referred to as an "instructor-made test." The primary problem with instructor-made tests, even though they may be valid and reliable, is the tendency to use the test over and over again. Soon the word gets out about the test content and answers, and everyone who takes the test knows the answers.

### Computer and Web-Based Testing

In recent years, the use of computers for training and testing purposes has grown by leaps and bounds. Distance learning over the Web is now available in conjunction with most colleges, universities, and two-year institutions. The fire and emergency medical services have made great strides embracing this technology, but more needs to be done.

**Teaching Tip**

All fire service instructors have witnessed personnel who have passed certification/promotional examinations but who we believed were not ready for the job. Conversely, we have seen people who we believed would be very good fire fighters, but who failed the examination. Although perceptual bias may be at work here, often those "gut feelings" are accurate. Poor tests are frequently to blame for these unexpected outcomes. Carefully screen each question for reliability and validity, as well as the examination as a whole for discrimination, to help prevent these errors.

According to the United States Distant Learning Association, the number of full-time distance learners increased from 483,113 students in 2002 to 1,488,482 students in 2006. *The Washington Post* estimated that the online learning population—including everyone from kindergarten students to enrollees in doctoral-level programs—would exceed 1.75 million students in 2007. According to *Distance Learning Today* newsletter, the Board of Regents for the University of Maryland recently mandated that its universities encourage students to take at least 12 hours of learning off campus, preferably online.

The fire and emergency service appears poised to increase its adoption of this new technology, in part because it has the potential to reduce costs for education and training for both the student and the department. Options available in terms of computer hardware and professionally developed test banks can put testing online 24 hours a day, 7 days a week. In fact, some fire and emergency medical service organizations are already conducting online and Web-based certification testing programs. Given the rising price of fuel, costs for hotels, worker salaries, and other costs, Web-based certification seems to be the wave of the future.

One of the primary benefits of online learning and testing is its convenience for personnel who work full-time jobs. An employer can also extend training and testing time into the available leisure time of willing employees.

As an instructor, you should collect information and look for successful applications of this technology for teaching and testing your students or employees. Find out what is going on in this fascinating technology and make recommendations to your supervisors. Remember, in the fire and emergency service you are dealing with situations that require rapid responses and high levels of technical knowledge. Providing your students with the best means for learning and performing is your primary challenge.

To begin, research the technology available: It is difficult to develop sound applications of technology without some sort of study. Once your research is complete, you will have a much better idea of what kind of technology would meet your specific requirements. There is so much going on in the realm of computers, networking, and online education/training that you can easily become quite confused about the various options and may even spend much more money on technology than is truly necessary. The diligent collection of data,

**Name:** ______________________ **Date:** ______________________

☐ **Pass** ☐ **Fail**

**SKILLS CHECKLIST: MAYDAY MESSAGE**

**OBJECTIVE:**
- Demonstrate the ability to call a MAYDAY message on a portable radio

**MATERIALS NEEDED:** Fire fighter in full protective clothing, full SCBA and portable radio equipment.

**STUDENT INSTRUCTIONS:** You will be asked to simulate firefighting operations in which you will encounter an emergency described to you by the instructor. After you are told what type of emergency you are facing, you will issue a MAYDAY report on your portable radio containing the **LUNAR** information. You will be given an opportunity to complete this skill by successfully completing all items on this skills checklist. If you are not successful you will not be re-tested until you have had additional instruction and practice.

**INSTRUCTOR INSTRUCTIONS:** Provide the student with instructions and describe the scenario. The student will only be permitted one attempt to complete all the items listed below. If they are not successful they will be required to receive additional instruction and practice before attempting the skill again. Place a check in the appropriate column as they complete the steps listed below. Any comments for improvement should be placed in the comments section. If a student does not complete a step correctly, in the comments section list the step number and comments on improvement.

| Steps | Pass | Fail |
|---|---|---|
| Don all PPE and attach regulator to face piece and begin breathing. | | |
| Receive scenario description from instructor on the emergency situation. | | |
| Grasp portable radio microphone and depress emergency button if equipped. | | |
| Initiate MAYDAY message in a calm, clear tone speaking into microphone.<br>State LUNAR report to I/C:<br>• Your **L**ocation, **U**nit, **N**ame, **A**ssignment, **R**esources needed | | |
| Await confirmation of your LUNAR report, and begin to take correct standard emergency actions. | | |
| Activate PASS device. | | |
| Provide updates on LUNAR as indicated by assigned scenario. | | |
| Notify I/C when self-rescue or RIT assistance is complete. | | |

**EVALUATOR COMMENTS:**

______________________

______________________

______________________

______________________

______________________

**FF SIGNATURE** **EVALUATOR'S SIGNATURE**

FIGURE 10.13 A skills performance checklist used in the application step of instruction.

thoughtful completion of an objective needs analysis, sober consideration of your budget, and thorough assessment of the software and equipment that will meet your most immediate needs are all steps that will help you realize the best possible return on your investment. Never forget that bookshelves and storage cabinets are filled with equipment and software that were purchased with good intentions but remain unused because they were not the right tools for the job.

Textbook publishers and other private organizations often provide test-item banks with their publications. This practice is good, but you must carefully analyze these test banks to make sure that they have been properly validated.

The most significant advantage to using valid and reliable computer-based test banks with large numbers of questions is the reduction in chances of test compromise. You can go even one step further by randomly generating any tests you administer every time you need to give a test. Large numbers of test items in a test bank help ensure that no two tests will ever be alike. For instance, if a test bank contains 1000 test items and you randomly generate a test each time you need one, the chance of getting the exact same test is almost nonexistent.

When choosing or creating a test, remember one thing if nothing else: Most testing in the fire and emergency service leads to certification, selection for a job, pay raises, and promotions. If you are testing for any of these reasons, do not use test banks that have not been rigorously validated and that are known to produce reliable tests. The question then becomes, "How do I know?" The checklist in FIGURE 10.14 should help you determine the quality of a test-item bank, no matter who is providing the product.

Of course, you can add many more questions to this checklist. Cost is one criterion that might potentially be added, but it should not be the primary concern with testing. In the fire and emergency medical services, instructors and training programs can face costly legal challenges because of the critical emergency tasks that trainees must perform. Training and testing programs are likely targets for litigation if a lawsuit making a liability claim is filed.

**Safety Tip**

As online learning occurs more frequently, it is important for students to be properly trained in basic ergonomics for keyboarding to prevent repetitive motion and other injuries.

| Questions to Be Answered by Test Bank Provider | Provider 1 | Provider 2 | Provider 3 |
| --- | --- | --- | --- |
| Do you claim to have valid and reliable test items in your test banks? | Yes __ No __<br>Notes: | Yes __ No __<br>Notes: | Yes __ No __<br>Notes: |
| Are there documentation and data available for determining test-bank validity and test reliability? | Yes __ No __<br>Notes: | Yes __ No __<br>Notes: | Yes __ No __<br>Notes: |
| Do your test banks comprehensively cover NFPA Professional Qualification Standards? If yes, do you have cross-reference tables to document the extent of coverage? | Yes __ No __<br>Notes: | Yes __ No __<br>Notes: | Yes __ No __<br>Notes: |
| How long have you been providing test banks? Is this your primary business focus? | Yes __ No __<br>Notes: | Yes __ No __<br>Notes: | Yes __ No __<br>Notes: |
| How many customers do you have, and can you provide user names and telephone numbers? | Yes __ No __<br>Notes: | Yes __ No __<br>Notes: | Yes __ No __<br>Notes: |
| Do you provide technical support for your test banks and software? If yes, then how? | Yes __ No __<br>Notes: | Yes __ No __<br>Notes: | Yes __ No __<br>Notes: |
| Is there regularly scheduled training for the test banks and software? If yes, how often and at what locations? | Yes __ No __<br>Notes: | Yes __ No __<br>Notes: | Yes __ No __<br>Notes: |
| Do you have technically competent staff members in testing technology? If yes, will you provide detailed résumés of their qualifications? | Yes __ No __<br>Notes: | Yes __ No __<br>Notes: | Yes __ No __<br>Notes: |
| What are your revision and updating strategies for your test bank? | Yes __ No __<br>Notes: | Yes __ No __<br>Notes: | Yes __ No __<br>Notes: |
| Do you provide qualified expert witnesses supporting your test bank in case of lawsuits? If yes, what are the costs for the service? | Yes __ No __<br>Notes: | Yes __ No __<br>Notes: | Yes __ No __<br>Notes: |

FIGURE 10.14 Test-bank validation checklist.

## Confidentiality of Test Scores

After the completion of the testing process, student scores must be maintained. These test results should be protected by strict security measures. Electronic results should be password-protected, and hard copies should be kept under lock and key. Test results should be released only with the permission of the student who has completed the testing process.

When you release test scores to a class, you should do so on an individual basis. This should be done in a private session one-on-one or in a personal letter to each student. The days of posting scores on a bulletin board, identified by students' Social Security numbers, are long gone.

**Safety Tip**

The proper handling of student information, including test grades, may be covered by laws that apply to both fire service instructors and their fire departments. Know your responsibility, and play it safe by maintaining strict confidentiality for all student information.

## Proctoring Tests

At this point, you should have attained both knowledge and skills in test-item writing and test development. Now, you should explore the other aspects of a sound testing program, such as proctoring of an exam. Proctoring tests involves much more than just being present in the testing environment; it requires specific skills to be performed professionally. Different types of tests require different proctoring skills and abilities.

### Proctoring Written Tests

Following are some suggested procedures for proctoring written tests:

- Arrive at least 30 to 45 minutes prior to the beginning time for the test.
- Make sure the testing environment is suitable in terms of lighting, temperature control, adequate space, and other related items.
- During the arrival of test takers, double-check those who should be in attendance by checking identification documents and record their presence. (You may develop a log or other means to document arrival time and departure time.)
- Maintain order in the testing facility. Discourage any activities that can be interpreted as cheating and do not provide any information to anyone that isn't also provided to the group as a whole.
- Maintain security of all testing materials at all times.
- Provide specific written test-taking rules and rules for behavior during the testing period, and give an oral review of all rules. Communicate all rules clearly.
- Remain objective with all test takers. Do not show any form of favoritism.
- Answer questions about the testing process and the test itself.
- Do not answer an individual question about the content of the test unless the information is shared with the entire group.
- Monitor test takers during the entire period of the test.
- Discipline anyone who becomes disruptive or violates the rules.
- Require all test-item challenges to be made prior to test takers leaving the room. Advise that their challenge will be noted and addressed after all of the tests are scored.
- Collect and double-check answer sheets or booklets to ensure all information is properly entered and any supporting materials are returned before allowing test takers to leave the room.
- Never leave the testing environment for any reason.
- Inventory all testing materials and return them to the designated person in the department.

### Proctoring Oral Tests

Oral tests are generally given in conjunction with performance tests. Even so, certain procedures need to be followed to maximize the effectiveness of the oral test:

- Determine the oral test items to be used and the type of oral question techniques that will be employed (e.g., overhead, direct, rhetorical).
- Make sure the oral test items are pertinent to the emergency task to be performed.
- Focus on the critical safety items and dangerous steps within the task to be performed.
- If the test taker misses the oral test item, redirect the question to another performer. If you are conducting a one-on-one oral test, provide the test taker with on-the-spot instruction. Safety knowledge and dangerous task performance steps must be clear and mastered before performance begins.
- Use a scoring guide for the oral test, and record any difficulties and lack of knowledge on the part of the test taker.
- If the group or individual has knowledge gaps regarding the critical safety items or dangerous performance steps, *do not* proceed to the performance test. More training and testing are required.
- Make a training record of all oral test results.

### Proctoring Performance Tests

Performance test proctoring requires considerably different skills than proctoring of written and oral tests. Specifically, proctors must have keen observational skills, technical competence for the tasks being performed, ability to record specific test observations including deficiency and outstanding performances, ability to foresee critical dangers that may lead to injury to the performer or others, and an objective relationship

## Performing a Horseshoe Hose Load

**Evaluator Instructions:** The candidate shall be provided with a fire apparatus, supply hose, and gloves.

**Task:** Peforming an horseshoe hose load.

**Performance Outcome:** The candidate shall demonstrate the ability to do a horseshoe hose load.

**Candidate Directive:** "Properly perform a horseshoe hose load."

| No. | Task Steps | First Test | | Retest | |
|---|---|---|---|---|---|
| | | P | F | P | F |
| 1. | To set up for a forward lay, the candidate places the male coupling in hose bed first at front corner. To set up for reverse lay, the candidate places the female hose coupling in hose bed first at front corner. | | | | |
| 2. | Lays first length of hose on edge against either wall of hose bed. | | | | |
| 3. | Makes a fold even with rear of hose bed. | | | | |
| 4. | Lays hose back to the front and around inside of hose bed so it approaches rear along opposite wall of hose bed. Makes a fold at rear of hose bed. Lays hose back around edge of hose bed inside first length of hose. | | | | |
| 5. | Continues to lay hose around edges of hose bed. Each fold of hose will take up decreasing amount of space in inside of horseshoe. | | | | |
| 6. | Once the center of horseshoe is filled in, the candidate begins second layer by bringing hose from front of hose bed and laying it around perimeter of hose bed. | | | | |
| 7. | Completes additional layers using same pattern as was done for first layer. | | | | |

**Retest Approved By:** **Retest Evaluator:**

FIGURE 10.15 A skills evaluation sheet used in the evaluation step of instruction can be a useful tool in evaluating and properly documenting a performance.

to the test takers FIGURE 10.15. Following are some suggested procedures including pre-test preparations:

- Arrive at the test site at least 1 hour prior to the start of the test.
- Check the test environment to make sure all needed tools, materials, and props are present and in good working order.
- Determine whether the performance test will require an oral test before actual performance. (Obtain oral test items if required.)
- Check the test takers' identification as they arrive and record their presence.
- Review the test procedures and ask for questions from the test takers.
- Verify that all test takers know what will be expected during performance of the task.
- Begin the performance test. Have the students tell you exactly what they will be doing *before* they do each step in the performance. (This critical step can help prevent an accident or improper performance before it happens.) In certification testing, missing safety-related steps or not knowing about critical performance issues that may cause injury or damage to equipment is cause to terminate testing and refer

the test taker for more training. The same is true for the final performance test in a training program that is expected to lead to licensure or certification.
- Remain silent and uninvolved in the task performance. Your job is to verify safe and competent performance. Be like a chair.
- Record test results. You may provide technical information and critique the performance if your institution or agency permits proctor interaction after the test.
- Provide test results to the person(s) designated to receive them.

## Cheating During an Exam

Student cheating during the examination process is an unfortunate by-product of the evaluation process. How this issue is addressed should be based on department policy developed by human resources personnel, the agency's legal advisor or attorney, and the chief of the department. Cheating by an individual reflects on the character of that person and, in many departments, can affect employment status, raises, or promotions.

In general, a policy that addresses cheating will explain what the instructor or proctor should do if a student has been observed cheating. In most cases, if a fire service instructor observes a student cheating, he or she should ask the student to leave the testing area. The student should be asked for an explanation of his or her conduct. If the fire service instructor feels that the student was, indeed, cheating, the student should be asked to leave the test site entirely or be given a verbal warning and be allowed to continue with testing.

The fire service instructor must document what was observed, what the student's response was, and what the fire service instructor allowed the student to do next. This information should be passed on to a supervisor, who will review what happened and then follow up as necessary with additional interviews of the fire service instructor and the student involved. The supervisor will determine what happens from this point forward. If it is determined that the student did cheat, then department policy will address the consequences. If it is determined that the student did not cheat, then arrangements should be made to allow the student to retake the test. Regardless of the outcome of any particular situation, the department should have a written policy that dictates how to handle possible cheating during an examination.

### Safety Tip

Determine whether your organization has a policy that addresses how to handle suspicions of cheating and follow it to the letter. Failure to do so is unfair to the student and can get you into hot water.

## Some Legal Considerations for Testing

Testing of persons for completion of training programs, job entry, certification, and licensure are governed by certain legal requirements and professional standards in the United States. These legal requirements and professional standards are readily available and should be important reference documents for anyone involved in the development of test items, construction of tests, administration of tests, analysis and improvement of tests, and maintenance of testing records. The three most important reference documents are *Uniform Guidelines for Employee Selection*, published by the Equal Employment Opportunity Commission; *Standards for Educational and Psychological Testing*, published by the American Psychological Association; and *Family Educational Rights and Privacy Act*, published by the U.S. Printing Office and other governmental organizations. Each of these references should be available in every division of training library for fire service instructors to read and become familiar with.

*Uniform Guidelines for Employee Selection* focuses primarily on employee selection/promotion procedures. Training and testing leading to hiring, promotion, demotion, membership in a group such as a union, referral to a job, retention on a job, licensing, and certification are all covered under these guidelines. According to this document, written and performance tests in such training programs should address job-related qualifications. It is up to the user of the testing materials to make sure that tests meet this criterion. If the tests are acquired from a professional publisher or distributor of tests, then that organization must be able to provide you with the necessary job-validity information. For example, if a question is based on the NFPA Professional Qualification Standards and its content is current, then the test item would be defined as containing job-validity information.

As a fire service instructor, it is important that you address job-content validity issues even for those test items that you personally develop and use on a day-to-day basis.

Almost all training in the fire service leads to some sort of certification, pay raise, or potential for promotion, so job-content–related testing should be paramount when using any form of testing.

## Providing Feedback to Students

As adult learners, we appreciate rapid feedback on our performance in the learning environment. When students are evaluated using written examinations, the instructor should make every attempt to provide the results of the evaluation in a timely fashion and when possible, take extra time to allow the student to review errors in their test. Simply giving the students their score really does not allow for a complete evaluation of the learning process, meaningful feedback means that the student has the opportunity to see where they have the opportunity to improve their knowledge or skill level and in the best case, the instructor can provide the student additional resources for additional learning or practice. Levels of acceptable performance can be set in percentage-based systems and relearning or additional training on missed objectives is typically required before the student is given an opportunity to retest on missed areas. Some state certification entities may require this step before an individual retests after a test failure.

In psychomotor or hands-on training situations, the instructor should have the opportunity to comment, correct or reinforce the students performance almost as it happens. If a step in the performance is missed that places the fire fighter in danger in any way, the instructor must immediately stop the skill demonstration and take corrective action. Other less critical errors provide the instructor an opportunity to have the student repeat the evaluation or for the instructor to demonstrate the proper steps in the procedure. Typically, in these evaluations, the student must perform all steps in the skills checklist in order to successfully complete the evaluation.

Instructors can collect data from the pool of students who have completed written examinations and review the success and failure rates of each question and in turn, evaluate any shortcomings in the delivery or learning process that may have occurred. An instructor's job is not just to present and evaluate, it is to complete the learning process by providing meaningful feedback to the students. Chapter 2 discusses the legal responsibility and the many *Ethics Tips* provided in this text relate to the responsibility of the instructor to take corrective action on many elements of student performance. Feedback must be applied consistently and in a standard fashion to be effective. Appendix C, Resources for Fire Service Instructors, provides several tools such as a Personal Improvement Agreement that can become a "contract" with a student to improve their performance if provided with the right types of coaching, counseling and resources. NFPA 1500, *Standard on Fire Department Occupational Safety and Health Program*, cites that "each member has the responsibility to maintain proficiency in their skills and knowledge and to avail themselves to the professional development provided to them" (5.1.9). The instructor should make sure that these opportunities to maintain their proficiency are available through feedback on training and job performance. If this is not done effectively, consider the possibility that information that a student did poorly on could be information or skills that in an emergency situation become the difference between safety and survival. Written feedback forms also serve as a record of the training and can be used to report test results to others in the department.

# Wrap-Up

## Chief Concepts

- Testing plays a vital role in training and educating fire and emergency service personnel.
- Tests should be developed for three purposes:
  - To measure student attainment of learning objectives
  - To determine weaknesses and gaps in the training program
  - To enhance and improve training programs by positively influencing the revision of training materials and the improvement of instructor performance
- Test development activities should adhere to a common set of procedures.
- Test analysis occurs after a test has been administered.
- Testing consists of two activities:
  - The preparation of test questions using uniform specifications
  - Quantitative/qualitative analysis to ensure that the test questions function properly as measurement devices for training
- Test specifications permit instructors and course designers to develop a uniform testing program made up of valid test items as a basis for constructing reliable tests.
- Eight types of written test questions are possible: multiple choice, matching, arrangement, identification, completion, true/false, short-answer essay, and long-answer essay.
- Performance testing is the single most important method for determining the competency of actual task performances.
- The confidentiality of test scores must be maintained.
- Proctoring tests involves much more than just being in the testing environment.
- A policy that addresses cheating will explain what the instructor should do if a student is observed cheating.
- It is important to provide the results of an evaluation to the student in a timely fashion.

## Hot Terms

**Essay test** A test that requires students to form a structured argument using materials presented in class.

**Face validity** Validity achieved when a test item is derived from an area of technical information by an experienced subject-matter expert who can attest to its technical accuracy.

**Job content/criterion-referenced validity** Validity achieved through the use of a technical committee of job incumbents who certify that the knowledge being measured is required on the job and referenced to known standards.

**Oral test** A test in which the answers are spoken either in response to an essay-type question or in conjunction with a presentation or demonstration.

**Performance test (skills evaluation)** A test that measures a student's ability to do a task under specified conditions and to a specific level of competence.

**Qualitative analysis** An in-depth research study performed to categorize data into patterns to help determine which test items are acceptable.

**Quantitative analysis** Use of statistics to determine the acceptability of a test.

**Reference blank** The place where the current job-relevant source of the test-item content is identified.

**Reliability** The characteristic of a test that measures what it is intended to measure on a consistent basis.

**Subject-matter expert (SME)** An individual who is technically competent and who works in the job for which test items are being developed.

**Systems Approach to Training (SAT) process** A training process that relies on learning objectives and outcomes-based learning.

**Technical-content validity** Validity achieved when a test item is developed by a subject-matter expert and is documented in current, job-relevant technical resources and training materials.

**Validity** The ability of a test item to measure what it is intended to measure.

**Written test** Any of several types of test items, such as multiple choice, true/false, matching, short-answer essay, long-answer essays, arrangement, completion, and identification test items. Answers are provided on the test or a scannable form used for machine scoring.

# Fire Service Instructor *in Action*

You have been charged with the task of developing the quizzes and final exam for your department's Fire Officer I course, which meets the criteria set forth in NFPA 1021, *Standard for Fire Officer Professional Qualifications*. The department wants to ensure that the evaluation tools cover all of the requirements of the standard.

**1.** To ensure the test is objective, which of the following test items will you not use?
  **A.** Written
  **B.** Completion
  **C.** True/false
  **D.** Essay

**2.** Laboratory exercises, scenarios, and job performance requirements (JPRs) that test the student's ability to perform tasks are all types of
  **A.** Written examinations.
  **B.** Oral examinations.
  **C.** Performance tests.
  **D.** Multiple-choice tests.

**3.** What is the most important reason for giving quizzes during the course?
  **A.** To average out the final exam scores
  **B.** To make students read their books as they go
  **C.** To identify areas that need more concentration
  **D.** To test areas not covered by the final exam

**4.** If students are able to pass the exam but are not capable of performing well as a company officer, the exam would not be considered
  **A.** Valid.
  **B.** Reliable.
  **C.** Consistent.
  **D.** Unusual.

**5.** If all students in one class score high, yet all students in another class score low, the test may not be
  **A.** Valid.
  **B.** Reliable.
  **C.** Consistent.
  **D.** Unusual.

**6.** What is the major problem with two-answer selection tests such as true/false questions?
  **A.** They tend to be ambiguous.
  **B.** They tend to be subjective.
  **C.** They are vulnerable to the "guessing" factor.
  **D.** They are difficult to construct.

**7.** If you decide to use a test-item bank,
  **A.** You can randomly generate the exam and it will be valid.
  **B.** You must carefully analyze the test bank to make sure the test items are properly validated.
  **C.** The students will probably not be able to pass the exam.
  **D.** You will not be able to provide a Web-based version of the exam.

**8.** As a fire instructor, you will need to ensure your testing process conforms to *Uniform Guidelines for Employee Selection*. Which organization publishes this document?
  **A.** Equal Employment Opportunity Commission
  **B.** Department of Labor
  **C.** International Association of Fire Chiefs
  **D.** American Psychological Association

# Evaluating the Fire Service Instructor

# CHAPTER 11

## NFPA 1041 Standard

**Instructor I**

NFPA 1041 contains no Instructor I Job Performance Requirements for this chapter.

**Instructor II**

**5.2.6** Evaluate instructors, given an evaluation form, department policy, and JPRs, so that the evaluation identifies areas of strengths and weaknesses, recommends changes in instructional style and communication methods, and provides opportunity for instructor feedback to the evaluator. [p. 192–202]

**(A) Requisite Knowledge.** Personnel evaluation methods, supervision techniques, department policy, and effective instructional methods and techniques. [p. 192–202]

**(B) Requisite Skills.** Coaching, observation techniques, and completion of evaluation forms. [p. 197–198]

**5.5.3** Develop a class evaluation instrument, given agency policy and evaluation goals, so that students have the ability to provide feedback to the instructor on instructional methods, communication techniques, learning environment, course content, and student materials. [p. 201–202]

**(A) Requisite Knowledge.** Evaluation methods and test validity. [p. 193–197]

**(B) Requisite Skills.** Development of evaluation forms. [p. 194–195]

## Knowledge Objectives

After studying this chapter, you will be able to:

- Describe the department's policy or procedure in respect to fire service instructor evaluations.
- Describe the methods for fire service instructor evaluation.
- Describe the evaluation process.
- Describe the role of feedback in the fire service instructor evaluation.
- Describe how to develop a class evaluation form.

## Skills Objectives

After studying this chapter, you will be able to:

- Perform an evaluation of a fire service instructor.

# You Are the Fire Service Instructor

You have been recently promoted into the training division. Many of your department officers are fire service instructors who are routinely assigned to prepare and deliver in-service classes. Such a class is being given at your department's training center to approximately 20 fire fighters and fire officers. Per your department's policies, you will be conducting a fire service instructor performance review during this class. The fire service instructor you will review is a popular senior captain who is known to be very good on the drill ground, but who sometimes struggles in the classroom. His presentation is on a subject with which he is very familiar. When he sees you enter the classroom, the instructor appears to be nervous. He begins the lecture before you can speak to him.

1. What should you consider when evaluating a fire service instructor?
2. By the first break you are concerned because, while the course outline is very good, the instructor's delivery comes off as halting and unprofessional. As the evaluator, what should you do?
3. By the end of the lecture, it is clear that the fire service instructor did not meet the standards for teaching. How will you resolve this dilemma?

## Introduction

At the historic Wingspread Conference in 1966, participants agreed on the need for the fire service to attain professionalism through education. In 1971, the Joint Council of National Fire Service Organizations (JCNFSO) created the National Professional Qualifications Board for the Fire Service (NPQB) to facilitate the development of nationally applicable performance standards for uniformed fire service personnel using the NFPA standards process. One of the four technical committees developed to facilitate this process was the Committee for Fire Service Instructor. In 1972, the NFPA established a committee charged with the development of a standard for the qualifications for a fire service instructor.

The first standard produced by the committee was issued in 1976 as NFPA 1041, *Standard for Fire Service Instructor Qualifications*. The goal of the standard was to establish:

> clear and concise job performance requirements that can be used to determine that an individual, when measured to the standard, possesses the skills and knowledge to perform as a fire service instructor.

NFPA 1041, *Standard for Fire Service Instructor Qualifications*, obligates the Instructor II to:

> evaluate instructors, given an evaluation form, department policy, and job performance requirements, so that the evaluation identifies areas of strengths and weaknesses, recommends changes in instructional style and communication methods, and provides opportunity for instructor feedback to the evaluator.

In reality, the fire service instructor is evaluated at all times during his or her performance, either formally or informally. The informal evaluation occurs during breaks or before or after class, when students compare their current fire service instructor to previous instructors from both inside and outside of the fire service. Additionally, formal evaluations are done for certification, as part of a fire department's improvement plan and as part of a self-improvement plan for the fire service instructor.

Fire fighter safety is a critical reason for fire service instructor evaluation. Since 1977, more than 240 fire fighters have died during training. From 2000 to 2005, fire fighter deaths during training averaged more than 10 per year. Some of these deaths occurred due to cardiac arrest or some other medical condition; others occurred as a result of trauma sustained during a training evolution. Approximately 50 percent of these deaths occurred on the drill ground. Given the risks inherent in training, fire service instructors need to be evaluated in each and every setting—not just the traditional classroom.

To become an evaluator, job performance requirements (JPRs) must be met. Often, meeting the JPRs includes demonstration of an evaluation of a fellow fire service instructor's presentation and/or reviewing the evaluation with the fire service instructor who was evaluated. The goal is to help the fire service instructor identify strengths and weaknesses, and become better through recommended improvements.

### Teaching Tip

NFPA 1041 requires that the Fire Service Instructor II be capable of identifying another fire service instructor's strong and weak points, and provide guidance for improvement.

## Fire Department Policy and Procedures

NFPA 1041 references departmental policies and procedures as one of the criteria for an evaluation. Many departments do not have an adopted standard for certifying fire service instructors; instead, they use state or national certification requirements for this purpose.

## Fire Department Requirements

NFPA 1041 requires that the department or authority having jurisdiction (AHJ) use qualified individuals as fire service instructors. The qualifications include more than just being knowledgeable on the subject: NFPA 1041 also expects that the fire service instructor will have the capability and skill level to appropriately demonstrate how to perform the skills included in the training session. The level of competency of the fire service instructor must be identified in a policy developed and enforced by the AHJ. In addition, this policy must include a method for verifying the qualifications and competency of the fire service instructor.

A department's policy or procedure for fire service instructor evaluation includes many components. The policy should identify who does the evaluation and ensure that the evaluations remain confidential. Confidentiality is important in that it allows the evaluator to be critical without embarrassing the fire service instructor being evaluated. A timetable needs to be established for evaluations, so that fire service instructors are evaluated initially and then periodically. The policy should establish an evaluation process for practical instruction as well as classroom presentations.

The fire service instructor evaluation policy should address the fire service instructor's preparations to deliver the material as well as his or her actual presentation of that material. Preparations include the reservation and preparation of the facility and the equipment required. The fire service instructor should assure that the classroom is appropriately arranged and that any necessary equipment is in place and in working order. He or she should have reviewed the lesson plan and adapted it for the target audience. Also, the fire service instructor needs to assure that the class starts on time.

Another issue that must be addressed is attire; the fire service instructor must be wearing attire appropriate for the setting. If the coursework will address practical evolutions, the fire service instructor should wear the same level of protection as that required of the students FIGURE 11.1. The fire service instructor should set a positive example for the students relative to personal protective equipment (PPE) on the drill ground. In the classroom, the fire service instructor should wear attire consistent with the attire expected of the students.

The evaluation policy also should include information on how the presentation is to be delivered. It should require that the fire service instructor follow the lesson outline, keep the class moving, use appropriate audio, video, or display equipment, finish material delivery in the allotted time frame, and deliver a summary. The policy must outline positive traits expected of the fire service instructor, including being unbiased, encouraging open discussion, being understandable, wearing neat attire, remaining flexible, and being honest. Undesirable traits, including the use of verbal fillers or undesirable mannerisms, should be identified and discouraged.

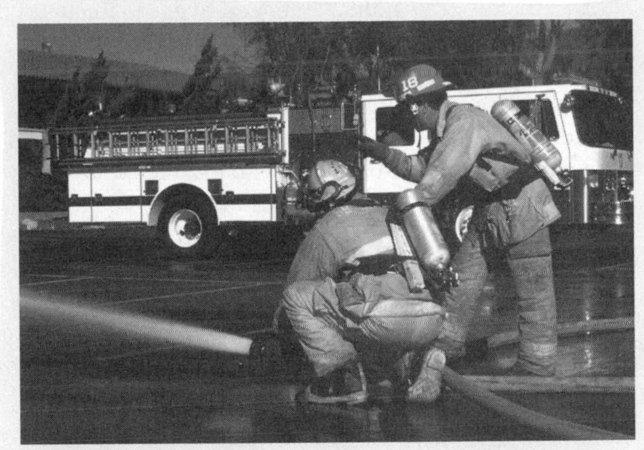

FIGURE 11.1 The fire service instructor should be in the same level of PPE as the students.

The department policy for evaluating a fire service instructor can be as simple as a short paragraph stating when evaluations will be done, by whom, and in what form; conversely, a detailed policy may enumerate the entire instructional process. In either case, the department's polices/procedures must be a part of the evaluation process and then be followed as outlined.

## Methods for Fire Service Instructor Evaluation

Fire service instructor evaluations may be a requirement for instructors' promotion. In some departments, you must obtain a basic level of fire service instructor certification to become a fire officer. Fire service instructor evaluations may also be required for state or national certification, as most states support a certification process for fire service instructors. The National Fire Academy and other institutions require initial and periodic evaluation. In some cases, fire service personnel may have to pass a fire service instructor evaluation process before they are allowed to instruct students. This evaluation may be done by a supervisor, a peer, or a student, or the instructor may do a self evaluation. Each evaluator puts his or her own knowledge and experience to work to complete the evaluation appropriately.

## Formative Evaluation

The formative evaluation process is typically conducted for the purpose of improving the fire service instructor's performance by identifying his or her strengths and weaknesses. For a student, this evaluation may take the form of a progress test or unit exam so that learning is measured and feedback can be given on the student's progress to date. For a fire service instructor, a formative evaluation may be done by either a supervisor or another fire service instructor for the purpose of professional development.

The formative evaluation is intended to refine the instructor's delivery and better prepare the fire service instructor to deliver the course. For this reason, it is often perceived as less threatening than other evaluations. A typical formative evaluation is designed to improve the fire service instructor's abilities by identifying individual skills or deficiencies and allowing for the development of improvement plans.

A formative evaluation contains criteria that may allow the evaluator to review and qualify the overall performance of the fire service instructor relating to the expected outcomes of the training session. The instructor's ability to meet these criteria may be measured by the instructional methods used during the training such as visual aids, student participation, and the application of testing instruments to the learning objectives. The evaluator may also participate in observation of student performance and take notes on the fire service instructor's interactions with the students. Much of formative evaluation centers on instructional technique, training content, and the delivery methods used by the fire service instructor. This information is then compared to student success rates.

## Summative Evaluation

A **summative evaluation** process measures the student's achievements, through testing or completion of evaluation forms, to determine the fire service instructor's strengths and weaknesses. Many applications of summative evaluations take place at the end of the training process. Such an evaluation is normally conducted for the purpose of making personnel decisions about fire service instructor certification, merit pay, promotion, and reassignment. Student evaluations of fire service instructors are commonly used for assessment, as part of the summative evaluation process. Because students may have different reasons for being in the class, precautions are necessary when using this type of evaluation. Summative evaluations usually are done by an administrator, such as the training director or training officer, rather than by a supervisor or another fire service instructor.

## Student Evaluation of the Fire Service Instructor

Students are in a position to rate the increased knowledge they have gained and their classroom experience as it relates to a particular fire service instructor. They may provide valuable information about other factors not readily available during an evaluation, such as the fire service instructor's punctuality, use of audiovisual equipment, and typical classroom demeanor.

Some departments include an evaluation form for students to complete at the conclusion of the course. Different delivery schedules and multiple fire service instructors delivering training over a long period of time may require evaluation forms to be distributed earlier so that students can complete the evaluation while specific instructor characteristics are still fresh in their minds. In general, a course evaluation typically includes a section that allows students to evaluate the fire service instructor, classroom setting, instructional material, handouts or audiovisual material, and ability of the material taught to meet their needs. All of these areas are usually rated on a scale of 1 to 5 or a scale of "strongly agree" to "strongly disagree" FIGURE 11.2.

This information can be viewed one evaluation at a time or by assigning a value to each response to provide statistical

University
Fire Fighter Training Instructor
EVALUATION

DIRECTIONS
- Use black lead pencil or pen.
- Fill the circles completely.
- Provide additional comments on back of form.

INSTRUCTOR ASSESSMENT

| | | Strongly Disagree | Disagree | Unsure | Agree | Strongly Agree |
|---|---|---|---|---|---|---|
| 1. Lead Instructor: | Related the course material to my needs | ○ | ○ | ○ | ○ | ○ |
| | Used course materials successfully | ○ | ○ | ○ | ○ | ○ |
| | Knew the subject thoroughly | ○ | ○ | ○ | ○ | ○ |
| | Encouraged student participation | ○ | ○ | ○ | ○ | ○ |
| | Answered students' questions | ○ | ○ | ○ | ○ | ○ |
| 2. Assisting Instructor: | Related the course material to my needs | ○ | ○ | ○ | ○ | ○ |
| | Used course materials successfully | ○ | ○ | ○ | ○ | ○ |
| | Knew the subject thoroughly | ○ | ○ | ○ | ○ | ○ |
| | Encouraged student participation | ○ | ○ | ○ | ○ | ○ |
| | Answered students' questions | ○ | ○ | ○ | ○ | ○ |
| | Was available outside of the classroom | ○ | ○ | ○ | ○ | ○ |

FIGURE 11.2 Sample of a bubble sheet evaluation form.

# Applying the JOB PERFORMANCE REQUIREMENTS (JPRs)

Evaluation of a fire service instructor and the content of a presentation serves as a form of quality assurance. These measurements of the fire service instructor's skills and the impressions of how the course material helped students understand and apply the learning objectives are an important part of a fire service instructor's professional development. A student review of the fire service instructor's attributes is one method of performing an evaluation. Another strategy is to have a senior fire service instructor conduct an in-person review of the course delivery.

## Instructor I

All instructor levels should develop an appreciation for how course and instructor evaluations benefit the instructional process. At the Instructor I level, you should consider all feedback provided in an evaluation as a measure of professional development and work to improve on any constructive criticisms provided while building on positive attributes.

## JPRs at Work

NFPA does not identify any JPRs that relate to this chapter. Nevertheless, the Instructor I should understand how to utilize information provided during the evaluation process.

## Instructor II

Evaluations of instructor performance are a type of personnel evaluation, and their construction and use should reflect the intent of any evaluation—namely, to improve the performance of the person being evaluated. The Instructor II will develop class and instructor evaluations and conduct evaluations of other instructors in an effort to improve both the instructor and the course elements.

## JPRs at Work

Develop evaluation instruments and evaluate instructors in an effort to improve instructor skills and the quality of the course materials by providing feedback on class presentation elements.

## Bridging the Gap Between Instructor I and Instructor II

Personnel who meet the Instructor I JPRs will be able to use their experiences in being evaluated when they transition to Instructor II by recalling the elements of the evaluations that they were able to put to use to improve themselves. When conducting evaluations, the Instructor II should be professional and constructive in identifying areas for improvement. Individual strengths and weaknesses should be highlighted, and both instructor levels should use this step in the instructional process in a positive manner.

## Instant Applications

1. Review the content of a course evaluation form and identify those elements that relate to fire service instructor qualities as well as those elements that relate to the features of the course.
2. Develop a list of desirable fire service instructor qualities. Rank these qualities in order based on which create the best learning environment.
3. Consult available standard personnel evaluations and identify any features that could be used to evaluate a fire service instructor during a training session.

averages. Optical scanning device readers can computerize these results and make the statistics easier to interpret. How effective this statistical information is depends on the effort that the department wishes to devote to its interpretation. If the department simply harvests the information, rather than using it for meaningful consideration, then the process is flawed. Another potential drawback is that student evaluation forms may ask questions that students cannot or do not judge reliably (i.e., many times these evaluations may indicate that the class should be shorter when the lesson plan actually calls for a longer presentation). The evaluation standards should be objective, identified prior to the evaluation, focused on the fire service instructor's performance, and part of an improvement program for the fire service instructor.

Some students may view the evaluation as an opportunity to take a shot at the fire service instructor. Others may just go through the motions of completing the evaluation, without giving it much thought or consideration. Some departments account for these problems by evaluating the students participating in the class—that is, students can be compared in terms of participation, knowledge, and skill level. This information can then be used when reviewing the students' evaluations for the fire service instructor and the course.

## Evaluation of Student Progress

To evaluate the fire service instructor, it may be necessary to evaluate student progress. Evaluations of student progress may be formal or informal depending on the situation. Students' progress may be evaluated for any of the following reasons related to fire service instructors:

- Certification
- Promotion
- Departmental requirements
- Personnel evaluation process
- Word of mouth
- Self-evaluation

Ultimately, the evaluation needs to provide an opportunity for the fire service instructor to identify student strengths and weaknesses. Once identified, the strengths can be promoted and the weaknesses addressed.

### Ethics Tip

Fire service instructors tend to naturally become acquaintances because they have a common passion and often attend the same seminars and training sessions. Over time, friendships will develop, and you may be put in the position of having to evaluate a friend. Evaluating a close friend can be difficult. If you find yourself in such a position, you will naturally want to point out the positives and avoid mention of any negatives. Do you have the fortitude to evaluate a friend honestly and fairly? As an evaluator, it is your responsibility to the department, to the students, and to the fire service as a whole to conduct evaluations fairly and honestly. Friendship is a personal relationship that cannot enter into the evaluation process.

### Certification

Many departments require that their fire service instructors have some form of certification, either state or national. The intent of this requirement is to establish a level of competency. Certification typically requires coursework in educational methodology and instructional technique and a demonstration of delivery techniques. NFPA 1500, *Standard on Fire Department Occupational Safety and Health Program*, requires that instruction be provided to fire fighters by a qualified fire service instructor who has met the minimum performance requirements of NFPA 1041 and the Fire Service Instructor I as well as allowing for local AHJ interpretations.

### Promotion

Achieving fire service instructor certification is one of the steps required before receiving a promotion in many departments. Departments that follow NFPA 1021, *Standard for Fire Officer Professional Qualifications*, require that all fire officers meet the requirements of Fire Service Instructor I prior to being eligible for certification at the Fire Officer I level. According to most fire service leaders, to be in charge of a group of people, you must have the knowledge, experience, composure, and skills to lead them in training as well as on the emergency scene. The inclusion of the requirement to be certified as a Fire Service Instructor I to become a Fire Officer I (usually the company officer level) is an acknowledgment of the importance of providing quality training to members of the fire company.

### Departmental Requirements

Fire department policies or procedures may require that fire service instructors be evaluated on a regular basis. This requirement indicates a desire by the leadership of the department to fulfill the requirements of NFPA 1021. The ability to instruct a fire company through in-service training sessions is often a key component of a company officer's job description and one of the reasons that fire service instructor certification is considered a co-requisite to officer certification—the two go hand in hand.

### Personnel Evaluation Process

Going through the personnel evaluation process is mandatory in many fire departments. Evaluating the fire service instructor on an annual or more frequent basis ensures that the fire service instructor doesn't become complacent or too reliant on outdated methods, materials, or technology. The ability to present up-to-date and relevant materials and general instructor abilities may be part of an evaluation. If the full-time job of the fire service instructor is to provide training and instruction, a more comprehensive evaluation should be completed based on those elements of the job description. This includes evaluation on presentation skills, program management, record and report keeping, and program delivery skills.

### Word of Mouth

If you have ever had the opportunity to be around a corner or on the other side of a wall during a break where the students are discussing the class, you are probably aware of the "word of mouth" phenomenon. Students are constantly evaluating you and your

**Teaching Tip**

Informal feedback from casual conversations by students can be an effective measure of fire service instructor performance, but caution should be used when interpreting those remarks. A single opinion may not necessarily represent the views of all students.

class, and their evaluation continues beyond the classroom. Just as in the collegiate setting, if students have a choice of instructors, then they will do their best to attend the classes presented by the instructors with the best reputations. The word of mouth method of evaluation isn't beneficial to the fire service instructor if that word of mouth never reaches him or her, however.

### Self-Evaluation

It can be beneficial for the instructor to periodically evaluate his or her own performance. These evaluations may be informal (a mental review) or formal (a supplied form). Whether informal or formal, this self-evaluation should be as objective as possible. Sometimes you can be too critical of yourself, which is just as detrimental as not identifying your faults. Experienced fire service instructors are often very critical of themselves. Like their students, instructors must try to improve their knowledge of a subject, their ability to keep students involved, and their ability to interact with the class.

# Evaluation Process

## Preparation

To make an evaluation of a new fire service instructor's performance a positive event, start by selecting a topic from the schedule with which you are familiar and enjoy teaching. Prior to conducting an evaluation, you need to review the evaluation criteria and its relevance to the job descriptions; review the lesson plans and supporting material; and then schedule the date, time, and location for the evaluation (if deemed appropriate by the AHJ) FIGURE 11.3.

If you plan to schedule your visit to the session with the fire service instructor, state the purpose of your visit and evaluation in positive terms. Identify the areas that you will be reviewing, including the delivery of the provided lesson plan material, the fire service instructor's ability to apply the material to the class, adherence to department and accepted safety practices, and the instructor's overall presentation skills. Encourage the fire service instructor to present normally and to not attempt to involve you in the delivery process. Be sure to arrive early enough to determine whether the fire service instructor is properly prepared. Discuss the preparation process with the fire service instructor prior to the start of class.

## Observations during the Evaluation

During the evaluation, look at the fire service instructor's entire skills set. By using all sources of information, you will give a more complete analysis of the fire service instructor's

FIGURE 11.3 Prior to the evaluation, review the lesson plans.

**Theory into Practice**

Find a standardized method of performing evaluations if one is not provided by your AHJ. A standardized method ensures a fair and consistent evaluation of all fire service instructors and ensures that learning opportunities are not missed. To perform an evaluation, follow these steps:

1. Identify the appropriate method for the evaluation.
2. Select or develop a form to be used to record the evaluation.
3. Familiarize yourself with the evaluation form.
4. Read the course outline lesson plan.
5. Schedule a time and location for the evaluation.
6. Meet with the fire service instructor prior to the class to explain the intent of the evaluation.
7. Make the atmosphere as friendly as possible. Try to alleviate the fire service instructor's anxiety.
8. Observe the fire service instructor delivering course content.
9. Take notes.
10. Complete the form.
11. Review the fire service instructor's performance with him or her.

performance. During classroom training evaluations, areas to evaluate include the instructor's dress: Is the instructor's attire professional, compliant with policy, and appropriate for the audience? The demeanor of the fire service instructor should be professional, outgoing, approachable, and appropriate for audience: Does the fire service instructor appear to be engaging the audience? Does he or she inspire confidence and respect?

When evaluating the presentation, note the fire service instructor's ability to relate the topic directly to the students' needs. Note any mannerisms, language usage, and comfort level with technology. Do you feel that you are watching a black-and-white, still-picture show or a full-color, interactive presentation?

**Safety Tip**

If you are evaluating a session that involves PPE, you should wear the same level of PPE as the students and the fire service instructor.

The use of lesson plan and use of the various components of the lesson plan should be evaluated part by part. The lesson plan components should be reviewed in terms of how they were used and whether the presentation was delivered in a consistent manner. Using a copy of the lesson plan being presented, you can also review how well the instructor followed the lesson plan.

During practical or drill-ground evaluations, the overall safety of the students is the major focus of the evaluation. Fire service instructors have the twin responsibilities of presenting training in a safe manner and of teaching safe practices. When reviewing a fire service instructor during practical sessions, consider whether he or she was able to anticipate problems in the delivery of instruction such as noise, weather, or equipment availability.

Certain types of training sessions, such as live fire exercises or complex rescue scenarios, may require the assignment of a safety officer and should always follow NFPA guidelines when applicable. Both the fire service instructor and the students must dress appropriately for activity, wear PPE correctly, and clean or inspect the PPE after heavy use. The best review of a practical training session is to determine how closely the fire service instructor's modeling during the training session matched the real fire-ground application of the skill.

**Safety Tip**

Closely scrutinizing a fire service instructor's adherence to student safety guidelines should be paramount.

## Lesson Plan

The lesson plan is the fire service instructor's road map. It is occasionally appropriate to stray off the road, providing that the side trip benefits the overall experience and enhances student knowledge. The fire service instructor must be familiar with the lesson plan and its learning objectives for the class—a key point to keep in mind during the evaluation.

## Forms

Because of the diverse reasons for which evaluations are undertaken, a variety of printed forms are used to document them. Fire service instructor evaluation forms vary widely, and some are better than others. When a department adopts a certain form, it must be used during your evaluation **FIGURE 11.4**. Before adopting an evaluation form, it is prudent to review several different versions of forms to determine which content and criteria best fit the needs of the department.

**Theory into Practice**

As the old saying goes, "The job's not over until the paperwork is done." Thorough and accurate records are essential to maintain credibility and help ensure fairness. Procrastination only makes the job harder. To complete a form for an evaluation, follow these steps:

1. Review the form.
2. Have appropriate writing materials.
3. Take notes and/or fill in the appropriate areas of the form.
4. Review the fire service instructor's performance with him or her.
5. Submit paperwork to AHJ as directed or required.

## Feedback to the Instructor

Following the evaluation, the fire service instructor should be given feedback on the strengths and weaknesses of the evaluated class. This step emphasizes the coaching aspect of the evaluation. Feedback helps improve performance and build confidence. Common examples of strengths cited usually relate to the instructor's expertise or knowledge of the subject matter. Most fire service instructors work very hard to know the material before presenting; indeed, they become content experts. Weaknesses often show up in the areas of eye contact, technology usage, and class participation skills. Each of these deficiencies can be built upon and improved through practice and experience. Of course, without an effective evaluation measure, the fire service instructor may never know which areas need improvement.

**Teaching Tip**

As an instructor, it is a good practice to retain copies of your evaluations and to periodically review them, especially in cases where you are teaching the same course. You can build upon your strengths and improve your class delivery based on previous comments.

## Evaluation Review

After the evaluation is complete, take some time to review the evaluation form or notes you have taken. Then review your notes in person with the fire service instructor **FIGURE 11.5**. Specifically consider the fire service instructor's strengths and weaknesses. During the review, strive to show professionalism by maintaining a level of formality appropriate for the situation.

# Southwest United Fire Districts Training Division Instructor Evaluation Form

Please shade the letter that indicates your evaluation of the course/instructor feature using #2 pencil.

| **Instructor Attributes** | **Excellent** | **Very Good** | **Good** | **Fair** | **Poor** |
|---|---|---|---|---|---|
| Ensure that the daily classroom and practical training evolutions were conducted in a safe and professional manner | (1A) | (1B) | (1C) | (1D) | (1E) |
| Ensure that all learning objectives were covered with more than adequate resource information | (2A) | (2B) | (2C) | (2D) | (2E) |
| Coordinate the activities of all students so that the class flow and pace was not interrupted | (3A) | (3B) | (3C) | (3D) | (3E) |
| The instructor was prepared to present the program | (4A) | (4B) | (4C) | (4D) | (4E) |
| Provide documentation to your progress in class and in general on a regular basis | (5A) | (5B) | (5C) | (5D) | (5E) |
| Monitor all class and practical activities and provide a structured learning environment | (6A) | (6B) | (6C) | (6D) | (6E) |
| Provide a positive role image to you at all times and conduct themselves in a professional manner | (7A) | (7B) | (7C) | (7D) | (7E) |
| Was accessible and responsive to your individual needs as a student in this program | (8A) | (8B) | (8C) | (8D) | (8E) |
| Was actively involved in your training and education throughout the program | (9A) | (9B) | (9C) | (9D) | (9E) |
| Worked to improve program content, policy and procedure | (10A) | (10B) | (10C) | (10D) | (10E) |
| The instructor is knowledgable about this program and acted accordingly | (11A) | (11B) | (11C) | (11D) | (11E) |
| The instructor was instrumental in your success in this program | (12A) | (12B) | (12C) | (12D) | (12E) |

**Instructor(s) Evaluation:** Please use the following scale to evaluate the instructor(s).

**A = Excellent;** clearly presented objectives, demonstrated thorough knowledge of subject, utilized time well, represented self as professional, encouraged questions and opinions, reinforced safety practices and standards

**B = Good;** presented objectives, knowledgeable in subject matter, presented material efficiently, added to lesson plan information

**C = Average;** met objectives, lacked enthusiasm for content or subject, lacked experience or background information, did not add to delivery of material

**D = Below Average;** did not meet student expectations, poor presentation skills, lack of knowledge of content or skills, did not interact well with students, did not represent self as professional

**E = Not Applicable;** did not instruct in section or course area, do not recall, no opinion

| **Instructor Name** | **Excellent** | **Good** | **Average** | **Below Average** | **Not Applicable** |
|---|---|---|---|---|---|
| | (13A) | (13B) | (13C) | (13D) | (13E) |

**FIGURE 11.4** A sample instructor evaluation form.

*Source:* Courtesy of Southwest United Fire Districts Training Division

# Voices of Experience

I had taken a chief's position with another department and had returned to my old neighborhood to do a safety officer's class. Several of the students knew me, and during the class we discussed some of the exploits of our mutual acquaintances. While these discussions were not part of the prepared course material, I felt that the informal discussions brought a personal side of the safety officer's requirements to the table. I was pleased with the class's participation and our interaction. I felt the class had gone well.

*"There was one bubble sheet that caught my attention."*

Upon returning home, I reviewed the evaluations. The form being used was a bubble sheet on one side, with room for written comments on the back. I was evaluated well on the bubble sheet and found several good comments on the reverse side. But there was one bubble sheet that caught my attention. This student was disappointed that I had allowed, and participated in, a conversation that had, apparently, taken the Lord's name in vain. I didn't recall this event, and I have a personal rule that I keep my language at the "G"-rated level.

I was disappointed in my evaluation. I believe that the classroom must be a safe environment where everyone can learn without feeling threatened or uncomfortable. I had obviously failed to do so for this student on this occasion. The evaluation was not signed. If it had been, I would have contacted the person and apologized. Since that day, I have made it a point to not become so involved in interacting with the students that I forget that it is my responsibility to keep the class at a level where everyone can learn comfortably.

*David Peterson*
Plainfield Fire Department
Kent County, Michigan

FIGURE 11.5 Review your notes in person with the fire service instructor.

Provide time for input from the fire service instructor before you give your feedback. This allows the fire service instructor to identify areas that he or she already is aware need improvement. You can then provide recommendations on how to make the improvements in those areas. This strategy allows weaknesses that you were going to address anyway to be covered in a nonthreatening manner.

All feedback should be delivered in a considerate manner, stressing the instructor's positive attributes, offering ideas, and suggesting changes to deal with weaknesses. Do not overwhelm the fire service instructor with criticism; instead, provide constructive feedback. Strive for a balance between positive and negative comments and set goals for fire service instructor improvement when appropriate. Provide examples from your experiences as a fire service instructor to provide direction. If the fire service instructor struggled with the use of computer equipment, offer time to review the equipment and provide practical teaching tips on that equipment. In this way, you can turn the evaluation session into a learning session. You may also want to include a follow-up review at a later date. Finally, always make yourself available to the new fire service instructor for advice and guidance.

To discuss an evaluation with a fire service instructor, follow these steps:

1. In a private area, meet with the fire service instructor to discuss his or her performance.
2. Once again explain the reason for the review.
3. Determine whether the fire service instructor has any comments.
4. Use notes and/or the evaluation form to identify the fire service instructor's strong points.
5. Use notes and/or the evaluation form to identify areas of the fire service instructor's performance that need improvement.
6. Review the overall outcome/final score of the evaluation.

**Teaching Tip**

Remind the fire service instructor that it is his or her obligation to remain a professional at all times. The fire service instructor is responsible for what occurs in the classroom.

## Student Comments

As mentioned previously, a valuable source for feedback is student comments on evaluation forms. When taken objectively, these comments can provide significant information. The challenge for the fire service instructor is to understand that all students are not taking this class to gain better skills; some students need the class or certificate of completion for other means. Read the written comments where provided. If a student took the time to write a remark, evaluate its weight when reviewing the class. Students who provide written feedback in addition to filling in check boxes show a genuine desire to improve the learning environment for the next student who takes that class.

**Teaching Tip**

You will typically see certain remarks on every student evaluation, such as one indicating that the class could have been shorter. Keep in mind that not all students are in the class to seek knowledge; for whatever reason, some students view themselves as prisoners of the system.

## Developing Class Evaluation Forms

One of the skills a Fire Service Instructor II needs to cultivate is the ability to develop a class evaluation form. Some departments may provide you with a class evaluation form, whereas others may have you develop your own. Whatever its source, the evaluation form should cover specific topics, such as the training environment and course material.

Evaluating the learning environment is important to student success. Was the room too hot or cold? Was the classroom lit properly? Did the audiovisual equipment work? Were the chairs and tables comfortable? Was the learning environment free from distractions? These are all questions that affect students' learning ability. When used properly, student evaluations of the learning environment can be used to justify additional funds for classroom improvements.

Evaluation forms should also include areas for evaluating the course material. Did the material meet the student needs? Did it meet the learning objectives? Did the training material match NFPA standards and the department's own mission statement?

The class evaluation form should also include questions regarding the fire service instructor:

- Did the class start and end on time?
- Did the fire service instructor know the course material?
- Did the fire service instructor identify the learning objectives?
- Was the fire service instructor dressed appropriately?
- How well did the fire service instructor meet the needs of the class?

The way in which the information gained from an evaluation form is used is just as important as the evaluation form itself. If a department wants to improve the quality of its programs, then it must act upon the information obtained from the evaluation form. If improvements to the classroom are necessary, then the department can budget for those improvements. If the course material is not meeting student needs, then revisions to the lesson plan may be necessary. If there are issues with the fire service instructor, those issues can be addressed as well.

# Wrap-Up

## Chief Concepts

- NFPA 1041 requires that Fire Service Instructor II be capable of identifying other fire service instructors' strong and weak points, and provide guidance for other instructors' improvement.
- Fire departments should establish policies or procedures for fire service instructor qualifications.
- A fire service instructor evaluation system should provide feedback on strengths and weaknesses demonstrated during the evaluation.
- Evaluators need procedures and standards that outline the evaluation process so as to promote fairness and ensure consistency during this process.
- The goal of the presentation observation is to evaluate the fire service instructor's ability. Evaluation visits may be either planned or unannounced. With both types of visit, the evaluation is intended to be an improvement tool.
- It is the responsibility of the evaluator to be prepared for the evaluation process, just as it is the fire service instructor's responsibility to be prepared to deliver the class.

## Hot Terms

**Formative evaluation** Process conducted to improve the fire service instructor's performance by identifying his or her strengths and weaknesses.

**Summative evaluation** Process that measures the students' achievements to determine the fire service instructor's strengths and weaknesses.

# Fire Service Instructor *in Action*

As the coordinator of a fire fighter recruit training program, you assign a new fire service instructor to assist in the delivery of the training program. Given that this individual is a new fire service instructor, you decide to sit in on an upcoming training session to evaluate her performance. Your attendance presents a set of possible distractions in her delivery of the material. In the past, some new fire service instructors have instructed to the evaluator instead of to the students.

1. Which type of evaluation process is designed to provide improvement recommendations to the fire service instructor?
   **A.** Formative evaluation process
   **B.** Personnel evaluation process
   **C.** Summative evaluation process
   **D.** Word of mouth evaluation process
2. Which type of evaluation process is normally used to determine certifications?
   **A.** Formative evaluation process
   **B.** Personnel evaluation process
   **C.** Summative evaluation process
   **D.** Word of mouth evaluation process
3. When conducting a fire service instructor evaluation, which of the following statements is true?
   **A.** The evaluation should always be scheduled prior to the class.
   **B.** The evaluation should never be scheduled prior to the class, but the fire service instructor should be told at the beginning of the class that he or she will be evaluated.
   **C.** The fire service instructor should not know until after class that he or she was evaluated.
   **D.** The decision to schedule the evaluation should be determined by department policy.
4. To conduct the evaluation of this fire service instructor, what is the minimum Fire Instructor certification level you must have achieved, according to NFPA 1041?
   **A.** Fire Service Instructor I
   **B.** Fire Service Instructor II
   **C.** Fire Service Instructor III
   **D.** Fire Service Instructor IV
5. If you decide to have students provide feedback for fire service instructor improvement, which of the following outcomes is more likely?
   **A.** The students always tend to go easier on the fire service instructor than they should.
   **B.** Students always tend to go harder on the fire service instructor than they should.
   **C.** Some students will not provide honest feedback to the fire service instructor.
   **D.** Student feedback is very accurate.
6. Which of the following points does departmental policy on evaluation of fire service instructors *not* need to include?
   **A.** The order in which the feedback is to be given
   **B.** Confidentiality requirements for the evaluation
   **C.** Frequency and timing of evaluations
   **D.** Who will evaluate the fire service instructor
7. Which of the following is *not* a reason to conduct an evaluation?
   **A.** Certification
   **B.** Promotion
   **C.** Tradition
   **D.** Personnel evaluation process
8. During the feedback session, what can you do to help lessen the impact of weaknesses that need to be addressed?
   **A.** Provide feedback on only the positive aspects of the instructor's performance to reinforce those skills.
   **B.** Give the worst problem first, so the rest of the weaknesses seem like less of an issue.
   **C.** Give the least worrisome problem first, so the fire service instructor is desensitized as the problems discussed get worse.
   **D.** Let the fire service instructor identify problems that he or she is aware of before you mention the problems that the instructor does not cover.

# Program Management

# PART V

# Managing the Training Team

# CHAPTER 12

## NFPA 1041 Standard

**Instructor I**

NFPA 1041 contains no Instructor I Job Performance Requirements for this chapter.

**Instructor II**

5.2 **Program Management.**

5.2.1 **Definition of Duty.** The management of instructional resources, staff, facilities, and records and reports. [p. 208–227]

5.2.2 Schedule instructional sessions, given department scheduling policy, instructional resources, staff, facilities and timeline for delivery, so that the specified sessions are delivered according to department policy. [p. 208–217]

- **(A) Requisite Knowledge.** Departmental policy, scheduling processes, supervision techniques, and resource management. [p. 213–215]
- **(B) Requisite Skills.** None required.

5.2.3 Formulate budget needs, given training goals, agency budget policy, and current resources, so that the resources required to meet training goals are identified and documented. [p. 217–223]

- **(A) Requisite Knowledge.** Agency budget policy, resource management, needs analysis, sources of instructional materials, and equipment. [p. 217–223]
- **(B) Requisite Skills.** Resource analysis and forms completion. [p. 220–223]

5.2.4 Acquire training resources, given an identified need, so that the resources are obtained within established timelines, budget constraints, and according to agency policy. [p. 220–223]

- **(A) Requisite Knowledge.** Agency policies, purchasing procedures, and budget management. [p. 217–223]
- **(B) Requisite Skills.** Forms completion. [p. 217–223]

5.4.3* Supervise other instructors and students during training, given a training scenario with increased hazard exposure, so that applicable safety standards and practices are followed, and instructional goals are met. [p. 223–227]

- **(A) Requisite Knowledge.** Safety rules, regulations and practices, the incident command system used by the agency; and leadership techniques. [p. 208–227]
- **(B) Requisite Skills.** Implementation of an incident management system used by the agency. [p. 208–227]

## Knowledge Objectives

After studying this chapter, you will be able to:

- Describe how to schedule instructional sessions.
- Describe the budget process, the creation of a bid, and budget management.
- Describe the process for acquiring and evaluating training resources.
- Describe how to safely supervise training.

## Skills Objectives

After studying this chapter, you will be able to:

- Demonstrate the ability to schedule training.
- Demonstrate the procedures for creating a training budget.

## You Are the Fire Service Instructor

You have identified a training need within your department; now you have to make several management decisions relating to the implementation of this training program. One decision revolves around the purchase of a curriculum support program for use in the delivery of the course. You have read that some of the best training sessions use a variety of teaching methods in combination to present the lesson plan. Determining which methods and which materials are suitable for your audience, plus which fit within your budget, requires some research. You have to choose between several professionally developed programs or develop the material yourself. Other management decisions focus on the scheduling of the instruction and the overall supervision of the delivery of the course to ensure quality, safety, and consistency.

1. How would you determine which of the professionally developed curriculum packages meet all current NFPA standards?
2. What types of curricula would you develop using your agency procedures or information?
3. How will you choose fire service instructors for this program?

## Introduction

Training and education are tools used by fire departments to improve their efficiency in operations and in fire fighter safety. Managing a training team in any type of fire department is a staff-level function, which is often carried out by a department training officer. This training officer may have a rank position within the department and must posses the managerial skills needed to accomplish the many job tasks assigned to the training team. A majority of the management and supervisory-level functions are performed by the Fire Service Instructor II or higher. If you wish to be a leader of a training team, you must improve your administrative and leadership skills, because instructional skills are only part of the requirements for filling this position.

The fire service training program is a critical part of the effort to reduce fire fighter injuries and line-of-duty deaths. Lack of training, failure to train, inadequate training, and improper training may all contribute to injuries and fatalities. Conversely, fire fighter injuries and fatalities are often prevented through effective training. Much is at stake and much relies on the quality of the training program. To become an effective manager, you must develop scheduling, budgetary, decision-making, and supervision skills.

Like any supervisor, you are responsible for making sure that department policies and SOP's are followed and referenced within your training classes with strict adherence to the use of safety-based policies such as the incident command system (ICS), incident management system, personnel accountability and the assignment of qualified instructors, especially to high-risk training events.

## Scheduling of Instruction

As a Fire Service Instructor II, you must take into consideration the many aspects of your department's operation when developing a training schedule. A comprehensive schedule includes all job areas and covers elements of both initial training and ongoing training. It also balances the various regulatory requirements with the training sessions needed to help fire fighters deliver safe and effective service. Regulatory requirements that affect the department's training schedule include everything from EMS recertification requirements to hazardous materials training. Ongoing skill and knowledge retention and refresher training programs are important parts of the scheduling process.

In part-time, paid-on-call, or volunteer organizations, the scheduling and delivery of training can be an extremely difficult task. Many times the availability of time to deliver training is compromised by members' occupations and family commitments.

Surveys of the quality of a training program often reveal that the delivery of training suffers owing to poor management of the training schedule. The dynamic nature of the fire service sometimes compounds these problems when incidents occur during training. The best training schedules, calendars, and organizational skills are tested daily when the alarm sounds FIGURE 12.1. Few fire departments have the luxury of taking companies completely out of service for training events and sessions. For both instructors and students alike, it can be frustrating when an elaborate training session comes to an abrupt end because of a call.

**FIGURE 12.1** Fire fighters leaving training for an emergency run.

## Teaching Tip

Review your department's job descriptions and categorize job functions into similar groups. Your department's training program should have initial and ongoing training content for every job description in the department.

In recognition of the fact that the role of the fire fighter is expanding to include more responsibilities, managing training so that it is delivered effectively is the primary function of the Fire Service Instructor II position. All fire fighter duties must be taken into account when developing the training scheduling. NFPA 1500, *Standard on Fire Department Occupational Health and Safety Programs*, states the importance of this point quite clearly:

> 5.1.3 The fire department shall establish training and education programs that provide new members initial training, proficiency opportunities, and a method of skill and knowledge evaluation for duties assigned to the member prior to engaging in emergency operations.

This mandate applies to all levels of the fire department, from the chief down to the newest fire fighter. Initial training focuses on meeting the job performance requirements (JPRs) that apply to each level of responsibility. As a fire service instructor, it is your duty to ensure that the training takes place at every level, before the fire fighter engages in emergency activities. Today it is no longer acceptable to throw an untrained fire fighter into a hostile situation without proper training and evaluation.

# Type of Training Schedules

Each department has specific issues that it must address in terms of organizing training schedules. Some fire service instructors take a cynical view, stating that a training schedule is not worth the paper that it is written on because of all of the things that can go wrong in the attempt to deliver the training. For example, outside training sessions at the drill tower may be compromised by poor weather conditions, or an emergency incident can delay a classroom session.

The two types of training that occur in the fire service are formal training courses and in-service training. At the most basic level, in-service training or on-duty training, often referred to as **in-service drills**, takes up the majority of a training schedule and are the most common delivery method of training **TABLE 12.1**. An in-service drill may consist of a single station drill, or it may involve all on-duty companies. Such in-service programs can be scheduled to run on specific days or be scheduled to take place only when a specific training need is identified. For the purposes of this chapter, in-service drills are considered to be conducted while fire fighters are on duty and available to respond to incidents, whereas formal training courses take place outside of the structured work day.

## In-Service Drills

In-service drills typically take place at a specified time and location. All types of fire departments use some form of in-service drills, because very little time is available to take units out of service for training exercises. With this kind of training, apparatus and personnel must be kept in some form of readiness in the event that an emergency occurs. On the training schedule, some departments indicate the topic and level of coverage that the drill will provide; others simply state, "to train on something" of the fire service instructor's choosing.

In-service training can be categorized into three levels:

- *Skill/knowledge development* occurs when new approaches to firefighting operations are introduced to the department **FIGURE 12.2**. These sessions expand on previously learned information or skill levels to increase the fire fighters' ability to do their job. This type of training is used when a new method, new piece of equipment, or new procedure is

**Table 12.1 In-Service Training Schedule for a Career Department 24/48 Work Schedule**

With adjustments, part-time or volunteer departments could use this schedule as well. Numbers represent days of the month.

| Sunday | Monday | Tuesday | Wednesday | Thursday | Friday | Saturday |
|---|---|---|---|---|---|---|
| 1 Make-up | 2 Hose | 3 Hose | 4 Hose | 5 EMS | 6 EMS | 7 EMS |
| 8 Make-up | 9 Ladders | 10 Ladders | 11 Ladders | 12 Company drill | 13 Company drill | 14 Company drill |
| 15 Make-up | 16 Hazardous materials | 17 Hazardous materials | 18 Hazardous materials | 19 Pump operator | 20 Pump operator | 21 Pump operator |
| 22 Make-up | 23 Tools and equipment | 24 Tools and equipment | 25 Tools and equipment | 26 Officer training | 27 Officer training | 28 Officer training |
| 29 Make-up | 30 Ropes | | | | | |

implemented. All fire fighters who will use a new piece of equipment should be trained on it prior to placing the equipment into service. This may take a considerable amount of time to accomplish, but the extra effort will pay off on the fire ground through better performance and use of the equipment.

- *Skill/knowledge maintenance* is one of the goals of a comprehensive training program FIGURE 12.3. Skill and knowledge degrade over time if not used regularly, and members of a life-safety profession cannot afford to not function at their highest level of ability. This type of training is intended to develop performance baselines for core duties and functions, which can then be used to establish company-level standards and to measure individual skills and weaknesses. All company members should be able to perform basic tasks within a reasonable amount of time. An incident commander (IC) makes incident assignments knowing how long it takes to stretch a line or ventilate a roof, for example; if the company is untrained or unprepared in these skills, other fire-ground assignments may suffer.
- *Skill/knowledge improvement* is necessary when individual weaknesses become apparent. These drills are used when errors or poor performance have occurred or when other undesirable outcomes have been observed. A company may be able to effectively stretch a hose line and place the nozzle in a ready position, but if the pump operator cannot have the desired gallons/minute (gpm) rate and nozzle pressure arrive at the nozzle, the entire company fails in its assignment. This is a signal for the company officer to train with the pump operator, using fire service instructors to improve the entire company's skill level.

FIGURE 12.2 Skill/knowledge development occurs when new approaches to firefighting operations are introduced to the department.

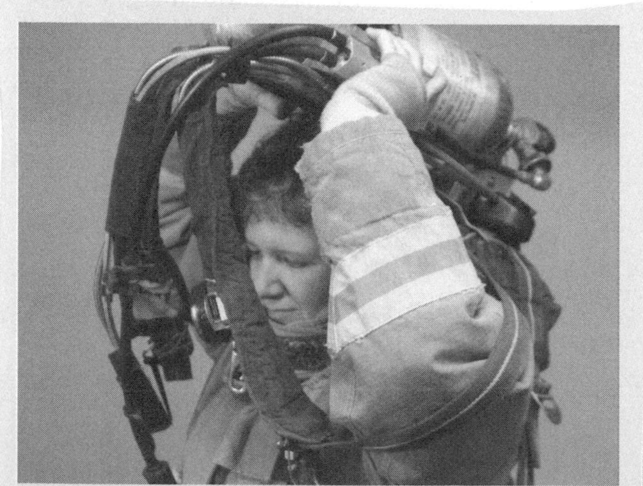

FIGURE 12.3 Skill/knowledge maintenance occurs in order to develop performance baselines for core duties and functions.

In-service training sessions are the most vulnerable to cancellation, delay, or reduction in participation owing to emergency response. Have a **supplemental training schedule** (a published program) available in the event that some mitigating factor requires a change in the original training session TABLE 12.2. The resources needed to accomplish the training should be on stand-by and ready to use when unexpected opportunities to conduct the in-service drill occur.

### Formal Training Courses

Certain types and levels of training require that the fire service instructor follow a set curriculum matched to a directed number of hours or evolutions. Training to NFPA Professional Qualification Standards using JPRs, for example, may require that students participating in a course attend a 40-hour program to meet all of the JPRs. This type of training schedule affects the use of the department classroom or training center and could possibly lead to overlapping use of the same facilities when in-service training is conducted at the same time. Formal training courses may also have specific facility-related needs such as audiovisual equipment, props, or appliances

**Table 12.2 Supplemental Training Schedule: Available Topics**

| Fire Suppression | Rescue Tools | Hazardous Materials | EMS |
|---|---|---|---|
| Nozzles and streams | Air bags | Emergency Response Guidebook (ERG) | Vital signs |
| Extinguishers | Glass removal tools | Communication | Patient movement |
| SCBA donning | Hydraulic tools | Decontamination | Documentation |
| Ladder carries | Hand tools | Spill containment | Initial patient surveys |

#### Teaching Tip

If your classroom or other training facilities are used to train outside groups or to hold department business meetings, make sure that their needs are covered and identified in one specific place so that no overlaps occur. Try to identify a central location where the training room schedule is posted and assign someone to be responsible for keeping the schedule up-to-date.

## Applying the JOB PERFORMANCE REQUIREMENTS (JPRs)

Most fire departments have a training division, which is often led by a training officer or other fire officer who is assigned to complete the management-related functions tied to training. Recordkeeping, scheduling, and compliance are some of the functions that must be attended to on a regular basis as part of training, regardless of the size of the department. Many resources must be managed, and budgets must be prepared and monitored as part of this endeavor. Fire service instructors need to be assigned and facilities secured to ensure that both instruction and student experiences are of the highest quality. Some of these functions are related to typical fire officer management and organizational principles. In the area of training program management, however, many special considerations must be addressed to manage the training team.

### Instructor I

Teamwork among the members of the training team is as important as teamwork is for the members of the fire company. As a division within the fire department, proper management requires the Instructor I to act as part of the team and to complete routine administrative tasks in addition to the delivery of training.

### JPRs at Work

NFPA does not identify any JPRs that relate to this chapter. As part of the training team, the Instructor I will function as part of the team by delivering training, managing resources, and ensuring the safety of students during training. Completing records, reports, and other proper documentation of the learning process will be a large part of the Instructor I's responsibilities.

### Instructor II

Juggling the resources, staff, facilities, and budget of a training program effectively requires sound management and leadership skills. Many fire officer JPRs will be put to work while completing these functions.

### JPRs at Work

Schedule and acquire resources necessary to conduct training sessions and assign qualified instructors while administering a budget for these functions. Ensure the safety of the students and instructors by administering department policies and applicable standards.

### Bridging the Gap Between Instructor I and Instructor II

The Instructor I can increase his or her ability to work as part of the training team by accepting additional responsibilities in management-related functions. The Instructor II can fill a mentoring position and also reduce his or her own workload by utilizing the services of the Instructor I in completing routine tasks. This prepares the Instructor I for increased management responsibilities and helps to fully develop the training team and utilize each member to his or her full potential.

### Instant Applications

1. Review your department's training policy and identify management-related responsibilities that must be completed by the Fire Service Instructor II.
2. Obtain a copy of the training budget to review expenditures and possible revenue sources.
3. Review your department's training schedules. Outline the process for determining training subjects, fire service instructor assignments, and the resources needed as part of the overall training program.

When I first accepted the position as a fire service instructor, I thought that I would be spending a majority of my time conducting live burn exercises, instructing auto extrication, and helping pump operators develop skills in fire hydraulics. It wasn't too long after I assumed this position that I found managing a training program was a significant undertaking. The many administrative and organizational details that it takes to make a successful training program work often draw you away from the classroom. The tasks of scheduling, planning instruction, and revising lesson plans took more time than actually teaching a class. The training schedule had to be adaptable, as new opportunities would appear for training classes or acquired burn structures would require a shuffling of the schedule.

You will find that your fellow fire service instructors are very willing to share their resources. Your network of peers is needed to help accomplish many of the management functions in training. They can assist in giving you ideas for training sessions that would have worked and those that have not along with providing you with ideas on the scheduling and delivery of training to diverse groups and types of departments. You shouldn't have to reinvent the wheel as many of the challenges we face have already been addressed by your peer network.

*"Your network of peers is needed to help accomplish many of the management functions in training."*

When training assignments are made to other fire service instructors, you must make every attempt to ensure that the training delivered is consistent and that any student who misses a training session is considered in make-up training opportunities. I have found that conducting periodic checks of fire fighter skills throughout the year allows you to evaluate how effective both your training system and your fire service instructors are.

Always work to keep your training program up-to-date with current and relevant information. The last thing a veteran fire fighter wants is to go through the same old rope and knot drill. Use case studies, and emphasize safety and survival, new technologies, and professional development components in your training plan. You are responsible for the development, delivery, and follow-through of each and every aspect of the training program. This takes hard work and dedication and a network of peer support to accomplish.

*Forest Reeder*
Pleasantview Fire Protection District
LaGrange Highlands, Illinois

that a department may have to budget for or acquire from other sources.

Other types of training also must be developed when constructing a comprehensive training schedule. For example, the schedule may need to accommodate any of these special areas of training:

- Technical rescue teams
- Special operations teams
- Hazardous materials teams
- Emergency medical services
- Vehicle operator training

The following areas of department operations may overlap with a training program and should be considered part of in-service training:

- Vehicle and equipment inspection and maintenance
- Hose testing
- Ladder testing
- Pump testing
- Daily/weekly/monthly equipment checks

Other duty areas within a department that must be considered within the framework of training program management include other staff-level functions within the organization. Ensuring that training is commensurate with duty means that the training needs of these personnel must be included in the development of your training plan. For example, you must consider the functions of fire cause determination personnel, fire inspection personnel, public education teams, dispatch and communication personnel, and apparatus mechanics when planning training. Each of these areas may take up substantial amounts of available time and require the coordination of resources and effective time management. These areas may also require specialized resources, fire service instructors, or equipment that must be considered when developing the schedule.

To facilitate planning, fire service instructors should develop a good working relationship with specialty team leaders. These team leaders have background and information that will help in the delivery of instruction. By working with these team leaders to prepare lesson plans and schedule sessions, this ensures that division objectives are incorporated into the course, and increases the likelihood that your training program will be successful.

## Scheduling for Success

To meet the needs of the department and its fire fighters, you must develop a training schedule that is consistent, easy to understand, and clear FIGURE 12.4. To accomplish this goal, you must have access to and understand the department's training policies. A department's training policy or standard operating procedure (SOP) that relates to training may identify who must train, how often, where, and who will provide the training. If this information is already known, the scheduling chore is almost complete.

Subject coverage in a curriculum-based training program should be sequential. Thus, when looking at your subjects, you must consider any prerequisite learning and skill levels needed. Ask yourself which topic needs to be covered before the next

### Teaching Tip

Review your department's SOPs relating to training participation and scheduling.

can be addressed. For example, it would be advisable to offer instruction on use of ladders before teaching a class on vertical ventilation to a group of new fire fighters. Students should learn and perform skills in a sequential order both to maximize their understanding and to improve their safety and survival.

The training policy or SOP that will help you develop a training schedule should include answers to the following questions:

- Who must attend the training session?
- Who will instruct the training session?
- Which resources are needed for the training session?

Information specific to the training session may also be used to help identify the scope and level of the training session. Students appreciate knowing what the training session will cover before they walk into the classroom. A training session notice will provide the necessary information to the students before the training takes place FIGURE 12.5. Highly motivated students may even complete reading assignments or research the topic before the training begins.

Fire officer development includes the ability to prepare a crew for a training session by practicing skill sets before the drill FIGURE 12.6. Advanced policies may also detail the responsibilities of those who attend, supervise, and participate in training. Likewise, recordkeeping processes are also included in this policy. Information about the training session itself includes the following details:

- What is the training subject?
- Is this a practical (hands-on) or classroom type of session?
- Are there any pre-training assignments?
- Where will the training take place?

## Developing a Training Schedule and Ensuring Regulatory Compliance

An **agency training needs assessment** should be performed at the direction of the department administration. The agency training needs assessment helps the Fire Service Instructor II to identify any regulatory compliance matters that must be included in the training schedule. It may also help identify the topics in which the department wants fire fighters to be trained. The following methods may be used to determine which topics need to be given priority over others:

- Identify the skills and knowledge necessary for safety and survival
- Review the skills and knowledge necessary for compliance with the department's mission statement
- Examine the subject areas tied directly to the job description of each fire fighter

**FIGURE 12.4** Training schedule.

*Source*: Courtesy of Southwest United Fire Districts Training Division

Although every department has specific regulatory issues that must be addressed in its training schedule, some issues have a universal impact. These areas require interpretation locally, but some type of overall training requirements is usually specified by many regulatory agencies.

Consider the Insurance Service Office (ISO): It has numerous training requirements that cover initial training for new fire fighters and vehicle operators and ongoing training for fire officers. Though technically not a regulatory agency for fire departments, ISO influences the operations of many departments because a lower property classification rating can be improved by implementation of an effective training plan. For this reason, it is a good idea to contact your local ISO representative and discuss the training requirements that are part of the ISO ratings survey.

Similarly, the Occupational Health and Safety Agency (OSHA) has specific training requirements that should be considered, covering topics ranging from respiratory protection to infectious disease protection. Because the enforcement power and authority of OSHA vary from state to state, contact your local OSHA representative to discuss which areas of your training program are subject to OSHA training requirements.

**Training Division**
**7550 Lyman Darien, IL 60561**
**Phone 630-910-2216 Fax 630-910-2085**
**www.sufd.org**

**Training Session Notice**

**Topic:** Communication

**Dates:**
January 22, 23, 24 2008 – Pleasantview Dispatch Center
January 29, 30, 31 2008 – Darien-Woodridge Dispatch Center

| Pleasantview Jan 22-23-24 | Darien-Woodridge Jan 29-30-31 |
|---|---|
| Time: 0830-0930 | Time: 0830-0930 |
| Instructor: Communications Chief Milam | Instructor: Communications Chief Milam |

**Equipment Needed:** All of the equipment needed is at the dispatch center(s).

*Each district should make appropriate coverage assignments for remaining companies during training. If companies are needed from training session, contact Dispatch for on-site dispatch.

FIGURE 12.5 Training session notice.
*Source*: Courtesy of Southwest United Fire Districts Training Division

Fines or other sanctions could be levied against your department if you fail to meet OSHA requirements.

Other regulatory commissions or agencies that may have training requirements include your local insurance carrier or risk management department, local EMS provider or licensing agent, the NFPA, International Fire Service Accreditation Congress (IFSAC), or State Fire Marshal or statewide certification entity. These agencies may have developed significant requirements that will affect entry-level and ongoing training for your students.

Knowledge of these and other regulatory agency requirements is a key component of the scheduling puzzle. Each agency may require training to take place at certain intervals or to last for a certain number of hours to maintain compliance with the agency's regulations. Become familiar with each of these areas and develop ways to cover as many areas as possible by combining resources and sessions to meet the various requirements.

A key component of the Fire Service Instructor II's job description requires you to keep current with the requirements of the regulatory agencies that affect your department. As new laws and standards are passed, your department training policy must evolve to keep up with the changing requirements. Best practices learned from case studies and other local requirements may also require the evolution of your department's training policy. As mentioned earlier in this book, high-risk areas of training, such as live fire training or special operations training, require a student to instructor ratio of 5:1, and the conduct of such drills should be outlined in your written training policy. Additional topics to be covered within the department training policy include specific safety rules during training, the scope of coverage of the training program, and many other areas of administrative responsibility. Many of these functions may fall within the duties of the Fire Service Instructor III or other fire officer.

FIGURE 12.6 Fire officer development includes the ability to prepare a crew for a training session by practicing skill sets before the drill.

## Scheduling of Training

A standardized process must be developed for scheduling training sessions. As stated earlier in this chapter, the schedule must incorporate some flexibility so that instructors and students can capitalize on unforeseen opportunities or work around obstacles in the delivery of training. For instance, you might have scheduled a block of training on hose lead-outs when a local developer calls you and offers the use of a structure scheduled for demolition. If the property is suitable for training exercises, it may present a great opportunity to enhance your training program through the use of the acquired structure. You may also want to consider covering your hose lead-out drills during the use of this structure along with the multitude of other areas you can train on.

Obstacles to training can range from weather complications to students' vacation plans. Also, new equipment may be purchased for which training is needed before the equipment can be put into service, taking precedent over other scheduled activities.

A process for the publication, distribution, and delivery of the training schedule must be developed as well. The goal is to keep all personnel who are affected by the training schedule informed about the training-related activities of the period FIGURE 12.7.

FIGURE 12.7 Training calendars and schedules should be posted.

## Master Training Schedule

A useful method to organize the multiple demands for schedule coverage is to use a system that aggregates all of the required training areas into one area. A simple tabular form, called a **master training schedule**, may suffice to begin this process TABLE 12.3. It serves as the starting point for the development of an in-service training program schedule. When a period of training (typically one month in the year) is planned, you can refer to this table to identify the topics that must be included into the schedule.

Following are the steps required to develop a master training schedule for a fire service department:

1. Complete an agency needs assessment of the regulatory agencies that require training on specific topics (e.g., OSHA, ISO, NFPA, local agency/authority).
   a. Determine the frequency of training needed.
   b. Determine the type of training needed:
      - Initially upon hire
      - Ongoing after hire
      - Classroom or hands-on
      - Individual, company, or multi-company
      - Mutual aid training
      - Training on special hazards or topics
2. Identify other areas of training needed to teach students the skills and knowledge necessary to keep them safe at emergencies.
   a. SCBA skills
   b. Fire fighter survival techniques
   c. Minimum company standards or other proficiency based training
3. Identify ongoing activities that will affect the training schedule, so that you do not overload the crews with too many activities within the same period.
   a. Equipment testing
   b. Fire prevention activities
   c. Apparatus testing
   d. Applicable recertification

**Table 12.3 Master Training Schedule**

| Subject | Jan. | Feb. | March | April | May | June | July | Aug. | Sept. | Oct. | Nov. | Dec. |
|---|---|---|---|---|---|---|---|---|---|---|---|---|
| **Mandatory Training Areas** | | | | | | | | | | | | |
| SCBA training | X | | | X | | | X | | | X | | |
| Driver training | | X | | | X | | | X | | | X | |
| Fire fighter survival skills | | | X | | | X | | | X | | | X |
| EMS continuing education | X | X | X | X | X | X | X | X | X | X | X | X |
| Protective clothing inspections | X | | | | | | X | | | | | |
| **Equipment Testing** | | | | | | | | | | | | |
| Hose service testing | | | | X | | | | | | | | |
| Annual pump testing | | | | | | | | | X | | | |
| Ground ladder testing | | | X | | | | | | | | | |
| Aerial ladder testing | | | | | | | | | | X | | |
| **Other Subject Training** | | | | | | | | | | | | |
| NFPA 1001 skills | X | | X | | X | | X | | X | | X | |
| NFPA 1002 skills | | X | | X | | X | | X | | X | | X |
| NFPA 1021 skills | X | | | X | | | X | | X | | | |
| NFPA 472 skills | | | X | | | X | | | X | | | |
| NFPA 1006 skills | | | | X | | | X | | | X | | |
| Other NFPA areas | | X | | | | | | | | | | |

4. Create a table that lists the information compiled in steps 1–3 in one column in individual rows, and create at least 12 additional columns to represent each month of the year.
5. Insert a check mark, X mark, or other notation indicating when each type of training needs to be completed into the number of monthly columns required to meet the standard.
   a. If quarterly SCBA training is needed, insert an X in one month of each quarter.
   b. If protective clothing needs to be donned, doffed, and cleaned semi-annually, insert an X in each of two six-month periods.
6. Once all elements have been added to the table, balance out the coverage to allow for an even distribution of the content to be covered each month.
7. Develop categories for other types of training by subject matter areas and distribute that throughout the calendar to give your program even coverage.
8. Confirm that all of the JPRs of a given job level have been covered by the planned training.
9. Review and revise the master training schedule on a regular basis, preferably every year.

## Budget Development and Administration

The budget process of a fire department can be very confusing, if not outright overwhelming. Understanding the budget development process and your role in purchasing, specifying resources and materials, and justifying the funding requested for a training project or program are all crucial to successful training program management.

## Introduction to Budgeting

A **budget** is an itemized summary of estimated or intended revenues and expenditures. **Revenues** are the income of a government from all sources appropriated for the payment of public expenses and are stated as estimates. **Expenditures** are the money spent for goods or services and are considered appropriations by the expenditure authority. Every fire department has some type of budget that defines the funds made available to operate the organization for a particular period of time, generally 1 year. The budget process is a cycle:

1. Identification of needs and required resources
2. Preparation of a budget request
3. Local government and public review of requested budget
4. Adoption of an approved budget
5. Administration of approved budget, with quarterly review and revision
6. Closeout of budget year

Budget preparation is both a technical and a political process. The funds that are allocated to the fire department define which services the department is able to provide for that year. The technical part relates to the task of calculating the funds that are required to achieve different objectives, whereas the political part is related to elected officials making the decisions on which programs should be funded among numerous alternatives.

### Budget Preparation

Successful budgeting requires justification of the amount of money being requested. A budget justification is useful in defending your position when request money is allocated for a specific line-item or category. A training division may be an example of a line-item in a budget document. When

**Line Item Budget Request 2009–2010**

| Category/Budget Code | Justification | 2009–2010 Requested | 2008–2009 Actual | 2008–2009 Requested |
|---|---|---|---|---|
| Contingency 10–04–00–460 | Unforeseen expenses related to training program | 1000.00 | 1000.00 | 1000.00 |
| Outside Seminars 10–04–00–510 | FF attendance at certification programs and seminars | 17,500.00 | 15,000.00 | 17,500.00 |
| Firefighter I Academy 10–04–00–513 | 1 new full-time member to attend academy | 2000.00 | 2000.00 | 2000.00 |
| Hosted Courses 10–04–00–514 | Expenses for in-house hosted classes; coffee, pop, snacks | 1000.00 | 1000.00 | 1000.00 |
| Text/Publications/Video 10–04–00–516 | Purchase of updated manuals, magazine subscriptions | 8500.00 | 7000.00 | 7610.00 |
| Drill Site Maintenance 10–04–00–518 | Upkeep of training center props and buildings | 30,500.00 | 25,000.00 | 30,500.00 |
| Misc. Supplies/Maintenance 10–04–00–519 | Smoke fluid, office supplies, repairs of equipment | 8,000.00 | 5375.00 | 5375.00 |
| Part-Time Instructors 10–04–00–521 | Part-time instructional staff wages | 35,000.00 | 25,000.00 | 31,000.00 |
| Instructor Seminars 10–04–00–522 | Attendance at local, state and national seminars, conf. | 15,500.00 | 7000.00 | 9000.00 |
| TOTALS | Support of Division of Training Mission and Assignments | 119,000 | 88,375 | 104,985.00 |

FIGURE 12.8 Training budget and justification columns.

developing the line-item for the training division, the justification can be based on previous expenses in this area, the agency training needs assessment, and the forecasting of the training needs for the upcoming period FIGURE 12.8.

## The Budget Cycle

Every department has a budget cycle, typically 12 months. The budget document describes where the revenue comes from (input) and where it goes (output) in terms of personnel, operating, and capital expenditures. Annual budgets usually apply to a fiscal year, such as the economic period starting on July 1 and ending on June 30 of the following year. With this type of schedule, the budget for fiscal year 2008, referred as FY08, would start on July 1, 2007, and end on June 30, 2008. In many cases, the process for developing the FY08 budget would start in 2006, a full year before the beginning of the fiscal year TABLE 12.4.

As the timeline in Table 12.4 shows, a long period separates the time at which you make a budget request and the time at which the money becomes available to spend. All training needs may not be apparent to you during the budget development process. Effective budget management requires you to anticipate as much as possible when preparing the budget and be able to develop contingency plans for unforeseen events. Some actions and purchases may have to be cancelled or delayed when others assume a higher priority.

Mandated continuing training, such as hazardous materials and cardiopulmonary resuscitation recertification (CPR) classes, must also be factored into the operating costs portion of the budget. That figure includes the cost of the training per fire fighter per class. The cost to pay another fire fighter overtime to cover the trainee's position during the mandated training could also be included in this amount. As new mandates are set, finding the money to cover these unexpected expenses can prove very difficult. Some departments have struggled with implementation of mandated training in the National Incident Management System (NIMS) area or in some aspects of hazardous materials response training, for example. The cost of this specialized training inflates the training budget request and often takes away from other vital programs in fire suppression training or fire officer training. This is truly one of the most difficult challenges for you to budget and justify.

**Table 12.4 Fiscal Year 2008 Timeline**

| | |
|---|---|
| August–September 2006 | Fire station commanders, section leaders, and program managers submit their FY08 requests to the fire chief. Their concentration is on proposed new programs, expensive new or replacement capital equipment, and physical plant repairs. Depending on how the fire department is organized, this is the point where a station commander would request funds for a new storage shed or a replacement dishwasher for the fire station. Documentation to support replacing a fire truck or purchasing new equipment, such as an infrared camera or a hydraulic rescue tool, would be submitted as well, and a proposal for a special training program would be prepared. In most fire departments, these requests are prepared by fire officers throughout the organization and submitted to an individual who is responsible for assembling the budget proposals. |
| September 2006 | The fire chief reviews the budget requests from fire stations and program proposals from staff to develop a prioritized budget wish list for FY08. At this time, larger department-wide initiatives or new program proposals are developed. For example, purchasing a new radio system or establishing a decontamination/weapons of mass destruction (WMD) unit might be part of the fire chief's wish list. |
| October 2006 | The city's budget office distributes FY08 budget preparation packages to all agency or division heads. Included in the package are the application forms for personnel, operating, and capital budget requests. The budget director includes specific instructions on any changes to the budget submission procedure from previous years. The senior local administrative official (e.g., mayor, city manager, or county executive) also provides specific submission guidelines. For example, if the economic indicators predict that tax revenues will not increase during FY08, the guidelines could restrict increases in the operating budget to 0.5 percent or freeze the number of full-time equivalent positions. |
| November 2006 | Deadline for agency heads to submit their FY08 budget requests to the budget director. |
| December 2006 | The budget director assembles the proposals from each agency or division head within the local government structure. Each submission is checked to ensure that it complies with the general directives provided by the senior local administrative official. Any variances from the guidelines must be the result of a legally binding agreement, settlement, or requirement or have the support of elected officials. |
| January 2007 | Local elected officials receive the preliminary proposed budget from the budget director. For some budget items or programs, two or three alternative proposals may be submitted that require a decision from the local elected officials. Once the elected officials make those decisions, the initial budget proposal is completed. It is known as the "proposed fiscal year 2008 budget." |
| February or March 2007 | The proposed budget is made available to the public for comment. Many cities make the proposed budget available on an Internet site, inviting public comment to the elected officials. In smaller communities, the proposed budget may show up in the local newspaper. |
| March or April 2007 | Local elected officials conduct a public hearing or town meeting to receive input from the public on the FY08 proposed budget. Based on the hearings and on additional information from staff and local government employees, the elected officials debate, revise, and amend the budget. During this process, the fire chief may be called on to make a presentation to explain the department's budget requests, particularly if large expenditures have been proposed. The department may be required to provide additional information or submit alternatives as result of the public budget review process. |
| May 2007 | Local leaders approve the amended budget. This becomes the "approved" or "adopted" FY08 budget for the municipality. |
| July 1, 2007 | FY08 begins. Some fire departments immediately begin the process of ordering expensive and durable capital equipment, particularly items that have long lead times for delivery. Many requests for proposals are issued in the early months of the fiscal year. |
| October 2007 | Informal first-quarter budget review. This review looks for trends of expenditures in an attempt to identify any problems. For example, if the cost of diesel fuel has increased and 60 percent of the motor fuel budget has been spent by the end of September, there will be no funds remaining to buy fuel in January. Now is the time to identify the problem and plan adjustments to the budget. Amounts that have been approved for one purpose may have to be diverted to a higher-priority account. |
| January 2008 | Formal midyear budget review. The approved budget may be revised to cover unplanned expenses or shortages. In some cases, this change occurs in response to a decrease in revenue, such as an unanticipated decrease in sales tax revenue. If the money is not coming in, expenditures for the remainder of the year might have to be reduced. At other times, revenues may exceed expectations, so that funds become available for an expenditure that was not approved at the beginning of the year. |
| April 2008 | Informal third- and fourth-quarter review. Review of activity within the third quarter and projections for the fourth quarter can be made at this point. Final adjustments in the budget are considered after 9 months of experience. Year-end figures can be projected with a high degree of accuracy. Some projects and activities may have to stop if they have exceeded their budgets, unless unexpended funds from another account can be reallocated to keep them going. |
| Mid-May 2008 | The finance office begins closing out the budget year and reconciling the accounts. No additional purchases are allowed from the FY08 budget. Purchases are deferred until the beginning of FY09. |

## Capital Expenditures and Training

Capital expenditures consist of the purchase of durable items that cost more than a threshold amount and last for more than one budget year. Local jurisdictions differ on the amount and the time period that these items are supposed to last. Hydraulic rescue tools, apparatus, and self-contained breathing apparatus (SCBA) are all examples of equipment purchased as capital items. Each of these items requires training when delivery takes place; thus additional costs may be incurred to prepare both fire service instructors and fire fighters to train on the new equipment. A Fire Officer III may be responsible for assigning values and depreciation rates to fixed assets and for developing policies that define them.

The training budget is really no different than any other part of the fire department's budget. All budgets must be prepared, justified, and managed throughout the life of the budget cycle. You must be able to understand the budget development and documentation process used by your department and be able to provide the necessary administrative review of your budget area. When budgets need to be cut, administrations may seek out areas that can be trimmed by reducing activities or purchasing of resources. In the training area, you might lose funding for new textbooks or curriculum packages, for class fees and seminar registrations, or for the purchase of that much needed training prop. In particular, capital expenditures for training towers or training sites are often early targets in a cash-strapped department.

**Theory into Practice**

Budgeting and justifications tend to go against the grain of most firefighting personnel. This activity is often associated with a "bean counter" attitude that misses the emotional reality of the fire fighter's job. Fire fighters recognize that there is no formula stating that "One life is worth *x* dollars," but rather that "Mrs. Smith died and her grandchildren will never get to play with her." While this is a perfectly understandable perspective from someone who sees the effects of the "bean counter's" decisions, it doesn't fix the situation.

Developing a budget is a great way to demonstrate what the training division does and what value it adds not only to the department, but to the community as a whole. Begin the process by viewing budget development as an opportunity rather than a necessary evil. Know the goals and recognize the direct relationship between attainment of those goals and your need for the requested resources. Be articulate in your justification by keeping to the point, but also be thorough. Don't fluff up the numbers to include a little "negotiation" room. Also, don't sell yourself short: Budget requests are a business decision. Demonstrate your professionalism by requesting enough funds to meet the needs of the department, recognizing that the department must live within its means, and make good decisions based on the outcome of the budget process.

## Training-Related Expenses

The following expense areas may be included in the training line-item:

- Fire service instructor salaries, benefits, and expenses
- Student expenses for class attendance
- Tuition, overtime if applicable, transportation, lodging, and meals
- Textbooks, publications, and DVDs
- Office supplies
- Equipment and training aid maintenance
- Course and seminar fees
- Facility maintenance and improvements
- Capital expenditures
- Subscriptions and memberships
- Contracts
- Educational assistance programs
- Tuition reimbursement
- Grant programs

Ongoing budget maintenance activities should monitor the expenditures made to date to ensure that the training program is fiscally sound. As a Fire Service Instructor II, your performance may be measured in part by your ability to operate within the financial constraints of your program. Good budget management starts with an understanding of the department's budget process and ends with you following the purchasing procedures and tracking expenses related to the training program on a regular basis.

## Acquiring and Evaluating Training Resources

Purchasing of resources and products used in the instructional area is a key management skill. In an era characterized by shrinking budgets, trial and error will not suffice as a means for selecting training resources. Instead, your purchasing responsibility and authority must be thoroughly understood and clearly documented as part of the training process.

Resource management of the equipment, materials, facilities, and personnel used for training purposes requires a full understanding of the training goals of the department. Purchasing new training equipment every year may not be an option for every department. Often, front-line equipment that has been replaced by newer equipment may be designated for training use. Although this practice may ease the resource allocation process, it can also require more complicated resource management.

## Training Resources

Training resources can be defined as any equipment, materials, or resources that are used to assist in the delivery of a training session. Such resources can be either consumable or reusable. Consumable items can be used only once, such as a student workbook that will be written in. Reusable items can be refilled, reset, or recharged, such as smoke fluid for a smoke-generating machine or roof cutout panels used in a ventilation prop simulator TABLE 12.5. Books and curriculum packages may also include reusable sections such as PowerPoint slides and reproducible handout materials. Prop structures may have certain durable components, whereas other parts may be disposable.

Each time you develop or prepare to instruct from a lesson plan, you should identify the types of resources needed to conduct that training session FIGURE 12.9. The following equipment and materials are typically identified in a lesson plan as resources:

- Materials that students need to complete learning objectives
- Materials that the fire service instructor needs to present, apply, and evaluate the learning objectives
- Props, aids, equipment, or other resources necessary to complete the session

Preparing for an effective training session requires you to review the availability, usability, and reliability of each needed resource. For example, all equipment must be checked for proper operation and safety and be ready to use for the duration of the training session. Enough fuel, batteries, blades, paper, flipchart paper, or whiteboard markers need to be available for the entire training session. As always, have a backup plan in case the resources you identified are not available or working properly.

**Table 12.5 Consumable and Reusable Training Resources**

| Consumable Resources (One-Time Use) |
|---|
| Pallets used for burning or chopping |
| Sheetrock used for wall breeching exercise |
| Straw or hay used for igniting fires |
| Flares used for igniting fires |
| Smoke fluid (used for a smoke machine) |
| Shingles or other roof coverings (used on a roof simulator) |
| Foam concentrate |
| Old window frames |
| Student handouts |
| **Reusable Resources (May be rebuilt, recharged or reused)** |
| Many curriculum items |
| Cutaways or cross section of pumps or ladder parts |
| Ropes used for knot tying practice |
| Prop structures such as wall props or roof simulators |
| Hose couplings or hose pieces |

### Ethics Tip

The fire service has long lived off of its "make do" attitude in getting the mission accomplished even in the face of inadequate resources. As a fire service instructor, you may have to use materials or equipment that serve as proxies for real-world items. This practice often requires some creative thinking. But what should you do when your training methods may lead to training automation on the fire ground? Training automation occurs when students drill until their actions become second nature. The problem with this is that some actions are used only for training purposes. For example, if the gate is opened only halfway during hose stream drills, fire fighters will not get the feel of a hose at normal pressure. As a fire service instructor, you have an obligation to know when substandard resources create risks.

## Hand-Me-Downs

When a power saw is taken out of service as front-line equipment and given to the training division to use in training, the fire service instructor is then able to teach a forcible entry session without worrying about using in-service equipment. The downside to this practice is that the power saw was likely taken off front-line service for a reason: Was it hard to start? Did it have broken parts? Did it leak fuel or oil? Or did it just become obsolete? If the answer to any of these questions is yes, then you must prepare for the impact of that issue. It may be a good idea to have in-service equipment ready to use at the training site in case the hand-me-down does not work; having spare equipment handy will ensure that the flow of the training session is not interrupted. For example, that old rotary cut-off saw that has been given to the training division should be backed up by the new one that was purchased to replace it. It is also important to highlight any operational differences between the two models.

Notify the proper company officers or shift commanders before removing in-service equipment from apparatus for use in training. You do not want to be in the uncomfortable position of justifying a delayed response to an incident because the company members had to repack hose before they were able to respond on a fire run.

Reserve apparatus may assist in delivery of the training session, but only if you identify any differences between the equipment students would normally use in the performance of their duties and the equipment being used for the training. Make sure that significant differences in safety equipment, SOPs, and other dynamic factors are spelled out and understood during the training session **FIGURE 12.10**. It would be great to allocate duplicate resources to the training division—that is, equipment that is the same as front-line equipment; unfortunately, few departments are able to afford such luxuries.

## Purchasing Training Resources

Resources used in the delivery of training (i.e., resources needed) should be identified during the construction of a lesson plan. In turn, books, projection equipment, training aids, computer resources, and even straw or hay used to generate smoke in a burn tower may need to be considered within the budget process.

When given budgetary approval to purchase training resources, you must follow all departmental guidelines in obtaining those items. Solicitation of bids for resources costing more than a stated amount of money is a requirement that many departments require their personnel to observe. Many departments are required to solicit bids from competitive vendors in an effort to get the best price for a product. Textbooks, smoke fluid for smoke generators, and training foams are all examples of materials that could have different price points from vendors. In contrast, other materials may be **single-source products**—products that are manufactured or distributed by only one vendor. In such a case, you may not have to go through the bid process.

## Evaluating Resources

Instructors who are responsible for evaluating resources for purchase must observe and follow all purchase policy requirements. Before the decision to purchase a resource is made, you must evaluate the quality, applicability, and compatibility of the resource for the department. Many high-quality resources have been developed in recent years, but some time-proven resources continue to dominate certain market areas. When considering and evaluating resources for purchase, consider the following issues:

1. How will this product/resource be used in your department?
   a. Answering this question involves a form of audience analysis that was discussed in Chapter 6.

## Lesson Plan

**Lesson Title:** Use of Fire Extinguishers

**Level of Instruction:** Firefighter I

**Method of Instruction:** Demonstration

**Learning Objective(s):**
7-1 The student shall demonstrate the ability to extinguish a Class A fire with a stored-pressure water-type fire extinguisher. (NFPA 1001, 5.3.16)
7-2 The student shall demonstrate the ability to return a stored pressure water-type fire extinguisher to service after use.

**References:** Fundamentals of Firefighter Skills, 2nd Edition, Chapter 7

**Time:** 50 Minutes

**Materials/Equipment Needed:** Portable water extinguishers, Class A combustible burn materials, Skills checklist, suitable area for hands-on demonstration, assigned PPE for skill demonstration, extinguisher recharging equipment

**Slides:** 73–78*

**Step #1 Lesson Preparation:**
- Fire extinguishers are first line of defense on incipient fires
- Civilians use for containment until FD arrives
- Must match extinguisher class with fire class
- FD personnel can use in certain situations, may limit water damage
- Review of fire behavior and fuel classifications
- Discuss types of extinguishers on apparatus
- Demonstrate methods for operation

**Step #2 Presentation**

**Step # 3 Application**

FIGURE 12.9 Lesson plan example showing resources needed for training evolution.

FIGURE 12.10 A reserve apparatus and an in-service apparatus.

### Safety Tip

When using reserve apparatus, ensure all safety features, such as pressure relief valves and pressure governors, are working properly.

### Teaching Tip

When purchasing training resources, make sure you understand your department's purchasing policy in relation to bid requirements.

**Table 12.6 Resource Evaluation Form**

| Criteria | Vendor/Product 1 Rating (1 = poor; 5 = excellent) | Vendor/Product 2 Rating (1 = poor; 5 = excellent) | Vendor/Product 3 Rating (1 = poor; 5 = excellent) |
|---|---|---|---|
| Ease of use within our department | | | |
| Instructor | | | |
| Student | | | |
| Amount of modifications needed for our department | | | |
| Standard compliance | | | |
| NFPA | | | |
| OSHA | | | |
| ANSI | | | |
| Other | | | |
| Cost of product/resource | | | |
| Similar products already in use by department | | | |
| Quantity discount available | | | |
| References | | | |
| Availability | | | |
| Vendor support | | | |
| Local alternatives | | | |
| Other | | | |

   b. Consider both the audience for the resource as well as the fire service instructors who will use it.
      - Will it require training for the fire service instructors who will use it?
      - Will it require new learning skills of the students?
2. With which standards is the product/resource developed in conjunction or compliance?
   a. This question may be very important when considering adoption or reference of a textbook or training package.
   b. Certain state training agencies use specific references for certification training.
3. Is this a "turnkey" product, meaning it is ready to use as is, or will the fire service instructor have to adapt the product to local conditions?
4. Is this product/resource compatible with your agency's procedures and methods?
   a. Consider the content specific issues of the product/resource.
   b. Does your department use these methods as part of its operations, or are there significant differences in the equipment or methods of the product/resource?
5. Does the product/resource fit within budgetary restrictions?
6. Are there advantages to purchasing larger quantities of the product/resource?
7. Can you get references from the vendors of the product/resource that you are considering, or must you do your own investigation within your local area or statewide instruction networks?
8. Can the product/resource be available for your use within the prescribed timeline?
9. Are there any local alternatives, such as your own design of a product/resource or identification of other local agencies that may be willing to share the use of the product/resource?
10. Is a manufacturer demonstration, product expert, or other developer assistance available to the instructor about the product/resource?
    a. On-site assistance
    b. Phone support
    c. Web site information
    d. Product literature
    e. Suggested user guidelines or instructions

Experienced fire service instructors often develop a resource evaluation form to help justify this process and validate the selection process TABLE 12.6. A simple checklist of these questions may assist in your decision making. Effective and efficient training program management requires the development and use of some standardized practice to evaluate products and resources. In today's budget-conscious departments, doing your homework ahead of budget requests is likely to pay off by getting the needed resources/products purchased quickly FIGURE 12.11.

## Managing the Delivery of Training

Managing fellow fire service instructors requires leadership skills and the ability to recognize training scenarios that require specific supervision relating to safety and standard practice. You must ensure that applicable standards, SOPs, and departmental practices are used in the development and execution of training sessions that involve increased hazards to students. Every fire service instructor has a different style of presenting materials, different background, and experiences that he or she brings to the training session. As a Fire Service Instructor II, you should know these facts when supervising training delivery. Allow for individual styles and experiences

**Training Division**
**7550 Lyman Ave. Darien, IL 60561**
**Phone 630-910-2216 Fax 630-910-2085**

**PURCHASE ORDER**

**Date:** April 1, 2008

**Purchase Order No.:** 040108-02

**Vendor:** Jones & Bartlett Publishers
**Address:** 40 Tall Pine Drive
Sudbury, MA 01776
**Sales Rep/Contact:** Lori Heyday
**Phone:** 1-800-526-1536
**Fax:** 1-781-662-3405

**Ship To:** Southwest United Fire Districts
7550 Lyman Avenue
Darien, IL 60561
Attn: Liz Javier
**Account #:** On File

| Date Required: 4/15/08 | Terms: On Account | Ship Via: Best Way | Internal Acct. No.: 10-460-4400 |
|---|---|---|---|

| Quantity | Description | Unit Price | Total Price |
|---|---|---|---|
| 4 | Fundamentals of Firefighter Skills Book/ Workbook bundle 0-7637-4914-1 | 79.50 | 318.00 |
| | | **Total Cost:** | $318.00 |

Approved by: ______________________________
(B/C Forest F. Reeder, Training Director)

Authorized Signature: ______________________________
(Chief Dan Hermes, Program Manager)

**Mail all invoices to:**
Southwest United Fire Districts
7550 Lyman Avenue
Darien, IL 60561
**Attn:** Liz Goldrick/Training

**Important:** Our purchase order number must appear on all invoices, packages, packing lists and correspondence. Southwest United Fire Districts is Illinois Sales Tax Exempt.
Tax Exempt # (Available Upon Request)

**FIGURE 12.11** Purchase request form.
*Source*: Courtesy of Southwest United Fire Districts Training Division

in the training sessions but make sure that all objectives, safety-related guidelines, department SOPs, and other performance expectations are covered adequately.

Like any other form of management, management of fire service instructors and students takes strong leadership and management skills. The quality of instructional delivery and the ability of the student to learn or do a new skill are two of the criteria that may be used to evaluate training success. Too many times in fire service history, training preparation and implementation has failed, with students and fire service instructors being injured or killed in poorly constructed or poorly supervised training sessions. Likewise, if the quality of the instruction is substandard, then the learning outcomes will be poor. These aftereffects can persist long after the student has left the classroom or drill ground and may lead to disastrous consequences on the fire ground.

For all these reasons, quality assurance is an important aspect of the management of a training program. Guard

**Illinois Fire Chiefs Foundation**
**End of Course Evaluation Form**

**Course Title:** ____________________ **Course Location:** ____________________

**Course Evaluation:** Please shade the letter that indicates your evaluation of the course/instructor feature using #2 pencil.

| Course Feature | Excellent | Very Good | Good | Fair | Poor |
|---|---|---|---|---|---|
| 1. AV Material / Course materials | (1A) | (1B) | (1C) | (1D) | (1E) |
| 2. Course organization / flow / delivery rate | (2A) | (2B) | (2C) | (2D) | (2E) |
| 3. Observance to safety procedures and practices | (3A) | (3B) | (3C) | (3D) | (3E) |
| 4. Hands on Experience *If applicable to course* | (4A) | (4B) | (4C) | (4D) | (4E) |
| 5. Time per subject *Add comments on specific areas on reverse* | (5A) | (5B) | (5C) | (5D) | (5E) |
| 6. Evaluation tools | (6A) | (6B) | (6C) | (6D) | (6E) |
| 7. Will this course help you with your professional development? | (7A) | (7B) | (7C) | (7D) | (7E) |
| 8. Did the course meet your expecations? | (8A) | (8B) | (8C) | (8D) | (8E) |
| 9. Would you recommend this course to others? | (9A) | (9B) | (9C) | (9D) | (9E) |
| 10. Overall impression of course | (10A) | (10B) | (10C) | (10D) | (10E) |

**Instructor(s) Evaluation:** Please use the following scale to evaluate the instructor(s).

**A = Excellent;** clearly presented objectives, demonstrated thorough knowledge of subject, utilized time well, represented self as professional, encouraged questions and opinions, reinforced safety practices and standards

**B = Good;** presented objectives, knowledgeable in subject matter, presented material efficiently, added to lesson plan information

**C = Average;** met objectives, lacked enthusiasm for content or subject, lacked experience or background information, did not add to delivery of material

**D = Below Average;** did not meet student expectations, poor presentation skills, lack of knowledge of content or skills, did not interact well with students, did not represent self as professional

**E = Not Applicable;** did not instruct in section or course area, do not recall, no opinion

| Instructor Name | Excellent | Good | Average | Below Average | Not Applicable |
|---|---|---|---|---|---|
| | (11A) | (11B) | (11C) | (11D) | (11E) |
| | (12A) | (12B) | (12C) | (12D) | (12E) |
| | (13A) | (13B) | (13C) | (13D) | (13E) |
| | (14A) | (14B) | (14C) | (14D) | (14E) |
| | (15A) | (15B) | (15C) | (15D) | (15E) |
| | (16A) | (16B) | (16C) | (16D) | (16E) |
| | (17A) | (17B) | (17C) | (17D) | (17E) |
| | (18A) | (18B) | (18C) | (18D) | (18E) |
| | (19A) | (19B) | (19C) | (19D) | (19E) |
| | (20A) | (20B) | (20C) | (20D) | (20E) |

**FIGURE 12.12** End-of-course survey.
*Source*: Courtesy of Illinois Fire Chiefs Foundation

against these factors, which are commonly encountered in poor training:

- Inconsistency of information between fire service instructors
- Failure to adhere or enforce safety practices
- Unprepared or unknowledgeable fire service instructors
- Not enough student involvement in the training session
- Unclear instructions or learning outcomes
- Poor presentation skills
- Repetitive and boring material

Most of these factors can be addressed through better oversight of the instruction process. End-of-course surveys **FIGURE 12.12** or peer review forms **FIGURE 12.13** that are completed by participants can provide valuable insight into the quality assurance part of the instructional process.

**Peer Review**
**Presentation Evaluation**

Student Name ______________________

Department ______________________

Topic ______________________

Start Time ____________ End Time ____________

**Communication Skills**
1: Poor; 2: Fair; 3: Average; 4: Good; 5: Excellent

| | | | | | |
|---|---|---|---|---|---|
| Voice Levels: | ☐ 1 | ☐ 2 | ☐ 3 | ☐ 4 | ☐ 5 |
| Eye Contact: | ☐ 1 | ☐ 2 | ☐ 3 | ☐ 4 | ☐ 5 |
| Mannerisms: | ☐ 1 | ☐ 2 | ☐ 3 | ☐ 4 | ☐ 5 |
| Appearance: | ☐ 1 | ☐ 2 | ☐ 3 | ☐ 4 | ☐ 5 |

**Presentation**

| | | | | | |
|---|---|---|---|---|---|
| Introduction: | ☐ 1 | ☐ 2 | ☐ 3 | ☐ 4 | ☐ 5 |
| Body Presentation: | ☐ 1 | ☐ 2 | ☐ 3 | ☐ 4 | ☐ 5 |
| Conclusion: | ☐ 1 | ☐ 2 | ☐ 3 | ☐ 4 | ☐ 5 |

**Comments:**

**FIGURE 12.13** Peer review form used for instant class feedback.

*Source*: Courtesy of Southwest United Fire Districts Training Division

**Safety Tip**

If safety is not adequately addressed, take immediate action. Ensure that students do not leave the training session without a thorough understanding of the safety aspects of the topic covered.

## Safety in Instruction

One criterion that outweighs all others in the evaluation of fire service instructors and supervision of training is safety. Making sure that the evolution accurately reflects the challenges of the real emergency scene requires that some exposure to hazards occur during training. Chapter 9 is devoted to safety during the learning process, and its key points should be a primary focus during every training session. The fire service instructor-in-charge is responsible for ensuring that applicable standards are enforced during the training session and local SOPs are followed at all times. When developing instruction in scenarios with increased hazard exposure, make sure that the staffing of safety-related positions—for example, instructor-in-charge, safety officer, IC, and assistant fire service instructors—is included in every training assignment.

As the use of an Incident Command System (ICS) is required for use during emergency scene operations, the best practice for instructors to follow is to staff all positions used by the local AHJ in its ICS policies during training events. This allows for compliance in this very important part of the standards process and also provides beneficial training for

### Safety Tip

Always provide adequate fire service instructor staffing levels commensurate with the risks of the evolutions and departmental policy.

those who function in those capacities, a part of the training delivery system that may often be overlooked. Remember, the best training scenarios are representations of the actual fire ground procedures used by the department.

Each of these roles must have key responsibilities and functions defined for each training session. Departmental SOPs and NFPA 1403 provide guidance in this respect. As the manager of the training program, you are responsible for ensuring that the implementation of the incident management system is practiced correctly at the live fire exercise: You must practice how you train. Adherence to safety rules and regulations includes wearing proper PPE and having appropriate backup resources in place. Prior to conducting training, hold briefings about expectations, outcomes, contingency plans (including mayday procedures and evacuation signals), and the communication process. Everyone must be aware of his or her individual assignment and role.

The live fire IC is responsible for monitoring of the communication process, use of the accountability tracking system, and appropriate rehabilitation of students and fire service instructors, just as the IC is on the fire ground. Thus this type of exercise provides training experience for the IC as well as the other participants. If an IC-in-training fills the IC role during training, a fire service instructor or experienced IC should also be assigned to make sure that all duties and responsibilities are performed and no safety or operational issues are missed.

Examples of increased hazard training exercises include live fire training, acquired structure use, training in simulated immediately dangerous to life and health (IDLH) conditions, and limited-visibility operations such as training where simulated smoke is used. Assign additional fire service instructors and safety officers at training sessions in which students will work from heights (e.g., on roofs) or in which students will operate equipment that can be pressurized or energized. Other training sessions that may increase the hazard exposure to students include evolutions that require heavy lifting and training on new equipment. Multiple-company evolutions in which multiple tasks are performed at the same time may require additional oversight to ensure that safety practices are followed and the expected outcomes of the training session are achieved.

# Wrap-Up

## Chief Concepts

- The Fire Service Instructor II must take into consideration many aspects of the department's operation when developing a training schedule.
- Each department has specific issues that must be addressed while organizing training schedules.
- Understanding the budget development process and the Fire Service Instructor II's role in purchasing, specifying resources and materials, and justifying requirements of a training project or program is crucial to successful program management.
- Purchasing of resources and products used in the instructional area is a management skill.
- The Fire Service Instructor II must ensure that the applicable standards, SOPs, and departmental practices are used in the development and execution of training sessions that involve increased hazards to students.

## Hot Terms

**Agency training needs assessment** An assessment performed at the direction of the department administration that helps to identify any regulatory compliance matters that must be included in the training schedule.

**Budget** An itemized summary of estimated or intended revenues and expenditures.

**Expenditures** Money spent for goods or services.

**In-service drill** A training session scheduled as part of a regular shift schedule.

**Master training schedule** A form used to identify and arrange training topics by the number of times they must be trained on or by the type of regulatory authority that requires the training to be completed.

**Revenues** The income of a government from all sources, which is then appropriated for the payment of public expenses.

**Single-source products** Products that are manufactured or distributed by only one vendor.

**Supplemental training schedule** An alternative training schedule that is used in the event that some mitigating factor requires a change in the original training session.

## Fire Service Instructor *in Action*

It is the fall of this year, and you are charged with developing the department's training program for the upcoming year. You are to identify budgetary requests, specify purchases of equipment and materials, and acquire training resources needed for the training program. Once these determinations are made, you must identify recognized standards and procedures on which to base your program and supervise the delivery of training and adherence to safety practices and procedures by the instructors.

**1.** Which type of budget is documented as a series of line-by-line categories of expenses or revenues?
   **A.** Line-item
   **B.** Capital
   **C.** Zero base
   **D.** Revenue forecasting

**2.** Which administrative process is used to determine the priorities and functions of the training program?
   **A.** Budget justification
   **B.** Training needs assessment
   **C.** Instructor meetings and discussions
   **D.** Task analysis

**3.** Which training activities may require assignment of additional instructors to observe safety-related details?
   **A.** A video presentation
   **B.** Any training session characterized by increased hazards
   **C.** A small-company discussion of fire-ground tactics
   **D.** A single-company drill reviewing water supply options

**4.** Which type of training do you need to consider when developing a training schedule for the department?
   **A.** Specialty area training
   **B.** In-service company training
   **C.** Training conducted by outside agencies that use the department's training facilities
   **D.** All of the above

**5.** While scheduling training for the next year, you want to reference professional qualification standards as put forth by NFPA. Which prefix number will these standards begin with?
   **A.** 10
   **B.** 14
   **C.** 15
   **D.** 19

**6.** You determine that a new facility will be required to meet the department's training needs within the next 5 years. Which type of expenditure would this be considered?
   **A.** Operational expense
   **B.** Capital expense
   **C.** Formal expense
   **D.** Supply expense

**7.** Which of the following statements about use of equipment during training is true?
   **A.** Hand-me-downs should not be used.
   **B.** Loaner equipment should not be used.
   **C.** Hand-me-downs may be used but their limitations must be understood.
   **D.** New equipment should not be used.

**8.** Training that occurs around the station and is often presented by the company officer is called
   **A.** Formal training.
   **B.** Informal training.
   **C.** Special operations training.
   **D.** In-service training.

# The Learning Process Never Stops

# CHAPTER 13

## NFPA 1041 Standard

NFPA 1041 contains no Instructor I or Instructor II Job Performance Requirements for this chapter.

## Knowledge Objectives

After studying this chapter, you will be able to:

- Identify and discuss the importance of continuing learning for the fire service instructor.
- Identify professional organizations that will help in the professional development of the fire service instructor.
- Identify and discuss the value and importance of coaching and mentoring the next generation of fire service instructors.

## Skills Objectives

There are no skill objectives for this chapter.

## You Are the Fire Service Instructor

At your annual department awards ceremony, the fire fighter receiving the award as rookie of the year comes to the microphone to give his acceptance speech. After the usual jokes and comments about his crew, he thanks his captain for setting a good example; he thanks his wife for her support during his year on the department; and then he thanks the training officers who helped him get through his initial training at the academy. He mentions your name specifically, emphasizing the support and effort you gave him. You reflect on this fire fighter and his classmates and realize how important your efforts are in training the next generation of fire service professionals.

1. How did your own professional development influence your skills as a fire service instructor?
2. What are you doing now to improve your skills as a fire service instructor?
3. How can you prepare the next generation of fire service instructors in your department?

## Introduction

The purpose of this book is to prepare you to be a fire service instructor. So far, you have learned the information and skills necessary to prepare a lesson plan, to teach from a lesson plan, to arrange for resources, and to evaluate both students and other fire service instructors. You have learned about budgeting techniques, supervisory skills, communication strategies, and ways to use today's technology in presenting material. So what's next? Have you reached the height of your career, or is there more to accomplish?

For some fire service personnel, teaching at a recruit school or being assigned to the training division is a momentary and brief assignment in their upward progression within a fire department. Some of those assigned to teach simply tolerate that duty, but for others teaching is exactly where they want to be—filling the role of a fire service instructor who is in a prime position to teach and influence current and future fire fighters. The role of a fire service instructor can be a formal designation or an informal one. Regardless of the designation, you are in a position to influence those around you through your actions, your words, and—most importantly—your example.

Once you are in the fire service instructor position and have received the training and certifications to prepare you for the job, you need to continue with your own growth and development. Reaching the position is just the beginning; it's the ongoing learning process that will make you a truly excellent fire service instructor. Learning is a lifelong process. Many fire service instructors nationwide go on to publish reports of their work in trade journals or books. When their ideas and lesson plans become publicly known through the literature, other fire service instructors have a starting point from which to build their own ideas and lesson plans. Publishing also adds to the creditability of the fire service instructor and helps in ensuring that the fire service is recognized as a profession.

**FIGURE 13.1** Reading is a great way to continue personal development.

## Lifelong Learning and Professional Development

Recall the fire service instructors who shaped and influenced your development. What is the common thread among these individuals? Most likely, they always had a magazine or an article in their hand or their nose in a book **FIGURE 13.1**. The

### Teaching Tip

When teaching a class, title an easel board "Resources." Over the course of the class, add resource information to this list and encourage students to do the same. Typically, this information will include Web site addresses, books, and names of other fire service instructors.

individuals who actively sought out opportunities to learn were often the ones called upon to teach at a moment's notice because they had a wealth of knowledge that others wanted to hear. The same individuals attended conferences to learn and focused on networking with their peers. Eventually, these individuals became presenters after obtaining a thorough knowledge base and developing their own ideas and theories to share with the fire service.

Another way to look at this idea is through an analogy. The speaker who is often heard at various fire service conferences did not start out as the keynote speaker. At one time he or she was the fire fighter sitting in the audience, taking notes, and listening to every word. This person took what was learned at the conference, returned to the firehouse, and looked for ways to put that new knowledge into practice. As he or she learned more and advanced into higher positions within the department, the fire service instructor had the opportunity to teach new concepts and ideas to the next generation. The ideas, concepts, or theories that this individual taught stood the test of time and became new policies and procedures followed within the department. Soon, word of mouth spread about these new ideas. Now the fire fighter who once sat in the audience taking notes is on the stand leading the discussion and is respected by his or her peers as a leader in the profession. All of this came about because of the individual's desire to learn and to develop—because the fire fighter was a true lifelong learner.

### Safety Tip

As a fire service instructor, you may feel that you have finally "arrived" and are at the top of your game. The fire service is constantly evolving, however. If you are not continuously learning, you are outdated. Become an avid lifelong learner to avoid this complacency.

An important part of becoming a great fire service instructor is having the basic desire to always improve and become a better fire service professional. Who benefits when you develop into a respected fire service instructor? Both you and the fire department, because a good fire service instructor supports the department's goals and the department recognizes the instructor's critical value to the fire service.

## Time Management

As grand and glorious as being a fire service instructor sounds, it also carries a great deal of responsibility. A department is often judged by the performance of its fire fighters. When a fire fighter is injured or killed in the line of duty, one of the first items reviewed and scrutinized is the training records and the worthiness of the fire service instructors. The need to keep up-to-date on all changes and improvements in the fire service is a difficult, but a critical task. How do you stay informed, do your job effectively, and still maintain your personal life?

As a fire service instructor, one of the first skills you need to learn and develop is time management. Several methods or techniques are available to help you make the best use of your time. You should use techniques that suit your personality and are easy and convenient to use. For example, setting a goal to read one magazine article per shift is a good start. Attending one fire service conference a year and earning a new certification are goals that will help in your development.

One helpful method is to set time aside each day to create a "to do" list. Prioritize the items on this list, and then work your way through the list, checking off what you have accomplished. Included on this list might be reading an article or reviewing a department SOP. Another method is to set goals for the year and break the goals down into months and weeks. Use whatever method works for you: The key is to do something that fits your needs.

## Staying Current

Subscribe to trade magazines to stay up-to-date on trends and new developments in the fire service. Also read other professional magazines in related fields such as homeland security or business. The benefit from reading trade journals is that it helps you stay current in the fire service. The benefit from reading material outside the fire service profession is that it helps you develop into a more well-rounded individual and professional.

### Safety Tip

Information is only as good as its source. Be sure to verify your source of information before you include that information in your lesson plans, particularly if it is potentially harmful to students.

A relatively new tool that is also available to you as a fire service instructor is the Internet. The World Wide Web has opened the door to resources that were difficult to obtain in earlier years, but now are accessible to anyone who has a computer and Internet access. The numerous Web sites hosted by colleges, professional journals, professional organizations, federal and state fire academies, and government agencies have volumes of information available. In fact, there is so much information available on the Internet that you may struggle with the task of limiting what you can read and use.

One drawback to the information found on the Internet is the sometimes questionable accuracy of the content presented. If you use material from the Internet, you need to verify the authenticity of the source. More than one fire service instructor has taken information from the Internet and used it in a lesson plan, only to find that it was inaccurate or completely wrong. This, in turn, hurts the reputation of the fire service instructor. The Internet is a tool and, just like with any other tool, you need to know how to properly use it and when to use it.

## Higher Education

Another method you can use is to pursue higher education. Most departments require fire fighters to have a high school diploma or a GED as a condition of employment. Many community or state colleges now offer degrees in fire science, emergency management, or communication. Although this coursework is in your discipline, colleges also require you to take general education courses such as history and English literature as part of the degree program, so that you will become a well-rounded thinker. Higher education has become more accessible today than ever before, with degrees being offered by traditional colleges as well as online programs. Both of these delivery methods provide the opportunity to earn an advanced degree and leave you the choice of which methodology you wish to pursue.

More departments have begun requiring an advanced degree for promotion, such as an associate degree for the captain's position and a bachelor's degree for the battalion chief's position. Other departments offer financial incentives for engaging in continuing education, paying for college or even giving fire fighters a raise for each college-level course they complete.

A major incentive for obtaining an associate or bachelor's degree is the requirement in place at the National Fire Academy that requires an advanced degree to be eligible to enter the Executive Fire Officer Program (EFOP). Currently an associate degree is required, but beginning in October 1, 2009, a bachelor's degree will be required for entrance into the EFOP. The EFOP is considered by many to be a key to promotion as a chief officer.

### Theory into Practice

Attending college can be a time-consuming and expensive proposition. You may be willing to make the time commitment, but your finances may not be flexible enough to handle the entire cost. If so, take some time to investigate the many possible funding sources for which you may qualify. For example, your fire department or city may have a reimbursement program. The federal government provides grants that are awarded to thousands of students every year. If you served in the armed forces, you may qualify for funding as a veteran. Many local charitable organizations also provide scholarships to students, including nontraditional students. Don't be locked into thinking that you can't attend college because of money concerns. If you are willing to make the commitment and expend the effort, funding sources are available.

## Conferences

Attendance at conferences, workshops, or seminars also offers you a chance to increase your skills and knowledge as a fire service instructor FIGURE 13.2. The many professional conferences available provide opportunities for growth and development at all levels. Some conferences provide hands-on training opportunities, whereas others are limited to classroom lectures or presentations. Regardless of which type of conference you choose to attend, take advantage of each and every opportunity you have to learn from your peers.

Take advantage of the opportunities at these conferences to network with other fire service instructors from around the country. Such conferences are great ways to learn new presentation skills that you can add to your own toolbox. Networking is a fundamental strategy used by great fire service instructors to learn from others and to bounce an idea off of a peer to determine if it is valid. Networking is a powerful tool in your toolbox that can open the door to sharing or collaborating on an idea and developing a new method.

### Theory into Practice

Although networking is often associated with politics and schmoozing, what you are really doing is creating a web of friends. The bigger the web, the more likely you are to have access to the precise resource you need when the time comes. Networking is designed as a give-and-take system.

The best way to create a large web is to be a giver. When someone needs something that you can provide, do it. If someone has a question, answer it. Helping others be successful creates a powerful network that can work to your benefit. By giving, you become the person who is sought out: The more sought out you are, the more contacts you will have. Develop your network not by schmoozing, but by being genuinely interested in helping others to succeed.

FIGURE 13.2 Conferences are an excellent self-improvement activity.

## Applying the JOB PERFORMANCE REQUIREMENTS (JPRs)

This chapter is the culmination of this entire book and sets the stage for the ongoing process of instructor professional development. As a professional fire service instructor, you have a responsibility to keep up-to-date on the educational trends, legalities, and professional organizations and resources available to you. Just as we hold fire fighters, driver/operators, and officers responsible for their professional development, fire service instructors are also required to continue to learn and practice their craft. Given the rapid rate of technological changes occurring and the increased workloads being placed on instructors, you should identify a peer work group and organizations that will be able to assist in your continuing educational missions. Part of your professional development will include the selection and mentoring of future instructors. This is an opportunity to pay back the fire service and improve it by sharing your experiences and guidance.

### Instructor I

The Instructor I must keep current on the latest instructional delivery techniques, media, and resources. Membership in state and national instructional organizations, if possible, is key to your professional development.

### Instructor II

The Instructor II will continue his or her learning by keeping up-to-date on recent legal decisions and methods of preparing instructional materials. Networking through trade shows, attending conferences, and taking courses designed to bring new information back to your department will help keep you motivated.

### JPRs at Work

NFPA does not identify any JPRs that relate to this chapter. Nevertheless, the concept of continuing educational and professional development in every job level is embraced by virtually every major fire service organization. All levels of instructors must continually refine their skills and knowledge throughout their entire career.

### JPRs at Work

NFPA does not identify any JPRs that relate to this chapter. Nevertheless, the concept of continuing educational and professional development in every job level is embraced by virtually every major fire service organization. All levels of instructors must continually refine their skills and knowledge throughout their entire career.

### Bridging the Gap Between Instructor I and Instructor II

Each level of instructor should now be prepared for the next level of instructional certification. The Instructor I has learned important skills in the area of course delivery, while the Instructor II has developed a series of skills in writing lesson plans, conducting evaluations, and developing media for many areas of coursework. The Instructor I should now begin taking the coursework necessary for certification at the Instructor II level and, with the help of a mentor or experienced instructor, be able to cross the gap to Instructor II successfully. The Instructor II is now ready to begin certification training for Instructor III responsibilities in curriculum design, budget and finance, and administrative areas of managing a training program.

In all cases, each level of instructor has a responsibility to the fire service and his or her organization to mentor the next generation of instructors. You can learn from them just as much as you will help them learn.

### Instant Applications

1. Identify at least three statewide organizations and three national organizations that are available to assist all levels of instructors in their professional development.
2. Review your local or state certification process to learn which affiliations or correlations exist in accordance with the Fire and Emergency Services Higher Education (FESHE) program, discussed later in this chapter; "The 16 Firefighter Life Safety Initiatives" developed by the National Fallen Firefighters Foundation; and NFPA standards relating to training and education.
3. Review the JPRs specific to your next level of instructor certification, and identify opportunities to experience learning in those areas.
4. Select an instructor or instructor candidate who has specific talents or background that can be used to improve a particular course. Urge that person to become involved in the training program.

## Lifelong Learning

Lifelong learning is an endless path with endless opportunities. For the one person who says he has seen it all, there are a hundred other individuals who ask for more. A major career killer is adopting the attitude that you can't teach students anything new. Your greatest challenge is to create the desire to learn in all of your students.

In your pursuit of becoming a lifelong learner, you need to remember the basics. Your teaching skills got you where you are, so continually brush up on your existing presentation skills and continue to learn new ones. Remember the teacher in high school whom no one liked because he taught as if he was still locked in the 1960s? Remember the teacher who constantly stayed current and fresh to students?

Keeping your teaching skills sharp and current is just as important as reading magazines, networking, or going to college. If you are the senior fire service instructor in the training division, make a point of teaching one subject once a month so that you keep your teaching skills up-to-date. As you have grown in your knowledge, have you stayed current with the types of students entering the fire service today? Are you still an effective fire service instructor for the new generation of fire fighters? Test yourself by teaching a class. Be honest with yourself about how effective you were. Great fire service instructors are lifelong teachers who love to teach.

**Teaching Tip**

Part of being an effective fire service instructor is learning to scan the horizon for industry trends and anticipate which will pan out. Part of the scanning process involves looking for new ways to add to your skills and abilities. Attending a one-day skills development program and reading a new book are two ways to stay on the cutting edge of your profession.

## Professional Organizations

There are many professional organizations that you can join to help in your professional development. Some organizations operate on the national level, but most are at the state, local, regional, or county level. Regardless of the level, these organizations provide many opportunities to network, share resources, and develop contacts. Some are very formal and highly structured, whereas others are more informal and relaxed.

The International Society of Fire Service Instructors (ISFSI) is an organization that is committed to the development of fire service instructors. ISFSI actively supports the professional development of the fire service through its push for qualified fire service instructors teaching quality material.

The National Fire Academy, which is part of the National Emergency Training Center in Emmitsburg, Maryland, is the nation's fire university. At the National Fire Academy, you can attend 6-day or 2-week courses that teach you how to develop lesson plans and build courses. In addition, other courses are available to help in all aspects of your professional and educational development. The National Fire Academy also supports two other programs that will help in your professional development: the Fire and Emergency Services Higher Education initiative and the Executive Fire Officer Program.

A key part of what the National Fire Academy is doing to professionalize the fire service can be seen in the Fire and Emergency Services Higher Education (FESHE) initiative. Across the United States, many different community and state colleges and universities offer degrees in different aspects of fire service management or engineering, but there are no standardized model degree programs. The FESHE initiative is intended to bring together representatives from colleges and universities to develop a standardized degree format and model curricula. If college degree programs follow the same set of rules, students will be able to transfer coursework from one school to another more readily. In addition, courses would present a standard curriculum, which increases the ability to transfer college credits and standardizes the information taught. Another purpose of the FESHE initiative is to develop a career path for those seeking to be chief officers.

Part of this development process has resulted in the National Professional Development Model. This model summarizes the combination of education, training, and experience that leads to the development of a well-rounded and fully developed fire officer who has the skills and ability to lead the fire service into the future.

The Executive Fire Officer Program (EFOP) is another program developed by the National Fire Academy that has helped to professionalize the fire service. Its purpose is to provide senior officers with the tools they need to become leaders of the future. Acceptance into the EFOP is by application process only.

The EFOP requires a major time commitment from both the fire officer and the fire department. The program takes four years to complete, with attendance at the National Fire Academy to four 2-week courses. At the end of each course, the student is required to submit an applied research project (ARP), which is then evaluated by staff members at the National Fire Academy. Once the ARP is approved, the student is allowed to register for the next course in the series.

## The Next Generation

A true mark of a great leader is his or her willingness to plan to be replaced. **Succession planning** is the process of identifying others within the department who could be mentored and coached to replace you and to carry out the functions that you perform.

The concept of training your replacement can be very difficult and challenging to deal with on an emotional level. Things may be easier if you understand that the process of training

**Teaching Tip**

Become certified by an accredited organization to give credibility to your competencies.

# Voices of Experience

We had been given an old building to use to for live fire training and we jumped on the opportunity to use it. The date of this training was before the development and adoption of NFPA 1403, *Standard on Live Fire Training Evolutions*. Our department had developed its own pre-burn skills checklist to use on acquired structures, so we felt that we had a good process by which to evaluate. Several weeks before the live fire drill was to be conducted, a group of fire service instructors preplanned the building, made drawings, checked the water supply, and did everything else we could think of to have a safe training activity.

The night of the drill, we divided the companies into teams, set up two water sources, pulled attack lines, and did a pre-burn walk-through of the structure. We used green hay in the back room to generate smoke without much heat so that everyone would have a chance to get in on the burns.

As the first fire was set, the interior fire service instructors awaited the first crews to make entry but nothing happened. As the fire service instructors waited in the front room, they could tell the fire was growing rapidly and became concerned because it was more fire than there should have been.

Just then, the safety officer called for everyone to exit as the structure began to "puff." At the same time, the fire service instructors in the front room noticed a sudden shift in the air movement. Before anyone inside could reach the exit, the back room flashed over and sent fire belching down the hallway, blowing the ceiling down onto the fire service instructors in the front room. The emergency evacuation signal sounded from the main engine and everyone on the outside scrambled to get to the fire service instructors inside. Everyone was okay and made it out without any injuries. After a full investigation, it was determined that fumes from glue that had been used to hold carpet on the walls of the room had enhanced the fire behavior in the room and caused a flashover. This is an example of the fact that "the learning process never stops." There are many incidents that are completely unpredictable, even for the skilled instructors.

*"Our department had developed its own pre-burn skills checklist to use on acquired structures, so we felt that we had a good process by which to evaluate."*

*Alan E. Joos*
Fire and Emergency Training Institute, Louisiana State University
Baton Rouge, Louisiana

FIGURE 13.3 Through succession planning, you ensure that your influence will continue through your successor.

another person to take your place usually means that you are also moving up and on to something new. Another benefit of training your replacement is that you can select someone who shares your own ideals and goals, thereby ensuring that your training program will continue on after you leave. Your influence could be felt for many decades in a department FIGURE 13.3.

How do you train your replacement and help him or her to be successful? Four basic concepts will aid you in teaching your replacement: identifying, mentoring, coaching, and sharing.

## Identifying

**Identifying** is the process of selecting those whom you would like to mentor, coach, and develop. The identification process almost requires a "crystal ball" approach of looking at individuals—that is, you need to see the person today and forecast who the person can be tomorrow. Look for these qualities:

- The quick learner. Look for those who truly understand how, why, and when (and when not).
- Individuals who expect 110 percent and inspire others by leading by example.
- Fire fighters with a passion for the fire service. Passion is contagious and can push a crew or a training division to move to the next level.

There are many important questions to ask the potential instructor. Oral interview questions may include:

1. Assume that you are hired to be the Training Officer and have been approached by members of a fire company who complain that their officer treats training as an unimportant issue and spends as little time as possible on drills. The members respect the officer as a good fire ground commander but he has no appreciation for training and does little to support the training program. What action would you take in this situation?
2. Why are you interested in becoming the Training Officer? What have you done to prepare for this position?
3. What personal attributes that you possess will be of the greatest benefit to you in fulfilling the duties of this position?
4. Do you have any plans for personal development which will benefit you in this position?
5. Throughout the course of the year, members of the department may need motivation, direction, and guidance regarding professional training and education needs. Do you see this as part of your role as Training Officer? Why or why not? If so, briefly describe your thoughts on how you can best provide this need.
6. What is your top priority or priorities that would have to be accomplished in the first three (3) months?
7. Describe specifically how you will gain the support, backing, and assistance of those company officers who do not consider training a priority.
8. One of the responsibilities of this position will be to function as the on scene safety officer at incidents. Have you had prior experience in this type of position? Also, explain your philosophy and approach to the position of scene safety officer.
9. How would you structure, document, and provide follow-up to ensure that all personnel complete required training?
10. How would you structure training sessions to accomplish training and maintain district coverage?
11. Assuming this department is a combination department of full time and part-time personnel, what additional challenges do you anticipate in developing and conducting training for this type of structure?
12. What is your philosophy regarding the role of the company officer in training?

Written/practical evaluation questions may include:

1. Prepare a written plan (not to exceed 3 pages—single-sided) describing how you will structure, conduct, and document the SCBA training program.
2. Full time personnel hired by the department are required to have prior certification as a Fire fighter II. These individuals once hired are placed directly on shift. A current training issue is to develop a forty (40) hour program to provide an orientation to the department and complete a basic assessment of skill levels. Prepare a written plan (not to exceed 3 pages—single-sided) of how you would structure, conduct, and document a new fire fighter orientation/assessment program.
3. An important aspect of department training programs will be the addition of certification courses that are in compliance with State Fire Marshal requirements. Prepare a written plan (not to exceed 3 pages—single-sided) detailing the steps that will be needed to implement for an in-house fire apparatus engineer program.

## Mentoring

**Mentoring** is a tool you can use to nurture and develop a fire fighter who has the basic skills and abilities to be a great fire service instructor. As a mentor, you select an individual to begin the process as either a formal or an informal protégé. Your role as a mentor is to demonstrate how to perform

**Ethics Tip**

Is it ethical to choose who gets your personal attention and support for success and who does not? This can be a painful question to answer. While you want everyone to have equal opportunities and chances of success, you also have limited resources. The reality is that you have an obligation to the fire service. Develop a process of screening to ensure that those who are most capable and willing to carry the torch to future generations are given additional support and mentoring.

functions or skills and to explain and answer questions. Most importantly, you act as a sounding board if your protégé has a problem. The following ideas form the foundation for a solid mentoring program:

- The mentor's job is to promote intentional learning and provide opportunities for the protégé to learn and grow. This does not mean you throw your protégé to the wolves and walk away; rather, it means that you give your protégé opportunities to teach or demonstrate a skill while you watch.
- As a mentor, you need to share your experiences, both good and bad. Learning comes from both positive and negative experiences.
- The mentoring process takes time, and patience needs to be shown on the part of both the mentor and the protégé.
- Mentoring is a joint venture and requires both parties to be actively involved in the process.

**Teaching Tip**

Sometimes the identification and selection process doesn't work. You might select a person whom you feel is ready for mentoring, but he or she does not respond to the process. If this happens, there still is an opportunity to learn and grow from the experience by parting on good terms and leaving the door open to future contact.

## Coaching

**Coaching** is a process usually associated with sports. The coach is a person who has experience and assists players in developing and improving their skills. Coaching is most often tied to manipulative skills development and physical training. The difference between mentoring and coaching is subtle. Coaching is the process of helping an individual see his or her abilities and coming up with solutions independently. A coach encourages through words but allows the player to develop his or her own skills. A good coach helps a player see his or her potential and then gives the player the freedom and support to achieve it.

As a fire service instructor, you must have expertise in numerous skills, such as firefighting skills for all levels you could be expected to teach. In addition to your technical skills, you have another set of skills at your disposal—namely, the skills used in the presentation of lessons, lesson and curriculum development, and use and development of evaluation tools. Part of the coaching process is allowing the fire fighter you are mentoring the opportunity to develop these skills as well.

Coaching also requires that you evaluate those persons whom you are grooming to replace you. These evaluations may take the form of either formal or informal evaluations or critiques. Just like teaching a child to ride a bike, you must help your protégé up to a point, and then allow him or her to go on alone. You might run alongside the bike for a while, but as the child gets the hang of it, you back off. So it is when coaching a new fire service instructor: You teach your protégé how to do a presentation during training, and then the fire fighter gets to do it for real in a recruit class while you stand in the back of the room and watch. Afterward, you give some advice, highlighting good points and offering suggestions for rough areas. This feedback helps your protégé to grow and improve. Coaching involves taking a "big brother" or "big sister" approach toward helping a person use the skills that he or she has and then teaching the person to do even better.

## Sharing

The last skill in succession planning is the basic concept of **sharing**. Sharing is a key part of mentoring and coaching. To be successful in succession planning, you must mentor and coach with a giving attitude. Not only do you need to take your protégé under your wing, but you need to do so with a sharing attitude. You must be willing to share everything that has made you such a successful fire service instructor. This includes your skills and knowledge, as well as all of the little nuances that have made your job easier and fun. All too often, those charged with the task of training their replacements hold back the little tricks of the trade. Some have the attitude that, "I had to learn it myself, so can you!" This attitude hurts both the department and your reputation. Share your knowledge and insight with your protégé as he or she takes on the responsibilities of training others. Give your protégé the extra step up rather than making him or her repeat all of your mistakes and waste all of that extra time and effort. When you share what you know with others and help others excel, you benefit both personally and professionally.

When you approach succession planning with this positive attitude and mentor the next generation of fire service instructors by teaching and coaching with a caring and sharing attitude, both you and your department move forward. Leave a legacy of professionalism that will outlive your time on the job.

# Wrap-Up

## Chief Concepts

- Professional development is a lifelong process.
- Professional development can take the form of training, certifications, and formal education.
- Professional development can also be achieved by participating in local, state, and national organizations and attending training conferences.
- Succession planning is the process of developing individuals who can be mentored and coached to take the lead in the future of the department.

## Hot Terms

**Coaching** The process of helping individuals develop skills and talents.

**Identifying** The process of selecting those whom the fire service instructor would like to mentor, coach, and develop.

**Mentoring** A relationship of trust between an experienced person and a person of less experience for the purpose of growth and career development.

**Sharing** The basic concept of giving to others with nothing expected in return.

**Succession planning** The act of ensuring the continuity of the organization by preparing its future leaders.

## Fire Service Instructor *in Action*

You have been promoted to deputy chief of operations. You are to begin your new assignment in 1 month and have been asked to identify your replacement from your staff in the training division. Several members of your staff are very qualified and capable of assuming your duties. You also are aware of several other candidates in the department who have worked for you at times during recruit classes. The chief asks for your suggestions before the end of the week.

**1.** What is the process of training your replacement called?
   **A.** Mentoring
   **B.** Coaching
   **C.** Succession planning
   **D.** Promotion

**2.** What is the process of helping a fire fighter to learn and develop a new skill by helping the fire fighter to set his or her own goals called?
   **A.** Mentoring
   **B.** Coaching
   **C.** Promotion
   **D.** Evaluation

**3.** What is the goal of a mentor?
   **A.** To provide intentional opportunities for his or her protégé to learn and grow
   **B.** To evaluate his or her protégé to determine if that individual is ready to take the helm
   **C.** To identify individuals who may be able to lead the department
   **D.** To develop every individual in the department

**4.** Which of the following is a vast new source of information that is available but may not always be reliable?
   **A.** Fire Emergency Training Network
   **B.** *Firehouse* magazine
   **C.** YouTube
   **D.** U.S. Fire Association Web site

**5.** Which of the following is *not* a method of professional development?
   **A.** Reading *The Wall Street Journal*
   **B.** Attending fire service conferences and seminars
   **C.** Reading fire-related material on the Internet
   **D.** All are methods of professional development

**6.** Which of the following is *not* a trait to look for when selecting a replacement?
   **A.** Someone who is willing to learn
   **B.** Someone who works for someone else
   **C.** Someone who is friendly
   **D.** Someone who has humility

# Analyzing Evaluation Instruments

## Introduction

A great deal of well-deserved attention has been given to the procedure for developing valid and reliable test items. However, it is equally important to follow up each administration of the test items with a post-test item analysis. This analysis will allow you to look at each item on the test in terms of its difficulty, or P+ value, and its ability to discriminate; this analysis also allows you to ascertain the reliability of the overall test. You can then use these data to identify potentially weak or faulty items, scrutinize each item to identify the possible cause of a weakness or fault, and make necessary improvements.

## Analyzing Student Evaluation Instruments

After a test has been administered, an analysis is conducted to find out how difficult the test items were, whether the test items discriminate between respondents with high or low scores, and if the results were consistent.

## Training Objective

Using data from one administration of a test and formulas for computing item difficulty (P+ value) and discrimination, conduct a post-test item analysis. This objective will be met when the P+ value and discrimination index have been computed for at least 10 items on the test, and each item has been identified as acceptable or needing review.

## Task Title

Conduct Post-Test Item Analysis

## Task Elements

1. Gather data.
2. Compute the P+ value.
3. Compute the discrimination index.
4. Compute the reliability index.
5. Identify items as acceptable or needing review.

## Gather Data

Immediately following the administration and grading of a test, it is important to collect information on the responses to each item. For each item, you need to know how many times each alternative was selected—that is, whether the alternative was the keyed (correct) response or not. On a tally sheet, tally each participant's response to each alternative on each test item.

To see how this process works, let's use a multiple-choice test item from a 10-item test that was administered to a group of 12 students. This particular test item had four alternatives—A through D—and B was the keyed response (i.e., correct answer). After tallying each participant's response to each test item, you are able to determine how many correct responses were marked and how many incorrect responses were marked for each distracter. Now, as you begin to analyze one test item at a time, isolate the data FIGURE A.1:

**Figure A.1 Isolation of Data**

| Item Number | Alternatives | | | | Responses | |
|---|---|---|---|---|---|---|
| | A | B | C | D | Right | Wrong |
| 1 | 1 | ***8** | 2 | 1 | 8 | 4 |

* Bold indicates keyed response.

Other types of test items will require different tally-sheet formats or provisions for more alternatives than the four shown here. FIGURE A.2, for example, provides space for 10 alternatives. Matching, arrangement, and identification test items require five alternatives; therefore, you will need to provide adequate space on your tally sheet.

**Figure A.2 Test-item analysis tally sheet**

| Item | Alternatives (Where * indicates the answer) | | | | | | | | | | Responses | |
|---|---|---|---|---|---|---|---|---|---|---|---|---|
| Number | A | B | C | D | E | F | G | H | I | J | Right | Wrong |
| | | | | | | | | | | | | |
| | | | | | | | | | Total | | | |

## Compute the P+ Value

The P+ value of a test item indicates how difficult that item was for the class taking the test. Your calculation of the P+ value will be useful later on as you revise a test item based on the results of your analysis. Here's the formula to calculate the P+ value of a test item:

$$\text{P+ value} = \frac{\text{Number right}}{\text{Number taking the test}}$$

$$\text{P+} = \frac{R}{N}$$

Now, let's apply that formula to our example test item. Eight of the 12 respondents answered the question correctly.

$$\begin{aligned} \text{P+} &= \frac{8}{12} \\ &= 0.666 \\ &= 0.67 \end{aligned}$$

The P+ value of that item is 0.67

The P+ value is merely a statistical indicator and should not be used by itself to judge the relative merit of a test item. Instead, it should be used in conjunction with the discrimination and reliability indexes.

## Compute the Discrimination Index

The discrimination index refers to a test item's ability to differentiate (or distinguish) between participants with high scores on the total test and those with low scores. Several formulas may be used to compute the discrimination index. Our example formula divides the group tested into three subgroups, as equal in number as possible, based on their grades on the overall test. The example test was administered to a group of 12 participants. We arrange their scores in descending order and divide them into three groups FIGURE A.3.

**Figure A.3 Group Determination**

| Score | Number of Respondents | Group (12/3 = 4) |
|---|---|---|
| 100 | 2 | High |
| 90 | 2 | High |
| 80 | 4 | Middle |
| 70 | 4 | Low |

Notes: When working with a group size that is not evenly divisible by 3, place equal numbers in the high and low groups and place the remaining numbers in the middle group.

It is important to have equal numbers in the high and low groups.

Once the groups have been determined, you can then calculate each test item's discrimination index. Another tally sheet, such as the one shown in FIGURE A.4, is useful for this purpose.

In this example, the four best grades constitute the high group and the lowest four grades constitute the low group. We now count the number of respondents in the high and low groups who answered the item correctly. In the high group, three respondents answered correctly; in the low group, two answered correctly. The number of respondents in the low group who answered correctly is subtracted from the number in the high group who also answered correctly. The difference (one) is then divided by one-third the total number of respondents (four).

The formula for computing the discrimination index follows:

$$D_s = \frac{H-L}{N/3} = \frac{3-2}{12/3} = \frac{1}{4} = 0.25$$

where:

H = number of respondents from the high group who answered the item correctly = 3

L = number of respondents from the low group who answered the item correctly = 2

N = number of respondents taking the test = 12

Number of groups = 3

The procedure follows these steps:

1. Sort test papers or answer sheets into high, middle, and low groups.
2. Tally all high-group and low-group responses to the item on the tally sheet in the space under "Test Item Alternatives."
3. Carry the number of correct responses for each group to the right column.
4. Insert the numbers representing the "respondents answering correctly" for the high and low groups into the formula to replace H and L, respectively.

$$\frac{H-L}{N/3} = \frac{3-2}{N/3}$$

5. Insert the total number of test respondents to replace N.

$$\frac{3-2}{N/3} = \frac{3-2}{12/3}$$

6. Complete the calculation to derive the discrimination index for that test item.

$$\frac{3-2}{12/3} = \frac{1}{4} = 0.25$$

The index of 0.25 indicates the degree to which the item discriminates between those respondents who obtained high total-test scores and those who obtained low total-test scores. Theoretically, the range of the discrimination index is from −1 to +1.

- An index of +1 results when all respondents in the high group and no respondents in the low group answer correctly.
- An index of −1 results when all respondents in the low group and no respondents in the high group answer correctly.
- If the same number of respondents from both groups answers either correctly or incorrectly, the index is 0 (the exact center of the range).

Here are some assumptions that can be made after studying discrimination-index trends:

- When more respondents from the high group than from the low group answer correctly, the item has positive discrimination.
- When more respondents from the low group than from the high group answer correctly, the item has negative discrimination.
- When the same number from both groups answers the item correctly or incorrectly, there is neutral or zero discrimination.

Test items that have very low positive discrimination, zero discrimination, or negative discrimination need to receive a quality review to determine their validity. If neutral or negative discrimination occurs with a test item of medium difficulty, you should determine the reason and correct that item before it is used again.

**Figure A.4 Test-item discrimination index tally sheet**

| Groups by Test Score | Test-Item Alternatives (Where indicates keyed response) | | | | | | | | | | Total Number of Respondents Answering Correctly |
|---|---|---|---|---|---|---|---|---|---|---|---|
| | A | B | C | D | E | F | G | H | I | J | |
| High third | 1 | 3 | 0 | 0 | | | | | | | 3 |
| Low third | 0 | 2 | 1 | 1 | | | | | | | 2 |

As mentioned earlier, the discrimination index has more meaning when it is used in conjunction with the P+ value (i.e., the measure of difficulty). Our example test item has a P+ value of 0.67 and a discrimination index of 0.25. Thus, although the P+ value is high, the difficulty level and the discrimination index are moderate to low. The answer to the question, "Is this good or bad?", depends on the purpose of the test item. Generally speaking, if both the difficulty level and the discrimination index are in the same range (low–low, medium–medium, or high–high), the test item is doing its job. An intentionally difficult yet valid test item should have a high discrimination index; an intentionally easy test item should have a low discrimination index.

## Compute the Reliability Index

The first two indexes are used to rate the difficulty and discrimination characteristics of a single test item. The reliability index, by comparison, rates the consistency of results for an overall test.

A test's reliability depends on three factors: the number of test items, the standard deviation, and the mean (average) of scores. As with the discrimination index, a number of formulas can be used to compute the reliability index. Two examples are the Kuder–Richardson Formula 21 and the Spearman–Brown Formula (also known as the split-half method).

**Kuder–Richardson Formula 21**

$$\text{Reliability} = \frac{ns^2 - M(n - M)}{ns^2}$$

where:
n = Number of test items
s = Standard deviation
M = mean

**Spearman–Brown Formula (Split-Half Method)**

$$r_{11} = 2\left(\frac{1\ S_a^{\,2} + S_b^{\,2}}{S_t^{\,2}}\right)$$

where:
$r_{11}$ = Reliability of total test
$S_a$ = Standard deviation of the first half of the test (odd-numbered items)
$S_b$ = Standard deviation of the second half of the test (even-numbered items)
$S_t$ = Standard deviation of the whole test

Given that both formulas use the standard deviation, this value must be calculated before proceeding with the reliability index. The formula for computing standard deviation is

$$\sqrt{\frac{\sum(x - x^2)}{n}}$$

where:
$\Sigma$ = Sum of
x = Each score
$x_2$ = Mean
n = Number of scores

The reliability index computation, using any formula, is a complex and time-consuming process that is best accomplished using a computer. Unfortunately, we do not have sufficient time to devote to this part of the post-test analysis, but your computer technician should have no trouble implementing the formula.

Ideally, the reliability index should be as close to 1.00 as possible. In reality, anything over 0.80 is generally acceptable. If the reliability index for the test does not reach 0.80, you need to look for possible causes, such as the following problems:

- Nonstandard instructions
- Scoring errors
- Non-uniform test conditions
- Respondent guessing
- Other chance fluctuations

For the following activities, we will assume that our example test has a reliability index of 0.87.

## Identify Items as Acceptable or Needing Review

To identify test items as acceptable or in need of review, the first step is to perform a quantitative analysis of the test item. This analysis uses difficulty (P+), discrimination, and reliability data to identify potentially weak or faulty test items. To accomplish the quantitative analysis, compare the test item's statistical data against the following screening criteria:

- Difficulty level (P+ value) between 0.25 and 0.75
- Discrimination index between +0.25 and +1.0
- Reliability of test 0.80 or greater

When a test item fails to meet these criteria, identify the item so that you can later conduct a qualitative analysis of it. During the qualitative analysis, you take a close look at the test item in terms of its technical content and adherence to the format guidelines for its type. Keep in mind that some test items identified as potentially weak or faulty during the quantitative analysis may *not* be found to be weak or faulty when reviewed more closely during the qualitative analysis. Some items on each test cover the "core" body of knowledge that everyone must possess. The instructor appropriately emphasizes that information during training, and the test item may then be intentionally written at a low enough difficulty level to ensure that the majority of the trainees "get it right." Therefore, when conducting the quantitative analysis, try to remain objective about the uses of statistical data.

## Revise Test Items Based on Post-Test Analysis

When you have determined the probable causes of your test-item faults, make changes accordingly. Even after a second and third revision, some items may remain problematic and require additional revision. The analytical item-analysis method does not guarantee immediate success in all cases, but it does promise gradual success over time.

# Resources for Conducting Live Fire Training Evolutions

## Contents



## Introduction

Live fire training is one of the most beneficial yet high-risk types of training that both an instructor and a fire fighter will participate in. NFPA 1403, *Standard on Live Fire Training Evolutions* (2007) was created as a guide to assist the instructor with resources and best practices when conducting these types of exercises. This text makes many references as to the responsibility of all fire service organizations and members to adhere to this standard. Legal precedent has been established along with the ethical responsibilities of the instructor to ensure that these high-risk training events yield the safest outcomes for all involved. Competent and professional instructors will have a full working knowledge and access to a complete edition of NFPA 1403 and also will have conducted research into live fire training related line-of-duty deaths available through a variety of resources such as the National Institute for Occupational Safety and Health (NIOSH) and the United States Fire Administration (USFA).

This appendix contains additional resources that may serve as templates that can be modified to fit the needs of individual departments when conducting live fire training evolutions. Formal training in the responsibilities of all positions identified in these resources is encouraged. As with any other type of training, defined objectives, specific learning outcomes, and predetermined activities must be conveyed to all participants before the first fire is ignited in an acquired structure or training tower.

## NFPA 1403, Standard on Live Fire Training Evolutions

Annex B, Annex C, and Annex D are taken directly from NFPA 1403, *Standard on Live Fire Training Evolutions*.

### NFPA 1403, Annex B, Live Fire Evolution Sample Checklist

*This annex is not a part of the requirements of this NFPA 1403 document but is included for informational purposes only.*

FIGURE B.1 provides a checklist for a live fire evolution.

# APPENDIX B

## LIVE FIRE EVOLUTION SAMPLE CHECKLIST

### PERMITS, DOCUMENTS, NOTIFICATIONS, INSURANCE.

_____ **1.** Written documentation received from owner:
- ☐ Permission to burn structure
- ☐ Proof of clear title
- ☐ Certificate of insurance cancellation
- ☐ Acknowledgment of post-burn property condition

_____ **2.** Local burn permit received

_____ **3.** Permission obtained to utilize fire hydrants

_____ **4.** Notification made to appropriate dispatch office of date, time, and location of burn

_____ **5.** Notification made to all affected police agencies:
- ☐ Received authority to block off roads
- ☐ Received assistance in traffic control

_____ **6.** Notification made to owners and users of adjacent property of date, time, and location of burn

_____ **7.** Liability insurance obtained covering damage to other property

_____ **8.** Written evidence of prerequisite training obtained from participating students from outside agencies

### PREBURN PLANNING.

_____ **1.** Preburn plans made, showing the following:
- ☐ Site plan drawing, including all exposures
- ☐ Floor plan detailing all rooms, hallways, and exterior openings
- ☐ Location of command post
- ☐ Position of all apparatus
- ☐ Position of all hoses, including backup lines
- ☐ Location of emergency escape routes
- ☐ Location of emergency evacuation assembly area
- ☐ Location of ingress and egress routes for emergency vehicles

_____ **2.** Available water supply determined

_____ **3.** Required fire flow determined for the acquired structure/live fire training structure/burn prop and exposure buildings

_____ **4.** Required reserve flow determined (50 percent of fire flow)

_____ **5.** Apparatus pumps obtained that meet or exceed the required fire flow for the building and exposures

_____ **6.** Separate water sources established for attack and backup hose lines

_____ **7.** Periodic weather reports obtained

_____ **8.** Parking areas designated and marked:
- ☐ Apparatus staging
- ☐ Ambulances
- ☐ Police vehicles
- ☐ Press vehicles
- ☐ Private vehicles

_____ **9.** Operations area established and perimeter marked

_____ **10.** Communications frequencies established, equipment obtained

### TRAINING STRUCTURE PREPARATION.

_____ **1.** Training structure inspected to determine structural integrity

_____ **2.** All utilities disconnected (acquired structures only)

_____ **3.** Highly combustible interior wall and ceiling coverings removed

_____ **4.** All holes in walls and ceilings patched

_____ **5.** Materials of exceptional weight removed from above training area (or area sealed from activity)

_____ **6.** Ventilation openings of adequate size precut for each separate roof area

_____ **7.** Windows checked and operated, openings closed

_____ **8.** Doors checked and operated, opened or closed, as needed

_____ **9.** Training structure components checked and operated:
- ☐ Roof scuttles
- ☐ Automatic ventilators
- ☐ Mechanical equipment
- ☐ Lighting equipment
- ☐ Manual or automatic sprinklers
- ☐ Standpipes

_____ **10.** Stairways made safe with railings in place

_____ **11.** Chimney checked for stability

**FIGURE B.1** Sample Checklist for Procedures for a Live Fire Evolution.

## LIVE FIRE EVOLUTION SAMPLE CHECKLIST (continued)

_____ **12.** Fuel tanks and closed vessels removed or adequately vented

_____ **13.** Unnecessary inside and outside debris removed

_____ **14.** Porches and outside steps made safe

_____ **15.** Cisterns, wells, cesspools, and other ground openings fenced or filled

_____ **16.** Hazards from toxic weeds, hives, and vermin eliminated

_____ **17.** Hazardous trees, brush, and surrounding vegetation removed

_____ **18.** Exposures such as buildings, trees, and utilities removed or protected

_____ **19.** All extraordinary exterior and interior hazards remedied

_____ **20.** Fire "sets" prepared:
- ☐ Class A materials only
- ☐ No flammable or combustible liquids
- ☐ No contaminated materials

### PREBURN PROCEDURES.

_____ **1.** All participants briefed:
- ☐ Training structure layout
- ☐ Crew and instructor assignments
- ☐ Safety rules
- ☐ Training structure evacuation procedure
- ☐ Evacuation signal (demonstrate)

_____ **2.** All hose lines checked:
- ☐ Sufficient size for the area of fire involvement
- ☐ Charged and test flowed
- ☐ Supervised by qualified instructors
- ☐ Adequate number of personnel

_____ **3.** Necessary tools and equipment positioned

_____ **4.** Participants checked:
- ☐ Approved full protective clothing
- ☐ Self-contained breathing apparatus (SCBA)
- ☐ Adequate SCBA air volume
- ☐ All equipment properly donned

### POSTBURN PROCEDURES.

_____ **1.** All personnel accounted for

_____ **2.** Remaining fires overhauled, as needed

_____ **3.** Training structure inspected for stability and hazards where more training is to follow *(see Training Structure Preparation)*

_____ **4.** Training critique conducted

_____ **5.** Records and reports prepared, as required:
- ☐ Account of activities conducted
- ☐ List of instructors and assignments
- ☐ List of other participants
- ☐ Documentation of unusual conditions or events
- ☐ Documentation of injuries incurred and treatment rendered
- ☐ Documentation of changes or deterioration of live fire training structure
- ☐ Acquired structure release
- ☐ Student training records
- ☐ Certificates of completion

_____ **6.** Building and property released to owner, release document signed

---

### RELEASE FORM

Having agreed with the Building Official, City of ______________________, that a structure owned by me and located at ______________________ is unfit for human habitation and is beyond rehabilitation, I further agree that the structure should be demolished. In order that demolition may be accomplished, I give my consent to the City of ______________________ to demolish, by burning or other means, the said structure.

I further release the City of ______________________ from any claim for loss resulting from such demolition.

Fire Department ______________________

Address ______________________

City, State ______________________

Date ______________________

Owner/Agent ______________________

Owner/Agent ______________________

Witness ______________________

**FIGURE B.1** Sample Checklist for Procedures for a Live Fire Evolution *(continued)*.

© 2007 National Fire Protection Association

## NFPA 1403, Annex C, Responsibilities of Personnel

*This annex is not a part of the requirements of this NFPA 1403 document but is included for informational purposes only.*

The lists in FIGURE C.1 outline the responsibilities of participants in a live fire training evolution.

### RESPONSIBILITIES OF PERSONNEL

**INSTRUCTOR-IN-CHARGE.**

_____ **1.** Plan and coordinate all training activities
_____ **2.** Monitor activities to ensure safe practices
_____ **3.** Inspect training structure integrity prior to each fire
_____ **4.** Assign instructors:
- ☐ Attack hose lines
- ☐ Backup hose lines
- ☐ Functional assignments
- ☐ Teaching assignments

_____ **5.** Brief instructors on responsibilities:
- ☐ Accounting for assigned students
- ☐ Assessing student performance
- ☐ Clothing and equipment inspection
- ☐ Monitoring safety
- ☐ Achieving tactical and training objectives

_____ **6.** Assign coordinating personnel, as needed:
- ☐ Emergency Medical Services
- ☐ Communications
- ☐ Water supply
- ☐ Apparatus staging
- ☐ Equipment staging
- ☐ Breathing apparatus
- ☐ Personnel welfare
- ☐ Public relations

_____ **7.** Ensure adherence to this standard by all persons within the training area

**INSTRUCTOR.**

_____ **1.** Monitor and supervise assigned students (no more than five per instructor)
_____ **2.** Inspect students' protective clothing and equipment
_____ **3.** Account for assigned students, both before and after evolutions

**SAFETY OFFICER.**

_____ **1.** Prevent unsafe acts
_____ **2.** Eliminate unsafe conditions
_____ **3.** Intervene and terminate unsafe acts
_____ **4.** Supervise additional safety personnel, as needed
_____ **5.** Coordinate lighting of fires with instructor-in-charge
_____ **6.** Ensure compliance of participants' personal equipment with applicable standards:
- ☐ Protective clothing
- ☐ Self-contained breathing apparatus (SCBA)
- ☐ Personal alarm devices, where used

_____ **7.** Ensure that all participants are accounted for, both before and after each evolution

**STUDENT.**

_____ **1.** Acquire prerequisite training
_____ **2.** Become familiar with building layout
_____ **3.** Wear approved full protective clothing
_____ **4.** Wear approved SCBA
_____ **5.** Obey all instructions and safety rules
_____ **6.** Provide documentation of prerequisite training, where from an outside agency

FIGURE C.1 Checklist for Responsibilities of Personnel.

© 2007 National Fire Protection Association

# NFPA 1403, Annex D, Heat Exhaustion and Heat Stroke in Training

*This annex is not a part of the requirements of this NFPA 1403 document but is included for informational purposes only.*

## D.1

The two most serious heat-related illnesses are heat exhaustion and heat stroke. The following material is excerpted from the NIOSH document *Occupational Exposure to Hot Environments, Revised Criteria*.

Symptoms of heat exhaustion include fatigue, nausea, headache, dizziness, pallor, weakness, and thirst. Factors that predispose a person to heat exhaustion include sustained exertion in the heat, failure to replace the water lost in sweat, and lack of acclimatization. Heat exhaustion responds readily to prompt treatments such as moving to a cooler environment, resting in a recumbent position, and taking fluids by mouth.

Heat stroke is the more serious of the heat-related illnesses and is considered a medical emergency. Symptoms of heat stroke include hot, red, dry skin, a rectal temperature of 40°C (104°F) or above, confusion, possible convulsions or loss of consciousness, or any combination of these symptoms. Factors that predispose a person to heat stroke include sustained exertion in the heat by unacclimatized workers, lack of physical fitness, obesity, recent alcohol intake, dehydration, individual susceptibility, and chronic cardiovascular disease. Heat stroke should be treated immediately. Treatments to reduce body temperature rapidly include immersing in chilled water, rinsing with alcohol, wrapping in a wet sheet, or fanning with cool, dry air, or any combination of these treatments. A physician's care is necessary to treat possible secondary disorders such as shock or kidney failure. While heat exhaustion cases greatly outnumber heat stroke cases, every case of heat exhaustion should be treated as having the potential to develop into heat stroke.

Acclimatization is a physiological adaptation to heat stress that occurs over a short period of time. After acclimatization has occurred, the body sweats more while losing less salt and can maintain a lower core temperature and lower cardiovascular demands. A person becomes acclimatized to a certain work intensity and temperature with repeated exposures to that work load and temperature. Formal acclimatization procedures might not be necessary for all fire fighters; however, training drills should be held outdoors regularly so that seasonal acclimatization can occur. For additional protection against heat stress, fire fighters might want to perform their regular aerobic training activities outdoors, especially during the spring and summer.

The metabolic demands of fire fighting range from 60 percent to 100 percent of maximum aerobic capacity. Tasks such as stair climbing, roof venting, and rescue operations, when performed in full gear, have an energy cost of 85 percent to 100 percent of maximum capacity and lead to near maximum heart rates.

It is clear from these estimates that a high level of cardiovascular fitness is an advantage in performing fire-fighting tasks. The higher level of fitness allows a longer work period and provides a greater reserve in case of an unexpected increase in work demands or in extreme environmental conditions. There are fire incidents during which even the fittest, most acclimatized fire fighter is exposed to significant heat stress. For this reason, many fire departments have adopted formal procedures for on-scene rehabilitation and have incorporated them into their manuals for standard operating procedures. The general goals of rehabilitation are as follows:

1. To provide physical and mental rest, allowing the fire fighter to recuperate from demands of emergency operations and adverse environmental conditions
2. To revitalize fire fighters by providing fluid replacement and food as needed
3. To provide medical monitoring, including treatment of injuries, to determine if and when fire fighters are able to return to action

Reproduced with permission from NFPA–1403-2007, *Live Fire Training*, Copyright © 2007, National Fire Protection Association. This reprinted material is not the complete and official position of the NFPA on the referenced subject, which is represented only by the standard in its entirety.

## Sample Live Fire Training Evolution and Safety Plan Forms

The remainder of this appendix provides sample forms that may serve as templates when conducting live fire training evolutions.

| **Any Fire Department**<br>**Building Floorplan/Evolution Setup** | Evolution # ______ Date ______ Time ______<br>Address ______<br>Fuel Load ______ |
|---|---|

| Any Fire Department<br>Site Plan / Vehicle Staging | Evolution # ______ Date ______ Time ______<br>Address ______<br>Fuel Load ______ |
| --- | --- |
| | |

**Any Fire Department**
**Evolution Objectives**

Evolution # ______ Date ______ Time ______

Name of Evolution ______

Address ______

**Objective 1** ______

**Objective 2** ______

**Objective 3** ______

**Objective 4** ______

**Objective 5** ______

**Objective 6** ______

**Objective 7** ______

**Any Fire Department**
**Evolution Assignments**

Evolution # ____________ Date ____________ Time ____________

Name of Evolution ____________________________

Address ____________________________

Evolution ____________________________

Incident Commander ____________________________

Safety Officer ____________________________

Ignition Officer ____________________________

Accountability Officer ____________________________

Engine Officer ____________________________

Truck Officer ____________________________

Squad Officer ____________________________

Rapid Intervention Officer ____________________________

**Engine Company**

Officer ____________________________

____________________________

____________________________

____________________________

Engineer ____________________________

**Truck Company**

Officer ____________________________

____________________________

____________________________

____________________________

Engineer ____________________________

**Squad Company**

Officer ____________________________

____________________________

____________________________

____________________________

**Rapid Intervention Company**

Officer ____________________________

____________________________

____________________________

____________________________

## Sample of a Completed Evolution Objectives Plan

Evolution **One Room Fire with Trapped Victims**

**Objective 1** Interior crews will have the experience of locating and extinguishing a fire in dense smoke conditions.

**Objective 2** Search crews will execute a search and rescue of victims under dense smoke and high heat conditions. Search crews should experience the value of ventilating windows as the search progresses.

**Objective 3** Truck crews will formulate and implement a plan for horizontal ventilation.

**Objective 4** Rapid intervention crews will observe exterior heat and smoke conditions. Shall note changes in color and density as fire progresses and is extinguished.

**Objective 5** Command will practice using the ICS system to direct crews on an interior firefight. Command will maintain personnel accountability.

## Sample of a Completed Evolution Assignments Listing

| | |
|---|---|
| **Evolution** | #2 ONE ROOM FIRE W/ RESCUE OF VICTIMS |
| **Lead Instructor** | DUTKIEWICZ |
| **Safety Officer** | MADDEN |
| **Ignition Officer** | HARTMANN |
| **Rapid Intervention Officer** | HYNES |
| **Accountability Officer** | D. SMITH |
| **Water Supply Officer** | CINQUEPALMI |

**Attack Company (Engine 1)**

| | |
|---|---|
| **Officer** | Simone |
| | Klekamp |
| | Mazurkiewicz |
| **Engineer** | Tums |

**Back-up Company (Engine 2)**

| | |
|---|---|
| **Officer** | Reeder |
| | Stolz |
| | E. Johnson |
| **Engineer** | Piper |

**Search 1 Company (Truck 1)**

| | |
|---|---|
| **Officer** | Breese |
| | Nagel |
| | Sterling |
| | Tufts |

**Search 2 Company (Squad 1)**

| | |
|---|---|
| **Officer** | Buhs |
| | Rivero |
| | Haran |
| | |

**Outside Safety Company**

| | |
|---|---|
| **Officer** | Ferro |
| | Olinski |
| | Smart |
| | |

**RIT**

| | |
|---|---|
| **Officer** | Hynes |
| | Exline |
| | Bohne |
| | |

# Resources for Fire Service Instructors

## Contents

# APPENDIX C

## Training Record
## Attendance Report

Date____________________ Station(s)____________ Description________________________________

Start Time_________________ End Time______________ Credit Hours (Total Time)______________

Method of Training: ☐ Classroom ☐ Practical ☐ Self-Directed Certification Credit: ☐ Yes ☐ No

Lead Instructor______________________________ Additional Instructor(s)____________________

| Print Name | Dept ID # | Signature | Hours Attended |
|---|---|---|---|
| | | | |
| | | | |
| | | | |
| | | | |
| | | | |
| | | | |
| | | | |
| | | | |
| | | | |
| | | | |
| | | | |
| | | | |
| | | | |
| | | | |
| | | | |
| | | | |
| | | | |

Objectives:_______________________________________________________________

Description of Training (Notes):_________________________________________

_________________________________________________________________________

_________________________________________________________________________

______________________ Instructor Signature

______________________ Training Officer Approval

| Type of Training | | | | |
|---|---|---|---|---|
| Company | Multi-Comp. | Officer | Mutual Aid | Night |
| Tower Burn | Classroom | Practical | Combo. | Driver |

| Equipment Used in Training Session | Feet of 1 ¾″ Hose Used | Feet of 2 ½″ Hose Used | Supply Hose Used / Ft. | Feet of Ladders | Number of Engines | Number of Trucks | Gallons of Water | Number of SCBA | Total Number of Firefighters |
|---|---|---|---|---|---|---|---|---|---|
| | | | | | | | | | |

# Training and Education Report

Name________________________ ID#_____________ Date of Birth_________________

S/S#________________________ Date Entered Fire Service_______ Date of Hire_________

Drivers License# _________________ Expiration Date__________ D/L Classification__________

## Certification Record

*Indicate date of Certification in Boxes Below*

| Basic Fire fighter | Advanced Fire fighter | Apparatus Operator (FAE) | Hazmat Awareness |
|---|---|---|---|
| Hazmat Operations | Hazmat Technician | Hazmat Specialist | Hazmat Incident Command |
| Technical Rescue Awareness | NFPA 1006–Trench Ops | NFPA 1006–Trench Tech | NFPA 1006–Rope |
| NFPA 1006–Rope 2 | NFPA 1006–Confined Space | Structural Collapse Operations | Structural Collapse Technician |
| Instructor 1 | Instructor 2 | Instructor 3 | Training Program Manager |
| Provisional Fire Officer 1 | Fire Officer 1 | Provisional Fire Officer 2 | Fire Officer 2 |
| Provisional Fire Officer 3 | Fire Officer 3 | Fire Prevention Officer | Vehicle Machinery Operations |
| Vehicle Machinery Technician | Water Rescue Operations | Watercraft Technician | IS 100 (IS100) |
| IS 200 (IS200) | IS300 (IS300) | IS400 (IS400) | IS 700 (IS700) |
| IS 800 (IS800) | | Emergency Medical Technician (EMT-B) | Paramedic (EMT-P) |

## Formal Education Record

High School Attended____________________ Year Graduated__________ GED__________

College Attended______________________ Credits Earned__________ Degree__________

Advanced Degree______________________ Credits Earned__________ Degree__________

Advanced Degree______________________ Credits Earned__________ Degree__________

# Training Make-Up Log

Month ____________ Shift ______________ Station ______ Officer ____________

INSTRUCTIONS: ANY FF MISSING A SCHEDULED DRILL WILL BE REQUIRED TO TAKE A MAKE-UP TRAINING SESSION. COMPANY OFFICER WILL LOG NAME/SUBJECT/DATE COVERED. WHEN FF RETURNS TO DUTY OR ON MAKE-UP DAY, OFFICER WILL FILL OUT REMAINDER OF FORM AS INDICATED. ALL MISSED DRILLS SHOULD BE MADE-UP BY END OF MONTH.

| Name | Subject | Date of Training | Date Make-up Complete | Training Record Y/N | F/F Initials | Company Officer (Training Completed) |
|---|---|---|---|---|---|---|
| | | | | | | |
| | | | | | | |
| | | | | | | |
| | | | | | | |
| | | | | | | |
| | | | | | | |
| | | | | | | |
| | | | | | | |
| | | | | | | |
| | | | | | | |
| | | | | | | |
| | | | | | | |
| | | | | | | |

______________________ Company Officer

______________________ Shift Commander Review

______________________ Approval

# Personal Improvement Agreement

Agreement Initiated by: ______________________ Date: __________
(Name and rank)

Fire Fighter Name: ______________________ ID#: __________

I. Concerns/Areas Needing Improvement: ______________________

II. Objective Number/JPR: ______________________
Other: ______________________
(Attach appropriate documents)

III. Shift Commander/Station Officer Action Plan: (What the officer will do to help improve the performance)
A. ______________________
B. ______________________

IV. Personal Action Plan: (What the member will do to improve themselves)
A. ______________________
B. ______________________

Drillmaster Follow-up Notes:

V. Document Action Plan Progress: (How are we doing?)
Dates: ______________________

Initial Follow-Up Date ________

VI. Has the area of concern been corrected? (Has improvement been seen in this area?)

Yes: ☐ ______________________

Cleared: ________

No: ☐ ______________________

Extended: ______

Fire Fighter Name (print): ______________________ Date: ________
Station Officer Name (print): ______________________ Initials: ________
Shift Commander Name (print): ______________________ Initials: ________

White Copy: Director of Training/file Yellow: Shift Commander Pink: Drillmaster Gold: Fire Fighter

**Director of Training Use**

☐ Suppression
☐ EMS
☐ Technical Rescue
☐ Hazardous Materials
☐ Officer Development
☐ Drillmaster Report

# Sample Training Policy

## PURPOSE AND SCOPE:

This directive provides definition as to the training responsibilities of each rank within the department. This directive also establishes the basic format of the department's training program, including training requirements and documentation.

## RESPONSIBILITIES:

### TRAINING OFFICER

The Training Officer is responsible for the overall administration and management of the department's Training Division. Through working with the staff and line officers, the Training Officer is to develop and implement a comprehensive yearly training plan. The Training Officer has overall responsibility and accountability to ensure that department training activities are current and consistent with applicable standards and practices.

### INSTRUCTOR-IN-CHARGE

The Instructor-in-Charge is the designated person who is responsible for the overall delivery of a specific lesson plan or training objective. All resources assigned to the training session are under the responsibility of the Instructor-in-Charge, including personnel, apparatus, and facilities and equipment. The use of all required and best-practice safety procedures throughout the training session will be the responsibility of the Instructor-in-Charge. All elements of documentation will also be overseen by the Instructor-in-Charge.

### INSTRUCTOR(S)

The instructor is responsible for assisting the Instructor-in-Charge at any high-risk training event or routine training session where it has been determined that additional instructors are needed. This may be for the purpose of safety, accountability, or reduction in the student-to-instructor ratio. The instructor is responsible for delivering lesson plans, evaluating performance, and providing feedback and documentation of the evolution or objectives as necessary. The instructor will be responsible for the safety of the members that are assigned to them during the evolution.

### SHIFT COMMANDER(S)

The Shift Commander is responsible for administering and monitoring the department training plan within his or her assigned shift. This includes assisting in the coordination, presentation, and evaluation of specific department level training sessions. The Shift Commander should make every attempt to attend high-risk training events, training events based on procedural operations where his or her presence is needed to simulate operations, and any other training event where he or she can assist in the evaluation of member performance.

Through periodic evaluation of companies and/or individuals during training and emergency operations, the shift Deputy Commander is responsible for identifying training deficiencies and providing recommendations to the Training Officer regarding the specific training needs of his or her shift.

**CAPTAIN(S)/OTHER MID-LEVEL SHIFT SUPERVISOR(S)**
The Captain is responsible for implementing and monitoring the department's training plan within his or her assigned area of authority (example—shift 2, west side). This includes reviewing monthly training reports to ensure company and individual compliance with training assignments. He or she also may be assigned by the Training Officer to assist in the coordination, presentation, and evaluation of specific department level training sessions. The Captain is responsible for completing the training responsibilities of a Company Officer.

**COMPANY OFFICER (S)/ACTING COMPANY OFFICER(S)**
The Company Officer is the key to the department's training program. He or she is the individual most responsible for the training and readiness of personnel. The Company Officer is required to complete monthly training assignments and submit all necessary documentation, including training reports and skills checklists. The Company Officer is to coordinate the various daily company activities so that training assignments are completed. He or she is responsible for coordinating company level training so that all members receive the training regardless of time off, vacations, Kelly days, etc.

The monthly training assignments represent the minimum of what must be done. The Company Officer is not limited to this, as each individual has strengths and weaknesses which must be addressed by the Company Officer. It is the responsibility of the Company Officer to improve the performance of the personnel assigned to him or her and to foster an environment that encourages the company toward continuous improvement.

The monthly Training Calendar and Company Officers' Training Packet lists the assigned training activities for the month. Some activities will have specific dates and/or time periods designated. For company level training assignments, the Company Officer has the authority to vary from the published schedule, if necessary for valid reasons. The Company Officer shall be responsible for scheduling and completing the training. The objective is for all training assignments to be completed by the end of each month.

The Company Officer is responsible for the safety of his or her personnel while training.

The Company Officer is responsible for maintaining the licensure required of the position. This includes EMT-B or Paramedic and appropriate drivers license classification.

**FIRE FIGHTERS**
Department fire fighters are expected to maintain a high level of preparedness through regular training and individual study. This includes keeping current on both fire service and departmental changes and notifying the Company Officer of training needs.

Fire fighters are responsible for participating in an aggressive, safe, and positive manner in all classroom and practical training.

Fire fighters are responsible for maintaining the licensure required of the position. This includes EMT-B or Paramedic and appropriate drivers license classification.

**MONTHLY TRAINING ASSIGNMENTS**

The Training Division is responsible for providing monthly training assignments through the Training Bulletin, Training Calendar, and the Company Officers' Training Packet issued at the beginning of each month. This shall specify the assigned and make-up dates for training (if applicable), who is required to attend, and other necessary information.

Monthly training will be classified as follows:

**MANDATORY**

Training that must be accomplished by all members. This may also include 40-hour personnel. Mandatory training will include Federal- and State-mandated courses and courses deemed as mandatory by the Anytown Fire Department. These training sessions and make-up sessions will be scheduled by the Training Division and coordinated through the Shift Commanders.

**REGULAR**

**Company level**—Training that is to be completed by each company during the month. Company training designated as part of the Essential Skills program shall be completed by each company member. Other types of company level training will specify whether make-up sessions are required for personnel who miss the training. Company Officers are responsible for monitoring this, and for scheduling and conducting any necessary make-up training.

**Department level**—Training that is primarily scheduled and conducted through the Training Division. Department level training that is designated as mandatory will have make-up sessions scheduled by the Training Division.

**DOCUMENTATION OF TRAINING**

The Training Officer shall be responsible for providing to the Shift Commanders and Captains a report as to the completion of training assignments from the previous month. The report shall be provided within the first 10 days of the next month. The Shift Commanders and Captains are responsible for following up with their Training Officers on the training that has not been completed.

The instructor of a specific training activity is responsible for completing and submitting the training report to the Training Division.

The Company Officer is responsible for ensuring that all skills checklists that may be required as part of a training assignment are completed and submitted to the Training Division.

For individual-type training activities (i.e. independent study, reviewing fire service publications, physical fitness, etc.) the individual is responsible for completing the report and submitting it to his or her Company Officer for signature.

Individuals attending classes outside of the fire department (i.e. fire officer classes, tactics seminars, etc.) are responsible for completing a training report upon returning from the course.

BY ORDER OF: ______________________________

DATE: ______________________________

# Instructor Training and Experience Validation

**Name:** ______________________ **Department:** ______________________

**Years of Fire Service Experience:** ______________ **Rank:** ______________________

## Certification Levels

☐ Instructor 1 ☐ Instructor 2 ☐ Instructor 3 ☐ Instructor 4 ☐ TPM ☐ Advanced Degree

☐ Fire Officer 1 ☐ Fire Officer 2 ☐ Fire Officer 3 ☐ EFO ☐ Other

- ➢ Copies of certifications must be available upon request

## Previous Live Fire Training Educational Experience

☐ NFPA 1403 course/class at state training facility, local academy, or agency

☐ Attended FDIC Live Fire Training programs

☐ Attended other national conference live fire training courses

- ➢ Copies of course/class completion certificates are available upon request

## Previous Live Fire Training Practical Experience

**Approximate # of Live Fire Training Exercises conducted at:**

☐ **Training Tower (Class A)** ☐ **Training Tower (Gas Fired)**

☐ **Acquired Structures** ☐ **Container/Simulator/Portable Unit**

☐ **Exterior Burn Prop** ☐ **Class B Fuel Burn**

☐ **Approximate # of Years of Live Fire Training Experience**

## Validation / Attestation Statement

*I have read and understand all components of NFPA 1403, Standard on Live Fire Training Evolutions, and agree to abide by all requirements specified for the positions/roles that I am assigned for this training event. I understand and accept responsibility for the position requirements for the activities I am assigned to complete and do so knowing the hazards and dangers associated with these assignments.*

______________________________

*Signature* *Date*

# Request for Travel Expenses to Attend Training/Conferences/Seminars

**Employee/Conf. Information:**

Date of Request: ____________________

Employee Name: ______________________________ Employee Number: ______

Class/Conference/Seminar Name: ________________________________________

Program Location: ________________ Conference/Seminar/Class Fee: ___________

District to be billed for conference fee: ☐ Yes ☐ No

Check made payable to conference name: ☐ Yes ☐ No

Purchase order to conference name: ☐ Yes ☐ No PO #_____________

Reason for attending: __________________________________________________________

Budget Code: ___________

Are any other members of the department attending?

☐ Yes # Attending: ☐ No

**Dates of Employee Attendance:**

Arrival Date: __________________ Departure Date: _____________________

**Category:**

**A:** ☐ Full coverage: Shift coverage, tuition, expenses paid by department

**B:** ☐ Partial coverage: Tuition, expenses as defined, employee to provide shift coverage

**C:** ☐ Sponsorship only: Attending on own with department approval

**Work coverage/time off during seminar**

| Date | Shift | Times |
| --- | --- | --- |
| | | |
| | | |
| | | |

**Lodging, Travel and Expense Information:**

**Hotel Information:**

Hotel Name:______________________________ Confirmation #:________________

Address:_________________________________ Check-in Date:________________

_________________________________________ Check-in Time:________________

Phone:___________________________________ Check-out Date:________________

Hotel Reservation Payment Method: ☐ Department ☐ Personal

**Transportation:**

Using own vehicle? ☐ Yes ☐ No If yes, number of miles roundtrip to conference?________

Department vehicle requested ☐ Yes ☐ No Vehicle # Assigned____________

Vehicle out date:______________ Return date: ____________________

Is rental car required? ☐ Yes ☐ No If yes, bring receipts to administration upon return for reimbursement.

Airfare required? ☐ Yes ☐ No

If yes, name of airline:________________ Cost of airfare:________________

Departing Flight #:________________ Returning Flight #:________________

Do you require reimbursement to you personally or the airline? ☐ Personally ☐ Airline

**Meals**

Number of meals included in conference fee:________________

Days/Number of meal expenses requested:________________

Rate of per diem:______________

Total expenses:______________

(Copy of seminar registration attached)

Date you wish to pick up any expense checks from Administration:________________

**Approval:**

Employee Signature: ______________________________ Date: __________

Battalion Chief Signature: ______________________________ Date: __________

Training Officer Signature: ______________________________ Date: __________

Deputy Chief Signature: ______________________________ Date: __________

Administration Signature: ______________________________ Date: __________

# Fire Fighter Skill Performance Ratings

- **Unskilled (0)**
  - Member failed evolution or skill (Mandatory re-evaluation will take place)
    - Exceeded time limit
    - Missed step in procedure
    - Created safety hazard to self or other member
    - Unable to perform task
    - Repeated failure of task attempt
  - Requires Personal Improvement Agreement and formal documentation on standard evaluation form
  - No credit is given for purpose of progress reporting or evaluation towards applicable certification

- **Moderately Skilled (1)**
  - Performance meets the minimum requirement for the task and is performed on first attempt. With additional practice, improved performance levels can be attained
  - Member's performance of skill may require supervision on fireground
  - Time performance near minimum requirement
  - All appropriate safety precautions are taken
  - Instructor/evaluator determines need for additional training or repeat of skill

- **Skilled (2)**
  - Performance is above the minimum level because:
    - The time was above average
    - Skill meets all performance levels and could be performed on fireground without supervision
    - No serious/critical errors were committed
    - All appropriate safety practices were observed and performed

- **Highly Skilled (3)**
  - Performance is at a high level of competence because:
    - It was error free
    - It was the fastest time
    - Member could supervise others doing same task and identify errors or suggest improvements
    - Member knows the role and importance of this task in relation to other fireground operations

**Example**

| Subject | Skill | Subset | Equipment | Performance Rating |
|---|---|---|---|---|
| Forcible Entry | Force an Inward-Swinging Door | Hand Tools | Halligan Bar/Flat Head Axe | 3 |

# Resources for Training Officers

| SCBA Training | |
|---|---|
| FF Near-miss incidents involving SCBA | www.firefighternearmiss.com |
| OSHA Standard 1910.134 | www.osha.gov |
| Model Respiratory Protection Program | www.state.nj.us/health/eoh/peoshweb/respromp.pdf |
| SCBA training ideas | www.rapidintervention.com and www.firehouse.com |
| **Driver Training** | |
| USFA "Emergency Vehicle Driver Training" program | www.usfa.fema.gov—publications section of Web site |
| Driver/Operator lesson plans | www.firehouse.com—training section of Web site |
| Fire apparatus and traffic safety related information | www.respondersafety.com |
| Driver safety information | www.drivetosurvive.org |
| IAFC 2006 National FF Safety Stand Down resources site | http://www.iafc.org/displaycommon.cfm?an=1&subarticlenbr=413 |
| Sample apparatus driving SOGs | www.vfis.com |
| San Diego FD driver training documents and manuals | www.sdfdtraining.com |
| **Company Officer Development** | |
| IAFC Officer Development Handbook | http://www.iafc.org/associations/4685/files/OffrsHdbkFINAL3.pdf *or* http://fire.state.nv.us/files_forms/IFACOfficersHdbk.pdf |
| Wildland Fire Leadership Development | http://www.fireleadership.gov |
| Long Beach (CA) Fire Department | http://www.lbfdtraining.com/index.html |
| **Miscellaneous Resources** | |
| National Fire Academy library | 1-800-638-1821 |
| Firefighter Close Calls | www.firefighterclosecalls.com |
| National FF Near Miss Reporting System | www.firefighternearmiss.com |
| Everyone Goes Home | www.everyonegoeshome.org |
| With the Command | www.withthecommand.com |
| T.R.A.D.E. (Training Resources and Data Exchange) | http://feti.lsu.edu/municipal/NFA/TRADE (14 CD's worth of training materials available for downloading) |

| Regulatory Agencies | |
|---|---|
| OSHA | www.osha.gov |
| U.S. EPA | www.epa.gov |
| NIOSH firefighter fatality reports | http://www.cdc.gov/niosh/fire |
| **NFPA Standards** | |
| NFPA Web site | www.nfpa.org |
| **Organizations** | |
| Illinois Society of Fire Service Instructors | www.ill-fireinstructors.org |
| International Society of Fire Service Instructors | www.isfsi.org |
| National Volunteer Fire Council | www.nvfc.org |
| International Association of Firefighters | www.iaff.org |
| International Association of Fire Chiefs | www.iafc.org |
| IFSTA | www.ifsta.org |
| **Fire Service Publications** | |
| Fire Engineering Magazine | www.fireengineering.com |
| FireHouse Magazine | www.firehouse.com |
| National Fire Rescue | www.nfrmag.com |
| Fire-Rescue Magazine | www.firerescuemag.com |
| **Online Magazines** | |
| Fire Nuggets | www.firenuggets.com |
| Vincent Dunn | www.vincentdunn.com |

# An Extract from: NFPA® 1041, Standard for Fire Service Instructor Professional Qualifications, 2007 Edition

## Chapter 4: Instructor I

**4.1 General.**

**4.1.1** The Fire Service Instructor I shall meet the JPRs defined in Sections 4.2 through 4.5 of this standard.

**4.2 Program Management.**

**4.2.1** Definition of Duty. The management of basic resources and the records and reports essential to the instructional process.

**4.2.2** Assemble course materials, given a specific topic, so that the lesson plan and all materials, resources, and equipment needed to deliver the lesson are obtained.

(A) **Requisite Knowledge.** Components of a lesson plan, policies and procedures for the procurement of materials and equipment, and resource availability.

(B) **Requisite Skills.** None required.

**4.2.3** Prepare training records and report forms, given policies and procedures and forms, so that required reports are accurately completed and submitted in accordance with the procedures.

(A) **Requisite Knowledge.** Types of records and reports required, and policies and procedures for processing records and reports.

(B) **Requisite Skills.** Basic report writing and record completion.

**4.3 Instructional Development.**

**4.3.1*** **Definition of Duty.** The review and adaptation of prepared instructional materials.

**4.3.2*** Review instructional materials, given the materials for a specific topic, target audience and learning environment, so that elements of the lesson plan, learning environment, and resources that need adaptation are identified.

(A) **Requisite Knowledge.** Recognition of student limitations, methods of instruction, types of resource materials, organization of the learning environment, and policies and procedures.

(B) **Requisite Skills.** Analysis of resources, facilities, and materials.

**4.3.3*** Adapt a prepared lesson plan, given course materials and an assignment, so that the needs of the student and the objectives of the lesson plan are achieved.

(A)* **Requisite Knowledge.** Elements of a lesson plan, selection of instructional aids and methods, and origination of learning environment.

(B) **Requisite Skills.** Instructor preparation and organizational skills.

**4.4 Instructional Delivery.**

**4.4.1** **Definition of Duty.** The delivery of instructional sessions utilizing prepared course materials.

**4.4.2** Organize the classroom, laboratory, or outdoor learning environment, given a facility and an assignment, so that lighting, distractions, climate control or weather, noise control, seating, audiovisual equipment, teaching aids, and safety are considered.

(A) **Requisite Knowledge.** Classroom management and safety, advantages and limitations of audiovisual equipment and teaching aids, classroom arrangement, and methods and techniques of instruction.

(B) **Requisite Skills.** Use of instructional media and materials.

**4.4.3** Present prepared lessons, given a prepared lesson plan that specifies the presentation method(s), so that the method(s) indicated in the plan are used and the stated objectives or learning outcomes are achieved.

(A) **Requisite Knowledge.** The laws and principles of learning, teaching methods and techniques, lesson plan components and elements of the communication process, and lesson plan terminology and definitions.

(B) **Requisite Skills.** Oral communication techniques, teaching methods and techniques, and utilization of lesson plans in the instructional setting.

**4.4.4*** Adjust presentation, given a lesson plan and changing circumstances in the class environment, so that class continuity and the objectives or learning outcomes are achieved.

(A) **Requisite Knowledge.** Methods of dealing with changing circumstances.

(B) **Requisite Skills.** None required.

# APPENDIX D

**4.4.5** Adjust to differences in learning styles, abilities, and behaviors, given the instructional environment, so that lesson objectives are accomplished, disruptive behavior is addressed, and a safe learning environment is maintained.

**(A)* Requisite Knowledge.** Motivation techniques, learning styles, types of learning disabilities and methods for dealing with them, and methods of dealing with disruptive and unsafe behavior.

**(B) Requisite Skills.** Basic coaching and motivational techniques, and adaptation of lesson plans or materials to specific instructional situations.

**4.4.6** Operate audiovisual equipment and demonstration devices, given a learning environment and equipment, so that the equipment functions properly.

**(A) Requisite Knowledge.** Components of audiovisual equipment.

**(B) Requisite Skills.** Use of audiovisual equipment, cleaning, and field level maintenance.

**4.4.7** Utilize audiovisual materials, given prepared topical media and equipment, so that the intended objectives are clearly presented, transitions between media and other parts of the presentation are smooth, and media are returned to storage.

**(A) Requisite Knowledge.** Media types, limitations, and selection criteria.

**(B) Requisite Skills.** Transition techniques within and between media.

**4.5** **Evaluation and Testing.**

**4.5.1*** **Definition of Duty.** The administration and grading of student evaluation instruments.

**4.5.2** Administer oral, written, and performance tests, given the lesson plan, evaluation instruments, and the evaluation procedures of the agency, so that the testing is conducted according to procedures and the security of the materials is maintained.

**(A) Requisite Knowledge.** Test administration, agency policies, laws affecting records and disclosure of training information, purposes of evaluation and testing, and performance skills evaluation.

**(B) Requisite Skills.** Use of skills checklists and oral questioning techniques.

**4.5.3** Grade student oral, written, or performance tests, given class answer sheets or skills checklists and appropriate answer keys, so the examinations are accurately graded and properly secured.

**(A) Requisite Knowledge.** Grading and maintaining confidentiality of scores.

**(B) Requisite Skills.** None required.

**4.5.4** Report test results, given a set of test answer sheets or skills checklists, a report form, and policies and procedures for reporting, so that the results are accurately recorded, the forms are forwarded according to procedure, and unusual circumstances are reported.

**(A) Requisite Knowledge.** Reporting procedures and the interpretation of test results.

**(B) Requisite Skills.** Communication skills and basic coaching.

**4.5.5*** Provide evaluation feedback to students, given evaluation data, so that the feedback is timely; specific enough for the student to make efforts to modify behavior; and objective, clear, and relevant; also include suggestions based on the data.

**(A) Requisite Knowledge.** Reporting procedures and the interpretation of test results.

**(B) Requisite Skills.** Communication skills and basic coaching

## Chapter 5: Instructor II

**5.1** **General.** The Fire Service Instructor II shall meet the requirements for Fire Service Instructor I and the JPRs defined in Sections 5.2 through 5.5 of this standard.

**5.2** **Program Management.**

**5.2.1** **Definition of Duty.** The management of instructional resources, staff, facilities, and records and reports.

**5.2.2** Schedule instructional sessions, given department scheduling policy, instructional resources, staff, facilities and timeline for delivery, so that the specified sessions are delivered according to department policy.

**(A) Requisite Knowledge.** Departmental policy, scheduling processes, supervision techniques, and resource management.

**(B) Requisite Skills.** None required.

**5.2.3** Formulate budget needs, given training goals, agency budget policy, and current resources, so that the resources required to meet training goals are identified and documented.

(A) **Requisite Knowledge.** Agency budget policy, resource management, needs analysis, sources of instructional materials, and equipment.

(B) **Requisite Skills.** Resource analysis and forms completion.

**5.2.4** Acquire training resources, given an identified need, so that the resources are obtained within established timelines, budget constraints, and according to agency policy.

(A) **Requisite Knowledge.** Agency policies, purchasing procedures, and budget management.

(B) **Requisite Skills.** Forms completion.

**5.2.5** Coordinate training record-keeping, given training forms, department policy, and training activity, so that all agency and legal requirements are met.

(A) **Requisite Knowledge.** Record-keeping processes, departmental policies, laws affecting records and disclosure of training information, professional standards applicable to training records, and databases used for record-keeping.

(B) **Requisite Skills.** Record auditing procedures.

**5.2.6** Evaluate instructors, given an evaluation form, department policy, and JPRs, so that the evaluation identifies areas of strengths and weaknesses, recommends changes in instructional style and communication methods, and provides opportunity for instructor feedback to the evaluator.

(A) **Requisite Knowledge.** Personnel evaluation methods, supervision techniques, department policy, and effective instructional methods and techniques.

(B) **Requisite Skills.** Coaching, observation techniques, and completion of evaluation forms.

**5.3 Instructional Development.**

**5.3.1 Definition of Duty.** The development of instructional materials for specific topics.

**5.3.2** Create a lesson plan, given a topic, audience characteristics, and a standard lesson plan format, so that the JPRs for the topic are achieved, and the plan includes learning objectives, a lesson outline, course materials, instructional aids, and an evaluation plan.

(A) **Requisite Knowledge.** Elements of a lesson plan, components of learning objectives, instructional methods and techniques, characteristics of adult learners, types and application of instructional media, evaluation techniques, and sources of references and materials.

(B) **Requisite Skills.** Basic research, using JPRs to develop behavioral objectives, student needs assessment, development of instructional media, outlining techniques, evaluation techniques, and resource needs analysis.

**5.3.3** Modify an existing lesson plan, given a topic, audience characteristics, and a lesson plan, so that the JPRs for the topic are achieved and the plan includes learning objectives, a lesson outline, course materials, instructional aids, and an evaluation plan.

(A) **Requisite Knowledge.** Elements of a lesson plan, components of learning objectives, instructional methods and techniques, characteristics of adult learners, types and application of instructional media, evaluation techniques, and sources of references and materials.

(B) **Requisite Skills.** Basic research, using JPRs to develop behavioral objectives, student needs assessment, development of instructional media, outlining techniques, evaluation techniques, and resource needs analysis.

**5.4 Instructional Delivery.**

**5.4.1 Definition of Duty.** Conducting classes using a lesson plan.

**5.4.2** Conduct a class using a lesson plan that the instructor has prepared and that involves the utilization of multiple teaching methods and techniques, given a topic and a target audience, so that the lesson objectives are achieved.

**(A) Requisite Knowledge.** Use and limitations of teaching methods and techniques.

**(B)* Requisite Skills.** Transition between different teaching methods.

**5.4.3*** Supervise other instructors and students during training, given a training scenario with increased hazard exposure, so that applicable safety standards and practices are followed, and instructional goals are met.

**(A) Requisite Knowledge.** Safety rules, regulations, and practices; the incident command system used by the agency; and leadership techniques.

**(B) Requisite Skills.** Implementation of an incident management system used by the agency.

**5.5 Evaluation and Testing.**

**5.5.1 Definition of Duty.** The development of student evaluation instruments to support instruction and the evaluation of test results.

**5.5.2** Develop student evaluation instruments, given learning objectives, audience characteristics, and training goals, so that the evaluation instrument determines if the student has achieved the learning objectives; the instrument evaluates performance in an objective, reliable, and verifiable manner; and the evaluation instrument is bias-free to any audience or group.

**(A) Requisite Knowledge.** Evaluation methods, development of forms, effective instructional methods, and techniques.

**(B) Requisite Skills.** Evaluation item construction and assembly of evaluation instruments.

**5.5.3** Develop a class evaluation instrument, given agency policy and evaluation goals, so that students have the ability to provide feedback to the instructor on instructional methods, communication techniques, learning environment, course content, and student materials.

**(A) Requisite Knowledge.** Evaluation methods and test validity.

**(B) Requisite Skills.** Development of evaluation forms.

**5.5.4** Analyze student evaluation instruments, given test data, objectives and agency policies, so that validity is determined and necessary changes are accomplished.

**(A) Requisite Knowledge.** Test validity, reliability, and item analysis.

**(B) Requisite Skills.** Item analysis techniques.

NOTICE: An asterisk (*) following the number or letter designating a paragraph indicates that explanatory material on the paragraph can be found in Annex A of NFPA 1041, *Standard For Fire Service Instructor Professional Qualifications, 2007 Edition.*

# NFPA® 1041, Standard for Fire Service Instructor Professional Qualifications, 2007 Edition, Correlation Guide

## Chapter 4: Instructor I

| NFPA 1041, *Standard for Fire Service Instructor Professional Qualifications*, 2007 Edition | Corresponding Textbook Chapter(s) |
|---|---|
| 4.1 | 1 |
| 4.1.1 | 1 |
| 4.2 | 1 |
| 4.2.1 | 1 |
| 4.2.2 | 6 |
| 4.2.2 (A) | 6 |
| 4.2.2 (B) | 6 |
| 4.2.3 | 1 |
| 4.2.3 (A) | 2 |
| 4.2.3 (B) | 2 |
| 4.3 | 6 |
| 4.3.1 | 6 |
| 4.3.2 | 6, 7 |
| 4.3.2 (A) | 4, 7 |
| 4.3.2 (B) | 7 |
| 4.3.3 | 6 |
| 4.3.3 (A) | 6, 8 |
| 4.3.3 (B) | 6 |
| 4.4 | 3 |
| 4.4.1 | 3 |
| 4.4.2 | 7, 9 |
| 4.4.2 (A) | 7, 8, 9 |
| 4.4.2 (B) | 8 |
| 4.4.3 | 3, 6 |
| 4.4.3 (A) | 3, 4, 5, 6 |
| 4.4.3 (B) | 3, 5 |
| 4.4.4 | 6 |
| 4.4.4 (A) | 6 |
| 4.4.4 (B) | 6 |
| 4.4.5 | 3, 4, 9 |
| 4.4.5 (A) | 3, 4 |
| 4.4.5 (B) | 3, 6 |
| 4.4.6 | 8 |
| 4.4.6 (A) | 8 |
| 4.4.6 (B) | 8 |
| 4.4.7 | 8 |

## Chapter 4: Instructor I

| NFPA 1041, *Standard for Fire Service Instructor Professional Qualifications*, 2007 Edition | Corresponding Textbook Chapter(s) |
|---|---|
| 4.4.7 (A) | 8 |
| 4.4.7 (B) | 8 |
| 4.5 | 10 |
| 4.5.1 | 10 |
| 4.5.2 | 10 |
| 4.5.2 (A) | 10 |
| 4.5.2 (B) | 10 |
| 4.5.3 | 10 |
| 4.5.3 (A) | 10 |
| 4.5.3 (B) | 10 |
| 4.5.4 | 10 |
| 4.5.4 (A) | 10 |
| 4.5.4 (B) | 10 |
| 4.5.5 | 10 |
| 4.5.5 (A) | 10 |
| 4.5.5 (B) | 10 |

## Chapter 5: Instructor II

| NFPA 1041, *Standard for Fire Service Instructor Professional Qualifications*, 2007 Edition | Corresponding Textbook Chapter(s) |
|---|---|
| 5.1 | 1 |
| 5.2 | 12 |
| 5.2.1 | 12 |
| 5.2.2 | 12 |
| 5.2.2 (A) | 12 |
| 5.2.2 (B) | 12 |
| 5.2.3 | 12 |
| 5.2.3 (A) | 12 |
| 5.2.3 (B) | 12 |
| 5.2.4 | 12 |
| 5.2.4 (A) | 12 |

# APPENDIX E

## Chapter 5: Instructor II

| NFPA 1041, *Standard for Fire Service Instructor Professional Qualifications*, 2007 Edition | Corresponding Textbook Chapter(s) |
| --- | --- |
| 5.2.4 (B) | 12 |
| 5.2.5 | 2 |
| 5.2.5 (A) | 2 |
| 5.2.5 (B) | 2 |
| 5.2.6 | 11 |
| 5.2.6 (A) | 11 |
| 5.2.6 (B) | 11 |
| 5.3 | 6 |
| 5.3.1 | 6 |
| 5.3.2 | 6 |
| 5.3.2 (A) | 6, 8 |
| 5.3.2 (B) | 6, 8 |
| 5.3.3 | 6 |
| 5.3.3 (A) | 6, 8 |
| 5.3.3 (B) | 6, 8 |
| 5.4 | 3 |
| 5.4.1 | 3 |
| 5.4.2 | 3 |

## Chapter 5: Instructor II

| NFPA 1041, *Standard for Fire Service Instructor Professional Qualifications*, 2007 Edition | Corresponding Textbook Chapter(s) |
| --- | --- |
| 5.4.2 (A) | 3 |
| 5.4.2 (B) | 3 |
| 5.4.3 | 9, 12 |
| 5.4.3 (A) | 9, 12 |
| 5.4.3 (B) | 12 |
| 5.5 | 10 |
| 5.5.1 | 10 |
| 5.5.2 | 10 |
| 5.5.2 (A) | 10 |
| 5.5.2 (B) | 10 |
| 5.5.3 | 11 |
| 5.5.3 (A) | 11 |
| 5.5.3 (B) | 11 |
| 5.5.4 | 10 |
| 5.5.4 (A) | 10 |
| 5.5.4 (B) | 10 |

# Glossary

**ABCD method** Process for writing lesson plan objectives that includes four components: audience, behavior, condition, and degree.

**Active listening** The process of hearing and understanding the communication sent; demonstrating that you are listening and have understood the message.

**Adapt** To make fit (as for a specific use or situation).

**Adult learning** The integration of new information into the values, beliefs, and behaviors of adults.

**Affective domain** The domain of learning that affects attitudes, emotions, or values. It may be associated with a student's perspective or belief being changed as a result of training in this domain.

**Agency training needs assessment** An assessment performed at the direction of the department administration that helps to identify any regulatory compliance matters that must be included in the training schedule.

**Ambient noise** The general level of background sound.

**Americans with Disabilities Act of 1990 (ADA)** A federal civil rights law that prohibits discrimination on the basis of disability.

**Andragogy** The identification of characteristics associated with adult learning.

**Application step** The third step of the four-step method of instruction, in which the student applies the information learned during the presentation step.

**Assignment** The part of the lesson plan that provides the student with opportunities for additional application or exploration of the lesson topic, often in the form of homework that is completed outside of the classroom.

**Attention-deficit/hyperactivity disorder (ADHD)** A disorder in which a person has a chronic level of inattention and an impulsive hyperactivity that affects daily functions.

**Baby boomer** The generation born after World War II (1946–1964).

**Behaviorist perspective** The theory that learning is a relatively permanent change in behavior that arises from experience.

**Bloom's taxonomy** A classification of the different objectives and skills that educators set for students (learning objectives).

**Budget** An itemized summary of estimated or intended revenues and expenditures.

**Certificate** Document given for the completion of a training course or event.

**Certification** Document awarded for the successful completion of a testing process based on a standard.

**Coaching** The process of helping individuals develop skills and talents.

**Cognitive domain** The domain of learning that effects a change in knowledge. It is most often associated with learning new information.

**Cognitive perspective** An intellectual process by which experience contributes to relatively permanent changes (learning). It may be associated by learning by experience.

**Communication process** The process of conveying an intended message from the sender to the receiver and getting feedback to ensure accuracy.

**Competency-based learning** Learning that is intended to create or improve professional competencies.

**Confidentiality** The requirement that, with very limited exceptions, employers must keep medical and other personal information about employees and applicants private.

**Continuing education** Education or training obtained to maintain skills, proficiency, or certification in a specific position.

**Degree** Document awarded by an institution for the completion of required coursework.

**Delegation** Transfer of authority and responsibility to another person for the purpose of teaching new job skills or as a means of time management. You can delegate authority but never responsibility.

**Demographics** Characteristics of a given population, possibly including such information as age, race, gender, education, marital status, family structure, and location of agency.

**Direct threat** A situation in which an individual's disability presents a serious risk to his or her own safety or the safety of his or her co-workers.

**Disability** A physical or mental condition that interferes with a major life activity.

**Dyscalculia** A learning disability in which students have difficulty with math and related subjects.

**Dysphasia** A learning disability in which students lack the ability to write, spell, or place words together to complete a sentence.

**Dyspraxia** Lack of physical coordination with motor skills.

**Essay test** A test that requires students to form a structured argument using materials presented in class.

**Ethics** Principles used to define behavior that is not specifically governed by rules of law but rather in many cases by public perceptions of right and wrong. Ethics are often defined on a regional or local level within the community.

**Evaluation step** The fourth step of the four-step method of instruction, in which the student is evaluated by the instructor.

**Expenditures** Money spent for goods or services.

**Face validity** Validity achieved when a test item is derived from an area of technical information by an experienced subject-matter expert who can attest to its technical accuracy.

**Feedback** The fifth and final link of the communication chain. Feedback allows the sender (the instructor) to determine whether the receiver (the student) understood the message.

**Firewalls** Software that acts as a virtual wall, separating a trusted environment from an untrusted environment by controlling and regulating the traffic between the two.

**Formative evaluation** Process conducted to improve the fire service instructor's performance by identifying his or her strengths and weaknesses.

**Four-step method of instruction** The most commonly used method of instruction in the fire service. The four steps are preparation, presentation, application, and evaluation.

**Freeware** Copyrighted software that is provided free by the author, who maintains the copyright.

**Generation X** People born after the baby boomers; they are today's adult learners.

**Generation Y** People born immediately after Generation X.

**Ghosting** When viewing text, faint shadows that appear to the right of each letter or number.

**Gross negligence** An act, or a failure to act, that is so reckless that it shows a conscious, voluntary disregard for the safety of others.

**Hostile work environment** A general work environment characterized by unwelcome physical or verbal sexual conduct that interferes with an employee's performance.

**Icon** A small, pictorial, on-screen representation of an object used to represent documents, file folders, and software.

**Identifying** The process of selecting those whom the fire service instructor would like to mentor, coach, and develop.

**In-service drill** A training session scheduled as part of a regular shift schedule.

**Instructor-in-charge** An individual who is qualified as an instructor and designated by the authority having jurisdiction to be in charge of fire fighter training.

**Intranet** A network in which only those belonging to the organization can access the network.

**Job content/criterion-referenced validity** Validity achieved through the use of a technical committee of job incumbents who certify that the knowledge being measured is required on the job and referenced to known standards.

**Job performance requirement (JPR)** A statement that describes a specific job task, lists the items necessary to complete the task, and defines measurable or observable outcomes and evaluation areas for the specific task.

**Kinesthetic learning** Learning that is based on doing or experiencing the information that is being taught.

**Learning** A relatively permanent change in behavior potential that is traceable to experience and practice.

**Learning domains** Categories that describe how learning takes place—specifically, the cognitive, psychomotor, and affective domains.

**Learning environment** A combination of the classroom's physical and emotional elements.

**Learning objective** A goal that is achieved through the attainment of a skill, knowledge, or both, and that can be measured or observed.

**Learning style** The way in which the individual prefers to learn.

**Lesson outline** The main body of the lesson plan. A chronological listing of the information presented in the lesson plan.

**Lesson plan** A detailed guide used by an instructor for preparing and delivering instruction.

**Lesson summary** The part of the lesson plan that briefly reviews the information from the presentation and application sections.

**Lesson title or topic** The part of the lesson plan that indicates the name or main subject of the lesson plan.

**Level of instruction** The part of the lesson plan that indicates the difficulty or appropriateness of the lesson for students.

**Liability** Responsibility; the assignment of blame. It often occurs after a breach of duty.

**Major life activity** Basic functions of an individual's daily life, including, but not limited to, caring for oneself, performing manual tasks, breathing, walking, learning, seeing, working, and hearing.

**Malfeasance** Dishonest, intentionally illegal or immoral actions.

**Master training schedule** A form used to identify and arrange training topics by the number of times they must be trained on or by the type of regulatory authority that requires the training to be completed.

**Medium** The third link of the communication chain. The medium describes how you convey the message.

**Mentoring** A relationship of trust between an experienced person and a person of less experience for the purpose of growth and career development.

**Message** The second link of the communication chain; the most complex link. The message describes what you are trying to convey to your students.

**Misfeasance** Mistaken, careless, or inadvertent actions that result in a violation of law.

**Modify** To make basic or fundamental changes.

**Motivation** The activator or energizer for an activity or behavior.

**Motivational factors** States of the person that are relatively temporary and reversible and that tend to energize or activate the behavior of the individual.

**Negligence** An unintentional breach of duty that is the proximate cause of harm.

**Oral test** A test in which the answers are spoken either in response to an essay-type question or in conjunction with a presentation or demonstration.

**Organizational chart** A graphic display of the fire department's chain of command and operational functions.

**Passive listening** Listening with your eyes and other senses without reacting to the message.

**Performance test (skills evaluation)** A test that measures a student's ability to do a task under specified conditions and to a specific level of competence.

**Power** The ability to influence the actions of others through organizational position, expertise, the ability to reward or punish, or a role modeling of oneself to a subordinate.

**Practical skills** Tasks or jobs that require the physical performance of a fire fighter to achieve.

**Preparation step** The first step of the four-step method of instruction, in which the instructor prepares to deliver the class and provides motivation for the students.

**Prerequisite** A condition that must be met before a student is allowed to receive the instruction contained within a lesson plan—often a certification, rank, or attendance of another class.

**Presentation step** The second step of the four-step method of instruction, in which the instructor delivers the class to the students.

**Psychomotor domain** The domain of learning that requires the physical use of knowledge. It represents the ability to physically manipulate an object or move the body to accomplish a task or use a skill. This domain is most often associated with hands-on training or drills.

**Qualitative analysis** An in-depth research study performed to categorize data into patterns to help determine which test items are acceptable.

**Quantitative analysis** Use of statistics to determine the acceptability of a test.

***Quid pro quo* sexual harassment** A situation in which an employee is forced to tolerate sexual harassment so as to keep a job, benefit, raise, or promotion.

**Reasonable accommodation** An employer's attempt to make its facilities, programs, policies, and other aspects of the work environment more accessible and usable for a person with a disability.

**Receiver** The fourth link of the communication chain. In the fire service classroom, the receiver is the student.

**Reference blank** The place where the current job-relevant source of the test-item content is identified.

**Regulation (code)** A law that can be established by legislative action, but is most commonly created by an administrative agency or a local entity.

**Reliability** The characteristic of a test that measures what it is intended to measure on a consistent basis.

**Revenues** The income of a government from all sources, which is then appropriated for the payment of public expenses.

**Safety officer** An individual who is appointed by the authority having jurisdiction and is qualified to maintain a safe working environment at all training evolutions.

**Sender** The first link of the communication chain. In the fire service classroom, the sender is the instructor.

**Sexual harassment** Unwelcome physical or verbal sexual conduct in the workplace that violates federal law.

**Sharing** The basic concept of giving to others with nothing expected in return.

**Single-source products** Products that are manufactured or distributed by only one vendor.

**Standards** A set of guidelines outlining behaviors or qualifications of positions or specifications for equipment or processes. Often developed by individuals within the regulated profession, they may be applied voluntarily or referenced within a rule or law.

**Statute** A law created by legislative action that embodies the law of the land at both the federal and state levels.

**Subject-matter expert (SME)** An individual who is technically competent and who works in the job for which test items are being developed.

**Succession planning** The act of ensuring the continuity of the organization by preparing its future leaders.

**Summative evaluation** Process that measures the students' achievements to determine the fire service instructor's strengths and weaknesses.

**Supplemental training schedule** An alternative training schedule that is used in the event that some mitigating factor requires a change in the original training session.

**Systems Approach to Training (SAT) process** A training process that relies on learning objectives and outcomes-based learning.

**Tailboard chat** An informal gathering where fire fighters discuss various issues.

**Technical-content validity** Validity achieved when a test item is developed by a subject-matter expert and is documented in current, job-relevant technical resources and training materials.

**Title VII** The section of the Civil Rights Act of 1964 that prohibits employment discrimination based on personal characteristics such as race, color, religion, sex, and national origin.

**Undue hardship** A situation in which accommodating an individual's disability would be too expensive or too difficult for the employer, given its size, resources, and the nature of its business.

**Validity** The ability of a test item to measure what it is intended to measure.

**VARK Preferences** A tool that measures a person's learning preferences along visual, aural, read/write, and kinesthetic sensory modalities.

**Vision** Having an alertness to the future, recognition of potential, and expectations of improvement.

**Visual, auditory, and kinesthetic (VAK) characteristics** Learning styles based on the idea that we all have a learning style preference based on sensory intake of information (visual, auditory, and kinesthetic).

**Willful and wanton conduct** An act that shows utter indifference or conscious disregard for the safety of others.

**Written test** Any of several types of test items, such as multiple choice, true/false, matching, short-answer essay, long-answer essays, arrangement, completion, and identification test items. Answers are provided on the test or a scannable form used for machine scoring.

# Index

## D

## E



## F

## G



## H



## I

## J



## K



## L

## M

## N

## O



## P

## U



## V

## W

# Photo Credits

3-2 © ClassicStock/Alamy Images; 3-3 © digitalskillet/ShutterStock, Inc.; 3-4 © SW Productions/Jupiterimages; 8-4 © Goygel-Sokol Dmitry/ShutterStock, Inc.; 8-8 © Amy Walters/ShutterStock, Inc.; 9-3 © Tom Carter/PhotoEdit, Inc.; 11-1 © Keith D. Cullom; 12-2 © Steven Townsend/Code 3 Images; 13-2 Photographed by Tony Greco, FDIC 2008. Used with permission of.PennWell Corporation.

Unless otherwise indicated, all photographs are under the copyright of Jones and Bartlett Publishers, courtesy of the Maryland Institute for Emergency Medical Services Systems, or were provided by the author.